A Review: Chill-Block Melt Spin Technique, Theories & Applications

Edited By

Mustafa Kamal and Usama S. Mohammad

Books End User License Agreement

DEDICATION

To Egypt; our motherland.

To our 25th January Revolution Martyrs; the Heroes.

Your children, Egypt, died, so you can live, and they disappeared for your people to stay.

Ahmed Shawqi (1869-1932)

To Humanity; we are all belonging to, and to Pol E. Duwez (1907-1984); the remarkable scientist and the pioneer of rapid solidification science and non-equilibrium processing of metals and alloys.

"Nobel Prize in Chemistry in 2011 on Quasicrystals by Chill-Block Melt Spin Technique"

Shechtman won the Nobel Prize in Chemistry in 2011 for his work on Quasicrystals using the technology of rapid solidification from the molten state of metallic alloys. His discovery of Quasicrystals revealed a new principle for packing of atoms and molecules.This led to a paradigm shift within Chemistry and the science of the symmetry of finite objects; the "Pattern and Symmetry in the field of Crystallography".

CONTENTS

CHAPTERS

FOREWORD

From first obtaining of amorphous materials, Au-Si alloys produced in 1959, in very small quantities by a technique of cooling of the melt with speeds of millions of degrees per second, this class of materials has been a great concern to materials science researchers.

The special application properties of this class of materials associated with the particular structure of metallic glasses have a particular interest of specialists in several fields (metallurgy, electronics, chemistry, aeronautics, and other). Since early 60's until now, the rapid solidification technology materials have made great progresses, so both amorphous, nanostructure and quasicrystalline materials can be obtained. These materials can be processed in a wide range of chemical compositions and under different forms, like wires, filaments, rods, and others both in labor and industrial quantities.

These papers provide to the specialists in the field of rapid solidification different theoretical data on physical and chemical phenomena that occur in rapid solidification process including the process parameters for the widely used chill-block melt spin technique. It also presents experimental data on mechanical, electrical, magnetic properties as basis elements for finding of new industrial applications of this type of material. The book edited by Prof. Dr. Mustafa Kamal and researcher Usama S. Mohammad is thoroughly documented and provides an extensive vision on the obtaining of new classes of materials by rapid solidification. The book can successfully serve as basis of study both for the renowned researchers in the field, for students specializing in the materials science and for specialists from companies that coordinate the activity of obtaining the metallic glasses on industrial scale.

Prof. Dr. Leontin Druga

Member of the Romanian Academy for Technical Sciences
Romania

PREFACE

Since the emergence of materials science, which denoted a new scientific concept, born out of physical metallurgy some time in the early 1950s, many advanced materials and techniques were developed and the interest in rapidly-solidified alloys has increased dramatically. The subject area has moved a long way since Pol Duwez reported his remarkable discoveries in 1960. It has become an integral part of the fabric of materials science and grown to be evident that there are often several alternative routes to a final outcome whether that is a particular microstructure, combination of properties or a product for engineering application. As a consequence, a number of metallic materials have emerged by simply the very rapid extraction of heat from the metallic melt, which not only are viewed as a challenging field for research but also, and due to their new properties, they are being considered for structural, functional and smart properties. Indeed, one of the most interesting applications of the technique is the synthesis of new metallic alloy phases which cannot be obtained either under equilibrium conditions or by normal quenching in the solid state.

The main product of the chill-block melt spin (CBMS) technique is a metastable metallic material. Metastability here is a state of equilibrium that is stable under working conditions, giving us the useful and valuable physical and chemical properties of the new metallic material. Amorphous, quasicrystalline, nano-, submicro- and fine micro-structured metallic materials have been effectively produced by CBMS route, with best reduced cost, making the uniqueness of this technique over the other rivals like melt atomization. Many advanced engineering fields require these products, like aerospace and robotics industries, as they have the best strength-to-weight ratios, especially when used for making advanced composite materials.

Metal physics requires a deep knowledge of the phase diagrams of elements and compounds as well as the understanding of phase transformations and the control of reaction kinetics. Thus, we tried in this study to simplify the subject of the industrial technique of rapid solidification processing (RSP) throughout the study of the CBMS technique which is considered world-widely a most promising means for producing new and advanced metallic materials, including metal matrix composites. The technique itself is simple and perfect and can be adapted to produce either low-cost or high-cost products by only one step, saving money and energy in comparison to other technologies.

This study implies three main topics; the first is the rapid solidification theory, the second is the CBMS technique and the third is a study of the effect of RSP on the structural and the physical properties of metals and alloys with applications. In our treatment, we used simple mathematics to make the topics clear as much as possible, followed by a very good deal of references for any further readings covering all sides of this review.

Mustafa Kamal and Usama S. Mohammad

Mansoura University,
Egypt

To the Reader

To our knowledge, this book contains information obtained from reliable and highly regarded sources. A wide variety of references are listed and mentioned sequentially in their exact places. Reasonable efforts have been made by the editors to publish reliable data and information, but the editors and the publisher cannot assume responsibility for the validity of all materials or for the consequences of their use.

2

CHAPTER 1

Rapid Solidification Processing

M. Kamal[a,*] and Usama S. Mohammad[b]

[a]*Chairman and Professor of Metal Physics, IUPAP-C5 Member, Department of Physics, Faculty of Science, Mansoura University, Al-Mansoura, Egypt and* [b]*Ph.D. Researcher in Metal Physics and Materials Science, Department of Physics, Faculty of Science in Damietta, Mansoura University, New Damietta City, Egypt*

Abstract: Rapid Solidification Processing (RSP) is an advanced technique for producing new materials. By applying high cooling rates larger than 10^5 °C, all thermal and diffusional processes are affected and new phases, new refined structures and new materials are emerged.

Key Words: Rapid solidification; chill block melt spin; undercooling; metastable phase; cooling rate; quasicrystal nucleation.

1.1 INTRODUCTION

Solidification is a phase transformation, which is familiar to every one; it is the process of formation of a solid. Solidification phenomena play an important role in many of the processes used in fields ranging from solid-state physics to production engineering. Solidification is generally combined by the formation of crystals, an event that is much rarer during the solidification of glasses or polymers. In metals, melting is accompanied by an enormous decrease in viscosity. If the properties of casting were easier to control, then solidification becomes most the important process. In this respect, solidification theory plays a vital role because it forms the basis for influencing the microstructure and hence improving the quality of cast product [1,2].

Some important processes, which involve solidification, are: *casting, welding, soldering, brazing, single crystal production* (or; *directional solidification*), and *rapid solidification processing*. Those aforementioned solidification processes involve extractions of heat from the melt in a more or less controlled manner. Solidification takes place if sufficient heat is extracted to (1) cool the superheated molten metal to or below its melt temperature (*i.e.; undercooling* or *supercooling*) and (2) to remove the latent heat from the *solidification front*. Heat extraction changes the energy of the phases in two ways:

1.   There is a decrease in the molar enthalpy of the liquid or solid, due to cooling, and it's given by:

$$\Delta H = \int c \, dT \; \ldots \tag{1-1}$$

Where c is the specific heat and dT is the temperature difference.

2.   There's a decrease in enthalpy, due to the transformation from liquid to solid, which is equal to the latent heat of fusion, where;

$$L = \Delta H_f = \Delta U + \Delta W \; \ldots \tag{1-2}$$

L is the latent heat of fusion, ΔU is the change in internal energy and ΔW is the work done by the system during phase change, ΔW is positive for expansion and negative for contraction and may be negligible in case of metals and metallic melts.

---

*Address correspondence to M. Kamal: Chairman and Professor of Metal Physics, IUPAP-C5 Member, Department of Physics, Faculty of Science, Mansoura University, Al-Mansoura, Egypt; E-mail: kamal422002@yahoo.com

3. To solidify the melt it must be undercooled by ΔT below the equilibrium melting temperature T_m with a decrease in *Gibbs molar free energy* ΔG (J mol^{-1}) where;

$$\Delta G = \Delta H - T \Delta S \; \ldots \tag{1-3}$$

In case of equilibrium; $\Delta G = dG = 0$, at melting point T_m ; $\Delta G_f = \Delta H_f - T_m \Delta S_f = 0$ for any phase α or β … *etc.* At any temperature T ; $\Delta G_f = L - T \Delta S_f \neq 0$ and

$$\Delta G_f = L \, (1 - T \,)/T_m \; \ldots \tag{1-4}$$

The form of a solidified microstructure depends on the system (*i.e.*; the alloy), composition and cooling conditions. It is important in this area to understand how the various microstructures are influenced by the alloy compositions and by the solidification conditions, using phase diagrams, thermodynamic treatments and modeling. These have been summarized in some interesting books [1, 3-13]. Nearly all of the solidification microstructures, which can be exhibited by a pure metal or an alloy, can be divided into two groups: single-phase primary crystals and polyphase structures (*e.g.*; eutectic, eutectoid, peritectic, bainitectic, cellular, spinoidal, dendritic …etc) [14].

1.2 RAPID SOLIDIFICATION

Hot steel has been quenched into water or oil, to harden it, for a millennium at least, and the technique has acquired many subtle variants. Numerous solutes, some of the ancient ones with magical overtones, have been added to quenching water to improve swords and armor. In recent years, polymer additions have enhanced control over quenching rates in heat treatment shops. In this process, 1000 °C s^{-1} would be reckoned an exceptionally high rate of cooling, dangerous because apt to cause cracking [15]. Attainable cooling rates depend primarily on the dimensions of the workpiece, and high cooling rates have been attained for research purposes by quenching thin foils by means of a fast-moving stream of a gas possessing high thermal conductivity. According to a 1951 paper by Duwez [16], with a metallic foil 50 μm thick quenched in a helium stream moving at several hundred feet per second, cooling rates of 10^5-10^8 °C s^{-1} can be achieved. That is probably close to the realistic limit for *solid-state quenching*. Continuous heating followed by immediate quenching of thin sheet just after cold-rolling (the so-called continuous annealing process for steel sheet) is still normal industrial practice in Japan [17] especially in traditional sword industry. Quenching of hot steel is of course only one of many industrial applications of solid-state quenching, all of them limited to *modest* cooling rates. All such applications have to, as their aim, preserve at room temperature a *metastable* phase structure, either for use "as is" or for subsequent modification in the direction of thermodynamic equilibrium without necessarily reaching it, as in the tempering of hardened steel. In addition, as Cahn said, there is no exaggeration to say that the artful control of *metastability* is the metallurgist's central skill [15]. Often, a solid-state quench preserves *metastably* the structure stable at the temperature from which the alloy is quenched, though under appropriate circumstances the alloy can transform to another non-equilibrium structure during the quench. Examples include the quenching of *austenite* to generate *martensite*, metastable carbides in engineering steels and cast irons, intermediate precipitates in structural age-hardening alloys, oxide glasses for structural and nonstructural applications as well as naturally-occurring films conferring corrosion protection on otherwise highly reactive metals. It is natural for a curious metallurgist to wonder what would happen if an alloy is quenched from the molten state: the preservation, metastably, of the structure stable at the starting temperature would yield a congealed liquid, that is to say a *metallic glass*. Alternatively, from analogy with the generation of martensite, one might expect the creation of a metastable crystalline structure during the quench. Such inquisitiveness on the part of an unusually adventurous metallurgist, Pol Duwez, gave rise to the subject of *Rapid Solidification Processing* (RSR) or the *Non-Equilibrium Processing* (NEP) as a new branch in metallurgy [18]. The status of rapid solidification is discussed in terms of the recent progress in modeling methods of achieving solidification at high cooling rates and its effects on alloy constitution and microstructures. Metallic materials can be solidified by electro- or electroless-deposition from salt solution, by vapor or sputter deposition, as well as by freezing of their melts [19]. Deposition from salt solutions is highly system-specific and slow (~ 0.1 mm h^{-1} is considered a high rate of electro-deposition). Vapor and sputter deposition, although they are very widely applicable, are slow (~ 20 μm h^{-1} for high rate sputtering) [19-21].

Solidification from the melt is widely applicable and also allows solid to form rather rapidly. Even the slowest steady state crystal growth from the melt is carried out at rates of 1 mm h^{-1} and a typical steel ingot will freeze in its cast iron mold at an average rate of ~ 100 mm h^{-1}. On this basis even normal solidification from the melt, is not slow. The aforementioned techniques are referred to as *conventional processing*, with cooling rates being about 10^{-1} to 10^{-2} °C s^{-1}, the slowest values being 10^{-5} to 10^{-6} °C s^{-1} for very large sand castings.

In rapid solidification, cooling rates must be at least of ~ 10^{5} to 10^{6} °C s^{-1}. For example, in the *gun technique* used by P. Duwez [22, 23] of splat cooling, in which a shock wave atomizes a small molten charge to form a *splat* of non-uniform thickness varying from 50 to < 0.1 μm on impact, with a rigid chill surface. Measurements indicated cooling rates of ~ 10^{5} to 10^{6} °C s^{-1} and 10^{10} °C s^{-1} for thinnest regions (recently, in a *sonochemical* apparatus, used for bulk amorphous metals production, the cooling rate is ~ 10^{10} °C s^{-1}; and 10^{14} °C s^{-1} for *in situ* laser surface glazing/melting) [11]. There are many other differences between conventional casting solidification and rapid solidification [19]. One difference is that it is possible to achieve relatively large (hundreds of degrees) supercooling of the melt before significant amounts of solid phase can form. This can result in constitutional changes in the retained equilibrium phases that can have compositions outside their equilibrium limits (*i.e.*; solid solubility extension), or stimulate non-equilibrium phases to form as a result of being favoured kinetically over equilibrium phases [15,18,19]. Such effects become the rule rather than the exception when cooling rates exceed 10^{6} °C s^{-1}. Resulting non-equilibrium phases may be crystalline or glassy (*i.e.*; amorphous) and maintenance of a high cooling rate following their formation helps to ensure their retention to ambient temperature. The availability of cooling rates > 10^{6} °C s^{-1} has facilitated glass formation even from readily-crystallizing metallic melts, establishing a new class of material; *the metallic glass* or *metglass*.

A second difference is that: shortening of diffusion distances and formation times results in substantial refinement and re-morphologizing of microstructure for both matrix and minor phases. This has important consequences for conventional alloys limited in processability and/or service performance by formation at ingot stage of non-uniform distributions of coarse subsidiary phases and segregates. The resulting possibilities for more effective use of alloy additions and of processing options are the origin of present arising efforts to establish *rapid solidification rate* (RSR) processing as a reliable means of manufacturing high quality, more economical products with improved combinations of properties.

1.3 SOME EVENTS IN THE EARLY DEVELOPMENT OF RSP AND NEP

Although much of the current activity on *rapid solidification* (RS) and NEP has its origin in Duwez seminal discoveries of 1960, the seeds of the technology and some of its effects were seen and known much earlier [15,18,19]. According to Jones [18], droplet RS technologies have their origin in the shot-casting process, said [24] to have been invented in 1650 by Prince Rupert, a nephew of Charles I and distinguished Royalist cavalry leader in the English Civil War. His process involved teeming molten lead through a preheated metal dish perforated by an array of equal sized holes. In a sufficient distance of free fall, droplets separate and round themselves under capillary forces into almost perfect spheres prior to solidification. The first shot tower, built by Watts in the late 1780s [25], allowed larger diameters of shot to be produced. Such towers were the forerunners of modern *drop tubes* used for microgravity research and containerless undercooling processing [15,18,26]. Although Agricola in 1556 [27] reported the use of a rotating disc to granulate liquid metals, the first patent on metal atomisation seems to have been granted to William Marriott of Huddersfield [28] in 1873. He used a steam ejector to generate and atomise the melt into a powder to be used for making salts and oxides of lead, presumably for paint manufacture. Cowing in 1906 [29] ejected melt from a rotating crucible, either collecting the product as solidified shot or intercepting the molten droplets on baffles where they spread and solidify as splats (rather than spheres), as in Fig. **1**.

Schoop in 1910 [30] formed a continuous adherent deposit of such impacted spray droplets, allowing repair of worn machine parts or application of wear or corrosion resistant coatings. The Duwez and Willens gun technique of 1963 [23] differed by using a shock wave to generate very small (~ 1 mm) droplets, impacting at sonic velocity onto a chill substrate (Fig. **3**). The idea of using spray deposition to generate preforms for subsequent densification by rolling was introduced by Singer in 1970 [31]. Long product RS technologies go back to Sir Henry Bessemer's patent [32], in 1846, of a system for direct production of tin foil and lead sheet from the melt, which adapted a successful practice for rolling semi-liquefied glass into plate glass. Ten years later, he verified his idea by casting a 1 m length of soft iron strip between twin rolls. In 1882, Small [33] patented an apparatus for making solder wire by melt-extrusion through an orifice (Fig. **4a**) followed in 1885 by Lyman's famous patent [34] for making printer's lead by casting into the channel between a steel belt and a grooved wheel. Strange and Pim [35] in 1908 and Staples in 1911 [36] filed patents embodying the principle of *chill-block* (or; *free-jet*) *melt-spinning* in which melt is discharged through an outlet or outlets on to a rotating chill surface on which it solidifies to form continuous ribbon or strip (Fig. **4b**).

It is notable that these early pioneers were motivated by an interest in developing better, or more direct, routes to manufacture specific products. The fact that RS occurred was incidental. Examples of the now well-known effects of RS, however, can be found prior to 1960. Dix in 1925 [37] showed a fully eutectic *micro-duplex* structure in a chill-cast *hypereutectic* Al-Fe alloy (Fig. **5a**), illustrating simultaneous suppression of primary intermetallic and replacement of an irregular eutectic structure by a regular one at sufficiently high solidification front velocities. Hofmann in 1938 [38] reported the first observation of nonequilibrium extension of solid solubility to 4 wt.% Mn in Al by chill casting into a tapered mould, compared with an equilibrium maximum solid solubility of 1.4 wt.% Mn (Fig. **5b**). The use of RS to much refine microstructure and to develop improved performance in engineering alloys dates back to 1950, when Busk and Leontis [39] reported consolidation by extrusion of gas jet or rotary atomised magnesium alloy powder to achieve finer grain size and higher strength than for the same alloys extruded from cast ingot. Production of amorphous Ni-P coatings by electro-deposition also dates from 1950 [40] and Buckel and Hilsch [41] established in 1956 that solid solubility of Cu in Sn and of Bi in Sn could be extended to beyond 20 and 45 at.%, respectively, by vapour quenching.

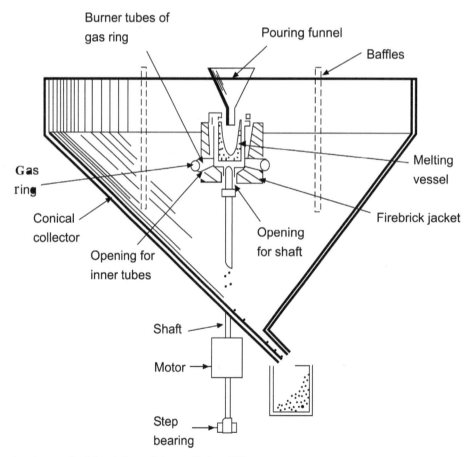

Fig. (1). Cowing's rotary method for shot or splat manufacture [18].

1.4 THEORY FOR RAPID SOLIDIFICATION PROCESSING

1.4.1 Empirical Treatment

Achievement of cooling rates in the range 10^2 to 10^{10} °C s^{-1} is dependent on rapid formation of a sufficient small dimension of cross section in good contact with an effective *heat sink*. For effectively perfect contact conditions between metallic melt and chill surface, standard heat flow analyses predict a cooling rate $\dot{T}$ at the uncooled boundary equal to (B/z^2). B is the *heat conduction factor* of the RS process and device; it is a function of relevant temperature intervals, materials properties, surface finish, cleanliness or purity of the contacting solids, substance (or lack of it) in the interstitial spaces (differentiating between the solid-solid conduction and the interstitial heat transfer caused by interfacial materials like air or gas pockets) and pressure. B has a value $\sim 10^4$ (mm^2 °C s^{-1}) [42] for Cu substrates. z is the unit section dimension, it must not exceed 10 mm, 0.1 mm and 10 µm if cooling rates of 10^2, 10^6 and 10^{10} °C s^{-1}, respectively, are to be achieved [19].

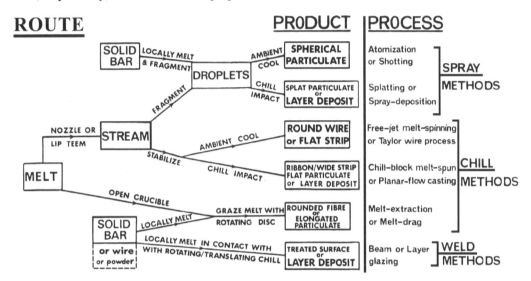

Fig. (2). Some production routes for RSP-technology and their products [19].

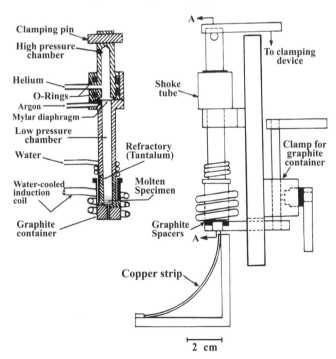

Fig. (3). Duwez gun technique of shockwave splat quenching/cooling from the melt by a copper *ski* slide/strip [15,23].

For the Chill-Block and Splat-Quenching melt spinning techniques, the measurements of $\dot{T}$ lies between the ideal cooling and a value some two orders of magnitude lower below which *Newtonian cooling* conditions satisfying the condition $(B_\infty/100) < \dot{T} z^2 < B_\infty$. B_∞ is the ideal heat conduction value of B (in vacuum with full solid-solid or melt-solid contact) where heat flow model predictions [42-45] give $3000 < B_\infty < 20000$ mm^2 °C s^{-1} for a range of metals (*e.g.*; Al, Ag, Cu, Fe and Sn) on Ag, Cu or Ni chill substrates with $B_\infty \sim 10^4$ mm^2 °C s^{-1} for Al or Fe on Cu due to the high heat conductivity of copper. Making $\varphi = B/B_\infty$, all measurements show that $0.1 < \varphi < 1$ for splat quenching and chill-casting, a characteristic for *near-ideal* cooling conditions, while $0.01 < \varphi < 0.1$ for the chill-block melt spinning, corresponding to *near-Newtonian* conditions, as we will see later [43-51].

The magnitude of *heat transfer coefficient* **h** at the *chilled boundary* required to ensure effectively perfect contact is $\underset{\sim}{>} (k/z)$, where

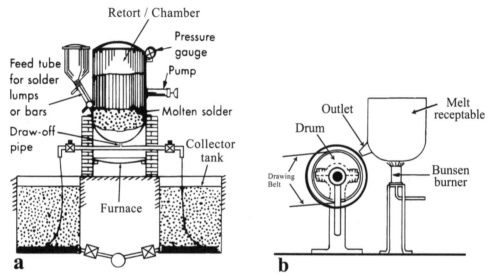

Fig. (4). (**a**) Small's method of direct casting of metal filament from the melt. (**b**) Strange and Pim's method of casting continuous thin sheet or ribbon by discharging melt on to a rotating chill surface [18].

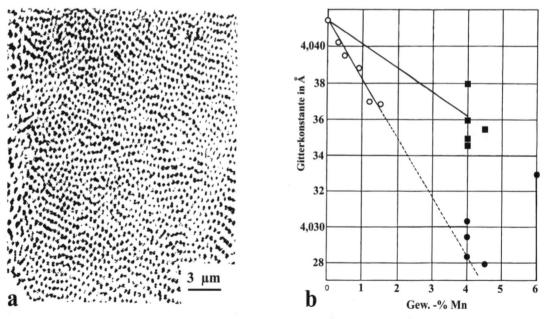

Fig. (5). (**a**) Microduplex eutectic structure in chill-cast Al–2.7 wt.% Fe. (**b**) Solid solubility extension of Mn in Al by chill casting. Key: (**O**) homogenised and quenched; (**●**) chill cast; (**■**) sand cast [15,18,37,38].

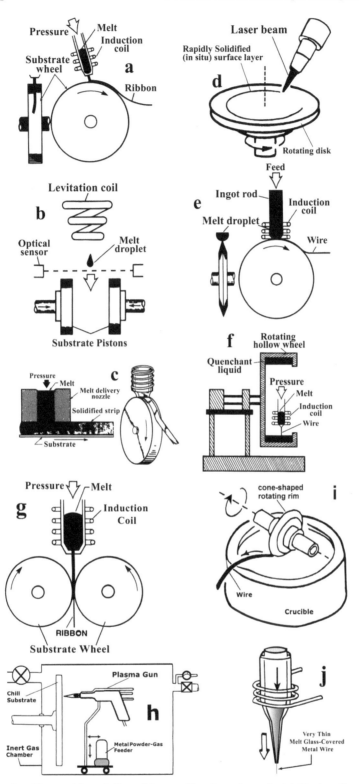

Fig. (6). Principal methods of rapid quenching of alloy ribbon or fiber from the melt. (**a**) Single roller chill-block melt-spinning (CBMS) [54-56]. (**b**) Anvil-and-piston technique or the drop smasher, using levitation melting by induction (pistons are pneumatically or electromagnetically accelerated) [57]. (**c**) Planar-flow casting (PFC) [56,58,59]. (**d**) Laser glazing by surface melting to produce *in situ* rapidly solidified surface layer of cooling rates reaching 10^{14} °C s^{-1}[60]. (**e**) Pendant-drop melt extractions for making tough metallic fibers and wires [61]. (**f**) In-rotating water spinning (INROWASP) of wire [62]. (**g**) Twin-roller quenching [63]. (**h**) Plasma jet technique [64]. (**i**) Melt drag [56,65]. (**j**) Taylor wire technique; a very fine wire filament is encased with a thin layer glass by the Taylor Cone and followed by quenching [66].

k is the thermal conductivity of solidified melt or chill surface material. So, the need to ensure good thermal contact becomes increasingly important with the increase of desired cooling rates and thus decreasing the required z, as achieved in the principal methods of rapid solidification techniques [2,15,18,52-66], shown in Figs. **6-9**. Alternative routes for achieving high cooling rates during solidification, as in Fig. **2**, have in common the basic requirements to rapidly produce a section of small z under conditions of sufficiently high h. Methods involving droplet formation (*spray methods*) which include the gun technique, are the most complex, usually involving three stages: stream formation, fragmentation and cooling. The rotary atomization technique of *rapid solidification rate* (RSR) powders [67] has been a major stimulus. Parallel developments are including *ultrasonic* [68,69] and *electrodynamic methods* [70,71] of droplet formation. Attention has also been given to the building-up of thick rapidly-solidified deposits by successive impact of incident droplets, thereby combining rapid solidification with an element of consolidation in a single operation [72-74]. Chill methods instead generally involve stabilizing any stream formed to generate a continuous filament/wire or thin strip/ribbon. Particular attention has been given to *chill-block melt spin(ing)* (CBMS) [56,75] and *planner-flow casting* (PFC) [58,76] methods developed as RSR processes. PFC and the related Battelle processes of *melt-drag* [56,65,77] and *melt-extraction* [61,78] from their products essentially without formation of a stream and, in the latter case, even without the need for melt to flow through an orifice, are among the most direct of available routes from melt to product. Weld methods of *directed-energy processing* [79] take this trend a stage further by *melting in situ* at the chill surface, rather than in a crucible, so giving an RSR-treated surface or a thick layer deposit on a suitable former. Modeling of this range of processing methods involves taking into account both *fluid dynamics* and *heat* and *mass transfer* topics. Fluid dynamics govern the size distribution and velocity of droplets produced by fluid-driven and rotary atomization. In addition, they control the amount of spreading and thinning of such droplets that occurs on impact with a chill surface. For spray methods, we will consider the relatively well-defined situation of direct drop formation at the rim of a rotating disc at low feed rates. Here, the droplet size is determined by the balance between the generating effect of centrifugal force ($= v\rho\omega 2\Delta/2$) and the restraining effect of surface tension force (directly proportional to γd), where (respectively); v and d are droplet volume ($4\pi r^3/3$ or $\pi d^3/6$) and diameter; ω and Δ are disc speed of rotation and diameter; ρ and γ are liquid density and surface tension. Giving [19,80]:

$$d = (a/\omega)(\gamma/\rho\Delta)^{1/2} \ldots \tag{1-5}$$

where a is estimated to be close to 4 as confirmed experimentally for a variety of liquids including some liquid metals, as in Fig. **7a**. Increased feed rate gives first ligament and then sheet formation, as in Figs. **7b** and **7c** [81] and prior breakdown into droplets. This is more complex situation and really is so difficult to model realistically, so it make us recourse to empirical relationships such that of Lubanska for fluid atomization of liquid metals and wax [82]:

$$d = (a/v)[\gamma\varphi\alpha(1+\beta)/\rho_0]^{1/2} \ldots \tag{1-6}$$

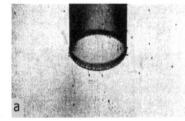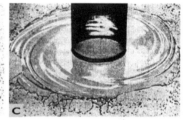

Fig. (7). The transition from (**a**) discrete droplets, is going direct to (**b**) prior ligament to (**c**) prior sheet formation of droplets with increasing feed rate (1 to 8 to 45 kg h⁻¹) [19,80].

where φ is melt orifice or stream diameter; ρ_0 and v are density and relative velocity of atomizing fluid; α and β are ratios of kinematic velocity and mass flow rate, respectively, of the melt of those of the atomizing fluid; and a is ~ 50, v must be known at the point of atomization. Eq. (1-6) is proved to show excellent agreement between measured mean droplet size and the predictions of it [19,83].

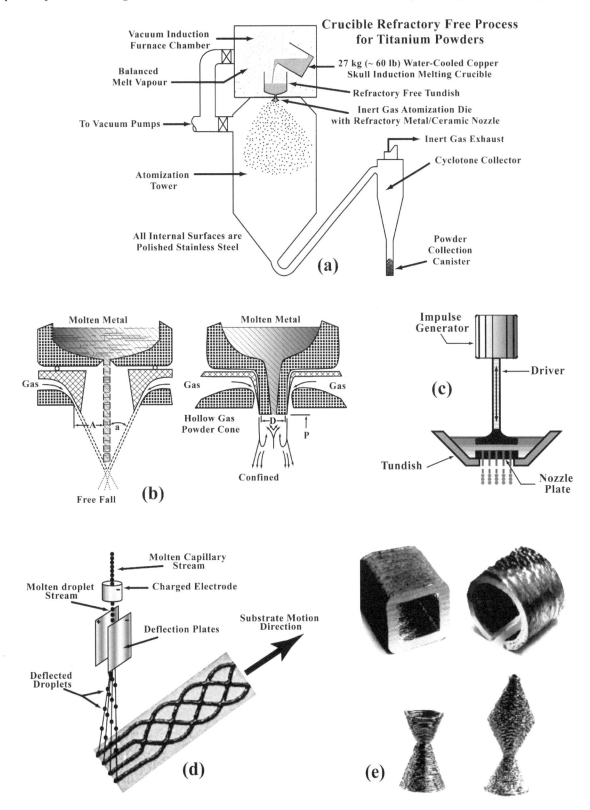

Fig. (8). RSR atomization techniques for metallic melts. (**a**) A schematic diagram of a common titanium powder gas tower atomizer [84]. (**b**) Two different atomizing dies with different controlling parameters for drop shape and attainable cooling rates [85]. (**c**) A piezoelectric transducer of an ultrasonic melt atomization nozzle/head for melt drop prototyping technology [86]. (**d**) and (**e**): forming complex and near net-shape profiles from atomized melt drops by the prototyping technology. Flying trajectory of metal droplets can be precisely computer-controlled by charging and electro-deflecting the droplets [87].

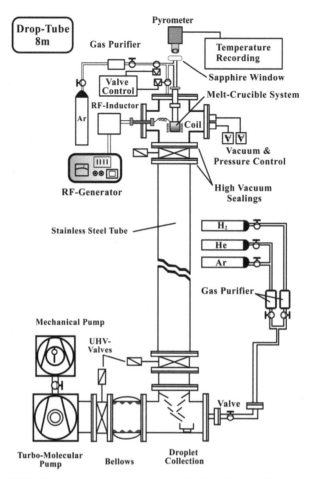

Fig. (9). Schematic view of the DLR (German Aerospace Center, in Cologne) short (8m) drop tube, the drop-tube technique combines rapid cooling of small particles (50-1000 μm) and reduction of volume/surface heterogeneous nucleation sites by containerless processing and by dispersion of the melt into a spray of small droplets. The drop tube is made of stainless-steel components all of which are compatible with the requirements of ultra-high vacuum (UHV) technique, ~ 10^{-7} mbar, and backfilling with purified He gas or He-3.5% H_2 mixture for high thermal conductivity (with very low content of oxygen gas; ~ 1 ppm). The drop-tube technique combines both principles of rapid cooling of small particles and reduction of heterogeneous nucleation, with cooling rates ~ 10^8 °C s^{-1} [88,89].

1.4.2 Theoretical Treatment for Rapid Solidification

1.4.2.1 Overview

There are serious practical difficulties with the accurate measurement of temperature in most methods of rapid liquid quenching. Besides the problems of measurement sensitivity and response, the relatively large latent heat of crystallization causes *recalescence* and an uncertain relationship between liquid cooling rate and temperature during solidification. As a result, many of the reports of liquid cooling rate evaluation are based on heat flow calculations. Alternatively, the size scale of microstructural features such as dendrites or cells are often related to the cooling rate or local solidification time. Although such relationships may be valid for secondary arm spacings in fully developed dendritic alloys, the extrapolation of size scale relationships determined at slow cooling rates to estimate high cooling rate values is of uncertain validity [15].

Much of the early rapid solidification processing work was focused on cooling rate because of its common usage as a variable in *solid-state transformations*, where heat evolution is usually negligible. At the same time, the application of *Time-Temperature-Transformation* (TTT) diagrams, representing the kinetics of crystallization at different melt undercoolings, was developed to estimate the critical cooling rates required to avoid crystallization and thus cause glass formation [90]. Because no heat is evolved when crystallization is avoided, the specification of a cooling rate is meaningful for *vitrification*. It is not, however, adequate for a complete analysis when crystallization occurs. The importance of the role of undercooling (= supercooling) becomes more apparent with the realization that metastable phases can form during slow cooling. Initially undercooled droplets [91,92], as well as controlled directional solidification studies [93,94], emphasize undercooling controlled kinetics. In the former case, the initial undercooling can control the *nucleation phase selection* and generate high solidification rate. In the latter case, the imposed solidification velocity controls the operating interfacial undercooling. In both cases, subsequent *competitive growth of phases* also contributes to phase selection and the evolution of the dominant microstructure. From this viewpoint, it becomes clear that rapid solidification should not always be equated to *rapid quenching* [95].

The present discussion focuses primarily on the principles underlying the general importance of the methods of melt processing: powders with cooling mainly by *convection* and melt streams with *conductive cooling* to a chill or quenching substrate. In spite of the appearance of seemingly different sample configurations and conditions, there is justification for treating many of the solidification characteristics involved in these processes on a common basis. Such a normalization of behavior is possible if the solidification kinetics involved in the various processing methods are viewed in terms of interface undercooling and solidification velocity. From this point of view, general principles governing the liquid-to-solid transformation at high rates can provide a unified framework to understand the variety of processing approaches used and the microstructures obtained in the rapid solidification. The central principles cover five areas: *heat flow, thermodynamics, nucleation kinetics, growth kinetics,* and *atomic clusters physics* [95]. To a certain extent, heat flow determines the relationship between externally controllable processing parameters and the internal fundamental solidification parameters. The thermodynamics of metastable and nonequilibrium solidification sets the possible range of solidification product phases while nucleation, growth kinetics (or; interface kinetics) and atomic clusters physics determine the detailed microstructural evolution. Within these five areas, specific features are examined including the role of melt subdivision, undercooling and liquid cooling rate on nucleation kinetics and the role of solute redistribution on growth kinetics [95].

1.4.2.2 Variables of Rapid Solidification Processes

Rapid solidification of crystalline alloys clearly involves the rapid removal of the latent heat of fusion by an appropriate *heat sink*. Two types of sinks are important: external and internal. External heat sinks are massive (relative to the casting size) metal substrates such as the wheel in melt spinning (MS), the underlying un-melted solid in surface (*in situ*) melting, or large volumes of high-velocity-cooling gas such as found in most atomization processes. The effectiveness of these heat sinks is characterized by a heat transfer coefficient h. For surface melting, infinite values of h are appropriate because the melt is in contact with its own solid, that is; the temperature is continuous across a melt-crystal interface. For atomization, the heat transfer coefficient is often dominated by convective-conductive cooling which increases dramatically with decreasing powder size. Jones [95,96] gives an expression for h which also depends on the properties of the quenching gas and the relative velocity between the droplet and the gas. Values for h typically range from 10^{-3} to 10^2 W cm^{-2} °C^{-1}. For quenching against a substrate, h values are difficult to measure and depend on the details of the surface of the quenching substrate. Estimates, often using dendrite arm spacings, vary widely (10^{-1} to 10^2 W cm^{-2} °C^{-1}). Clearly, such large uncertainties make detailed predictions difficult for substrate quenching. A second type of heat sink is internal in the sense that the casting is its own heat sink. This occurs when the alloy is *undercooled* below its *liquidus* temperature prior to the initiation of solidification. The release of latent heat can then be absorbed by the sample itself. According to Boettinger and Perepezko [95], the undercooling of the liquid prior to solidification, ΔT, is spatially uniform and the effectiveness of this heat sink is expressed by the *dimensionless undercooling*, $\Delta\theta$, given by:

$$\Delta\theta = \Delta T \frac{C}{L} \dots \qquad\qquad (1\text{-}7)$$

where C is the heat capacity per unit volume of liquid and L is the latent heat per unit volume ($= \Delta H_f$; the enthalpy of fusion, L/C for Al is 91 °C). *Hypercooling* is defined by the condition where $\Delta\theta \geq 1$, the hypercooling limit is $\Delta T_{hyp} = \Delta H_f / C_P^L$ and complete *isenthalpic rapid solidification* is possible [95,97,98] (*i.e.*; the heat transfer is rapid enough to permit hypercooling). Thus, nucleation is suppressed until the sample has undercooled so far that the release of the entire latent heat is insufficient to reheat the sample to the equilibrium freezing point T_m and freezing can then go to completion without any further heat extraction [94,99-102]. The cooling rate varies of course as a function of droplet or ribbon diameter and depth within the droplet or ribbon, and is perturbed by the release of latent heat in a way that depends on the undercooling that preceded the start of freezing. If $\Delta\theta > 1$, all of the latent heat can be absorbed by the casting without *reheating* to the liquidus temperature, T_L and without need for any external heat extraction during the actual solidification process. This type of heat sink is very effective in causing high solidification velocities especially in long drop tubes. Unfortunately, an a priori knowledge of the nucleation temperature of an alloy is as difficult to determine as the heat transfer coefficient. One important general trend, however, is that: the increase in initial undercooling will completely decrease particle/grain size [92,103].

1.4.2.3 Liquid Undercooling

The physical shape of rapidly solidified samples is clearly of importance in evaluating the overall thermal conditions for the different processing methods and sample configurations. However, it is of perhaps equal importance to consider the effect of sample geometry on the extent of melt undercooling at the onset *temperature of nucleation T_N*; the temperature at which a *drastical increase* in nucleation rate occurs, with rapid growth. For example, with powders the production method and sample shape are known to have an important role in allowing for the isolation of the most active internal nucleation sites to a small fraction of the powder population [92,102]. In this case, regardless of the imposed cooling rate, the powder configuration and associated melt subdivision are conducive to high melt undercooling. For a melt stream or surface melt layer configuration, the geometric shape does not appear to have a direct influence on physical nucleant isolation by melt subdivision. However, the *thermal transport* and *solute redistribution conditions*, accompanying the rapid progression of a solidification front under a high rate of heat extraction to the underlying substrate, may be shown to lead to a localization of nucleant influence, so that melt undercooling is obtained for a sufficient time interval to allow for the competitive development of structural modifications. Based upon experimental experience, an effective approach that may be applied to obtain a large melt undercooling involves the cooling of a collection of fine liquid metal droplets [104]. By dispersing a liquid sample into a large number of small independent drops with sizes from 5-40 μm, only a small fraction of these drops will contain *potent nucleants* so that the majority of the drops can display a large undercooling. Past experience with the droplet method has identified a number of processing parameters that govern the optimization of undercooling in powder samples. These processing variables include droplet size refinement, droplet surface coating catalysis, uniformity of coating, melt superheat, cooling rate [92,104-107], alloy composition [104], and applied pressure [108]. The key responses to changes in these processing variables are highlighted in Table **1**. Even when these processing conditions are arranged to produce maximum undercooling, most experience [72,91,109] suggests that solidification is still initiated by a *heterogeneous nucleation* site(s) associated with the sample surface, even in the extremely evacuated experimental systems. Therefore, it appears that a close attention to the nature of the powder surface coating is of prime importance in achieving reproducible, large undercooling values in fine powders [102].

1.4.2.4 Heat Transport During Atomization and Melt Spinning

The interplay of initial undercooling and external heat transfer is clearly shown in the schematic representations of the two melt-spinning conditions given in Fig. **10**. In (**a**), the liquid-solid interface is shown to begin very close to the point where the liquid first comes into contact with the wheel. In (**b**), liquid exists for some distance or time in contact with the wheel without nucleation [95,102]. During this time the liquid can become undercooled and the liquid-solid interface, once nucleated at the wheel surface, will move at a higher rate than would the interface shown in (**a**). That is: a combination of high initial undercooling and high heat transfer coefficient promotes the *most* rapid solidification [110,111]. Some authors [112] argue that when the ribbon thickness depends inversely on the square root of the wheel speed, the cooling is nearly *ideal* ($h \rightarrow \infty$). However, several other mechanisms, which can contribute to ribbon formation including growth into an undercooled melt or growth of a momentum boundary layer, would also result in the same square root relation between ribbon thickness and wheel speed. Theoretical details of the relative roles of the magnitudes of the heat transfer coefficient and of the initial undercooling have been examined for powders by Levi and Mehrabian [113], and for quenching against a substrate by Clyne [101,102]. A major complication occurs for heat flow analysis, even for pure materials when high solidification velocities are involved. The temperature of the interface T_I cannot be treated as a constant equal to the melting point T_m, but rather is a function of the interface velocity and nature. As a result, the heat flow analysis depends on the details of this function that can be very complicated, especially for dendritic growth and/or for alloys. Levi and Mehrabian as well as Clyne treated the case of a pure metal freezing with a smooth (*nondendritic*) liquid-solid interface governed by a kinetic law for the interface velocity, V, given by $V = \mu(T_m - T_I)$, where μ is the *linear interface attachment coefficient*. Even for pure metals, the value of the coefficient is not certain. For aluminum, Levi and Mehrabian have used a range of values for μ between 2 and 50 cm s^{-1} °C^{-1} while Clyne has used 4 cm s^{-1} °C^{-1} [95,101,102].

The model of *collision-limited growth* [114] would suggest a value for μ near 300 cm s^{-1} °C^{-1} for Al. Despite this uncertainty, these calculations show the importance of the initial undercooling, ΔT, on the development of high solidification velocities. Fig. **11**, adapted by Levi and Mehrabian, shows the interface temperature and interface velocity as solidification proceeds from one side of a powder particle to the other (an increasing *fraction solid*). The curves show the case of an initial dimensionless undercooling of 0.5 (~ 182 °C for Al) for various values of the heat transfer coefficient h. The velocity starts at a high value (> 3 m s^{-1}) and slows as the interface crosses the particle. This slowing is due to the evolution of the latent heat at the liquid-solid interface and the resultant reduction in the interface undercooling. The effect of changing the heat transfer coefficient by two orders of magnitude affects primarily the velocity after the fraction solid exceeds the dimensionless initial undercooling (0.5 in this case). Growth for small fraction solid is controlled by internal heat flow, while growth for large fraction solid is controlled by external heat flow.

If no initial undercooling occurred, the growth velocities, for the entire particle, would be near those seen at high fraction solid (Fig. **11**), which are typically around 10 cm s^{-1}. Similar effects of initial undercooling are shown by Clyne [101] for quenching against a substrate (as in melt-spinning quenching). Fig. **12** shows temperature-time plots at two positions inside a 50-µm-thick layer of an Al melt in contact with a substrate with $h = 10^2$ W cm^{-2} °C^{-1}. For example, at a position 5 µm from the substrate, the temperature in the liquid *drops* until nucleation occurs at the substrate at $\Delta\theta = 0.38$, which is calculated from the *homogeneous nucleation theory* [8,10,102]. The temperature at this position then rises rapidly as the liquid-solid interface proceeds from the substrate to the 5-µm position. Thereafter, the temperature at this position falls slowly until the interface reaches the 50-µm position (end of solidification) and then falls more rapidly. A similar plot for the 50-µm position shows a small initial drop in temperature followed by a rise as the interface approaches. The higher temperature, when the interface passes the 50-µm position compared to the 5-µm position, shows the *slow down* of the liquid-solid interface across the sample thickness. As described above for an Al-melt, much slower solidification rates occur in samples initially undercooled by the same amount.

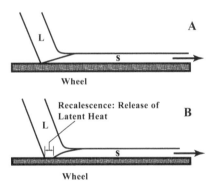

Fig. (10). Schematic representation of the liquid-solid interface position during melt spinning if nucleation occurs near the wheel surface. In (**a**); nucleation occurs at a temperature near the alloy liquidus. In (**b**); nucleation occurs at large undercooling (that is, solid-liquid interface is displaced downstream). Vertical distances are exaggerated for clarity [95].

1.4.3 Thermodynamics of Rapid Solidification Processing

Although rapid solidification is widely considered as a *nonequilibrium* process, it is clear that different degrees of departure from *full equilibrium* occur and constitute a *hierarchy*, which is followed with increasing solidification rate. This hierarchy is shown in Table **2** [95,102]. The conditions required for global equilibrium (I) are usually obtained only after *long-time annealing*. *Chemical potentials* and temperature are uniform throughout the system.

Under such conditions, no changes occur with time. During solidification or other phase transitions, gradients of temperature and composition must exist within the phases. However, one may often accurately describe the overall kinetics using diffusion equations to describe temperature and composition within the phases and using the phase diagram to give the possible temperature and compositions for boundaries between the phases.

The *Gibbs-Thomson effect* is included to determine shifts in equilibrium grain shape due to the grain boundary curvature caused by the solute supersaturation, volume-dependant and concentration-dependant chemical potential. This is called local equilibrium (II). The approach is never strictly valid, but it is based on the notion that interfaces will equilibrate much more quickly than will bulk phases. The conditions present in (II) are widely used to model the majority of solidification and phase transformations excluding massive (or; *partitionless*) and *martensitic* transformations. For example, under the assumptions of fast diffusion in the liquid phase, very slow diffusion in the solid phase and local equilibrium at the interface, the *Scheil equation* describes quite accurately the *non-equilibrium coring* or *microsegregation* in conventional castings [8,10]. Metastable equilibrium (III) can also be used locally at interfaces and is important in ordinary metallurgical practice. The change of cast iron from the stable gray form (*austenite* and *graphite*) to the metastable white form (*austenite* and *cementite*) with increasing solidification rate (and interface undercooling) is a familiar example [115]. The eutectic temperature and composition for white cast iron is a measurable thermodynamic transformation temperature as is any stable eutectic temperature. *Metastable equilibrium occurs when the free energy is a minimum (but not an absolute minimum) with negative change in free energy* [10,102]. It is often thought of as a *constrained equilibrium*. The constraint for the existence of this metastable equilibrium between liquid, austenite and cementite is the absence of graphite. Graphite can be absent because its nucleation or growth is difficult. Subject to this constraint, the free energy is minimized and a large fluctuation (or; nucleation) is necessary to reach the stable equilibrium. When solidification is complete, a two-phase mixture of austenite and cementite can exist in a global metastable equilibrium. In general, local metastable equilibrium is important during rapid solidification because many equilibrium phases, especially those with complex crystal structures, seem to have difficulties with nucleation and/or growth. Hence, *metastable phase diagrams* are important in describing interface conditions for many rapid solidification processes [8,95,116]. Significant loss of *interfacial equilibrium* (IV in Table **2**), whether for a stable or a metastable phase, is thought to become important for simple crystalline phases when the crystal growth rate exceeds the *diffusive speed* of solute atoms exchanging between the liquid and solid; V_D. An upper bound on this diffusive speed is (D/a_o), where D is the liquid diffusion coefficient and a_o is the *interatomic dimension; the atomic jump,* with a jumping frequency $\nu_{SL} = D/a_o^2$. Experiments on doped Si [117] and on metallic alloys [118] have shown that, for crystal growth rates > 1 m s^{-1} (rapid growth rates), significant *interfacial nonequilibrium* effects exist and solute is trapped into the freezing solid at levels exceeding the existing equilibrium solubility. Thus, the chemical potential of the solute increases upon being incorporated in the freezing solid in a process called *solute trapping*. This increase in chemical potential of the solute across the interface must be balanced by a decrease in chemical potential of the solvent in order for crystallization to occur; *i.e.*, to yield a net decrease in free energy. The free energy change during solidification, ΔG, is given by:

Table 1. Undercooled melt processing: powders [95].

Parameter	Undercooling response	Remarks
Powder size	Increased ΔT with size refinement at constant $\dot{T}$	Nucleant isolation follows
Poisson statistics		
Powder coating	Function of coating structure and chemistry; major effect in limiting undercooling of the finest powders	Most effective coating is catalytically inert; ΔT in nucleation is usually heterogeneous
Cooling rate	ΔT generally increases with increasing $\dot{T}$	Changing $\dot{T}$ can alter the operative nucleation kinetics
Melt superheat	System specific	Appears to be related to coating catalysis
Alloy composition	T_N follows trend of T_L	Melt purity not usually critical; near glass transition T_N decreases rapidly
Pressure	T_N parallels melting curves trend	Change in response can signal alternate phase formation

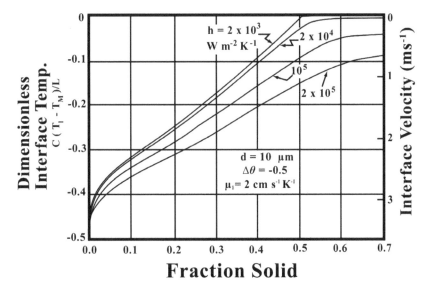

Fig. (11). Calculated interface temperature [113] and interface velocity for the solidification of a powder particle initially undercooled by 1/2(L/C). The temperature rises and the velocity falls as growth proceeds from the point of nucleation on the powder surface across the powder particle. The effect of various values of the heat transfer coefficient *h* is also shown [95,102].

Table 2. Hierarchy of Thermodynamic Equilibrium [95,102].

I.	**Full Diffusional (Global) Equilibrium** 1. No *chemical potential* gradients (compositions of phases are uniform). 2. No temperature gradients. 3. Lever rule applicable.
II.	**Local Interfacial Equilibrium** 1. Chemical potential for each component continuous across the interface. 2. Phase diagram gives compositions and temperatures only at Solid-Liquid interface. 3. Correction made for interface curvature (*Gibbs-Thomson Effect*).
III.	**Metastable Local Interfacial Equilibrium** 1. Important when stable phase cannot nucleate or grow fast enough. 2. Metastable phase diagram (a true thermodynamic phase diagram missing the stable phase or phases) gives the interface conditions.
IV.	**Interfacial Non-Equilibrium** 1. Phase diagram fails to give temperature and compositions at interface. 2. Chemical potentials are not equal at interface. 3. Free energy functions of phases still lead to criteria for the *impossible* [a].

(a) See: Baker and Cahn [119].

$$\Delta G = [\, (\mu_S^A - \mu_L^A)(1 - C_S^*) \; - \; (\mu_S^B - \mu_L^B) C_S^* \,] \ldots \tag{1-8}$$

where μ_S^A and μ_S^B are the chemical potentials for species A and B in the solid, μ_L^A and μ_L^B are the chemical potentials in the liquid, respectively. These potentials are functions of the temperature and solid or liquid composition (C_S^* or C_L^*) *at the interface during solidification*. Despite the loss of interface equilibrium during rapid solidification, the free-energy functions of the solid and liquid phases (and their associated chemical potentials) can be used to restrict the range of compositions that can exist at the interface at various temperatures, as shown by Baker and Cahn [102,119]. This restriction is obtained by the requirement that: only those processes that reduce the free energy of the system are possible (ΔG must be negative). Fig. **13** shows the regions of allowable solid compositions at the interface for a fixed interfacial liquid composition and for various interfacial temperatures [120]. Such allowable regions can be calculated from a thermodynamic model of the interesting system, as we will see later.

1.4.3.1 T_o Curves

Metastable equilibrium is a true thermodynamic equilibrium in the sense that the free energy is minimized [119,121]. Chemical potentials of the components for each phase involved in the

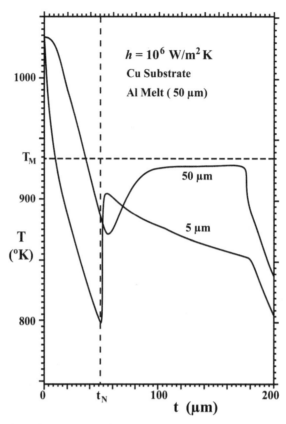

Fig. (12). Calculated temperature time histories [101] for two positions within a liquid layer 5 and 50 μm from a chilling substrate. The nucleation is assumed to occur at the substrate surface and is estimated to occur at an undercooling of ~ 0.4(L/C). The recalescence after the passing of the liquid-solid interface is evident at both positions [95].

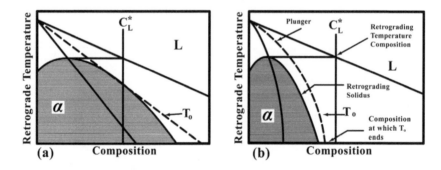

Fig. (13). In (**a**); Regions of thermodynamically allowed solid composition that may be formed from liquid of composition C_L^* at various temperatures. The value of T_o is the highest temperature at which partitionless solidification can occur. In (**b**); the T_o temperature plunges and partitionless solidification ($C_L^* = C_S^*$) is impossible for liquid of composition C_L^* [120].

metastable equilibrium are equal. In Table **2**, however, another situation is described in (IV), and relates to a situation where chemical potentials are not equal across an interface growing at a high rate and large undercooling. These rapid growth rates can trap the solute into the freezing solid at levels exceeding the equilibrium value for the corresponding liquid composition present at the interface [10,95,102].

The increase in chemical potential of the solute across the interface must be balanced by the decrease in chemical potential of the solvent in order for crystallization to occur; that is, to yield a net decrease in free energy [119]. To achieve this, the interface temperature must lie significantly below the liquidus temperature, whether stable or metastable. For any selected pair of liquid and solid compositions, a thermodynamic temperature can be described that is the highest temperature where crystallization can occur as shown in Fig. **13**. However, one often considers a limiting case, called *partitionless solidification*, which is favored at very high solidification rate, where the composition of the solid formed at the interface, C_S^*, equals the composition of the liquid at the interface, C_L^*. The T_0 temperature is the highest temperature where this can occur [122,123]. This is the temperature where the molar free energies of the liquid and solid phases are equal for the composition of interest; *i.e.*, the temperature where ΔG = 0 for $C_S^* = C_L^*$ in Eq. (1-8). As illustrated in Fig. **13**, a T_0 curve represents only part of the thermodynamic information available when solidification occurs without local equilibrium. T_0 curves exist for the liquid with stable or metastable phases, and lie between the liquidus and solidus for those phases. In fact, for dilute alloys, the slope of the T_0 curve is $m_L[(\ln k_o)/(k_o-1)]$, m_L is the *liquidus slop* and k_o is the *equilibrium partition coefficient*. Fig. **14** shows schematically, possible T_0 curves for three eutectic phase diagrams [120]. An important use of these curves is to determine whether a bound exists for the extension of solubility by rapid melt quenching or solubility is bound. If the T_0 curves plunge to very low temperatures as in Fig. **14a**, single phase α or crystals with composition beyond their respective T_0 curves cannot be formed from the melt. In fact, for phases with a *retrograde solidus*, the T_0 curve plunges to absolute zero at a composition no greater than the liquidus composition at the *retrograde temperature*, thus *placing a bound on solubility extension* [97]. Experiments on laser-melted doped Si alloys seem to confirm this bound [98]. Eutectic systems with *plunging* T_0 curves are good candidates for *easy metallic glass formation*. An alloy in the center of such a phase diagram can only crystallize into a mixture of solid phases with different compositions, regardless of the departure from equilibrium. The diffusional kinetics of this separation from the liquid phase frequently depresses the solidification temperature to near the glass transition T_g where an increased liquid viscosity effectively halts crystallization.

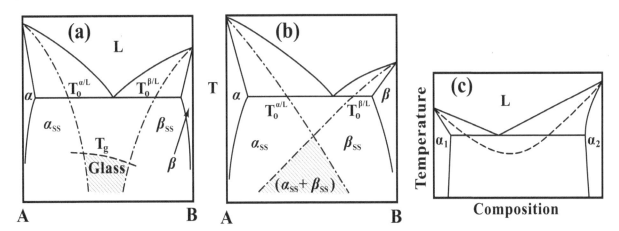

Fig. (14). Schematic representation of T_0 curves (dashed lines) for liquid to solid transformations in three types of eutectic systems [110,120]. The subscript (*ss*) refers to the *supersaturated* solid solutions.

In contrast, alloys with T_0 curves which are only slightly depressed below the stable liquidus curves, as in Fig. **14b** and **14c**, make good candidates for solubility extension and unlikely ones for glass formation. In Fig. **14b**, the crystal structures of α and β are different and the T_0 curve crosses, whereas in Fig. **14c** the crystal structures are the same and the T_0 curve is continuous across the diagram. At temperatures and liquid compositions below the T_0 curves, partitionless solidification is thermodynamically possible. Ni-Cr and Ag-Cu are examples of the behavior in Fig. **14b** and **14c** [95,102].

1.4.3.2 Nucleation of Metastable Phases

The definition of equilibrium is given by $dG = 0$ and can be illustrated graphically as in Fig. **15**. If it were possible to evaluate the free energy of a given system for all conceivable configurations, stable equilibrium configuration would be found to have the lowest free energy. As illustrated in Fig. **15**, various atomic configurations can be represented by points along the abscissa. Configuration **A** would be the stable equilibrium state. At this point, small changes in the arrangement of atoms to a first approximation produce no change in *G*. However, there will always be other configurations, **B**, which lie at a *local minimum* in free energy and therefore satisfy the condition d*G* = 0 as well, but which do not have the lowest possible value of *G*. Such configurations are called *metastable equilibrium states* to distinguish them from

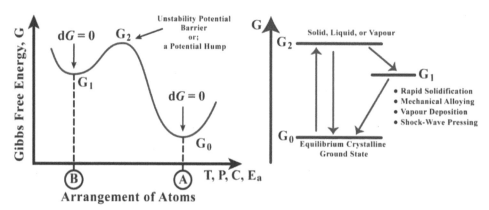

Fig. (15). The general schematic variation diagram of Gibbs free energy with the arrangement of atoms. Configuration *A* is the lowest free energy and is therefore the arrangement when the system is at stable equilibrium. Configuration *B* is a metastable equilibrium. The abscissa may be *T*, *P*, *C* or E_a. *T* is the temperature, *P* is the pressure, *C* is the concentration and E_a is the acquired energy (*e.g.*; thermal, radioactive, mechanical … *etc*). Transformation of a material from its specific ground state to a metastable equilibrium state is the "*energize and quench*" route for synthesizing metastable materials. G_0 and G_1 are potential wells while G_2 is a potential barrier or *hump* [1,6,126,127].

the stable equilibrium states. The intermediate states for which d*G* ≠ 0 are unstable and they are only ever realized momentarily in practice. If, as the result of thermal fluctuations, the atoms become arranged in an intermediate state, they will rapidly arrange into one of the free energy minima. If by a change of temperature or pressure, for example, a system is moved from a stable to a metastable state, it will, given time, transform to the new stable equilibrium state. Taking the definition of a metastable state to be *any state which has a free energy in excess of that of stable equilibrium*, it is then natural to quantify the degree of metastability by the value of that excess. Any transformation that results in a decrease in Gibbs free energy is thermodynamically possible such that the *activation energy* (per mole) of the reaction $\Delta G = G_2 - G_1$ < 0, where G_1 and G_2 are the free energies of the initial and final states respectively [1,6,10]. A state of unstable equilibrium is destroyed by any fluctuation, however small. An example would be a precipitate dispersion in which all the precipitates have exactly the same size and there are unstable states and they are not in equilibrium at all. The transformation need not go directly to the stable equilibrium state but can pass through a whole series of intermediate metastable states. Sometimes metastable states can be very short-lived, at other times they can exist almost indefinitely as in the case of diamond at room temperature and pressure. The reason for these differences is the presence of the *free energy hump* between the metastable and stable states, as in Fig. **15**.

The study of transformation rates in physical metallurgy belongs to the realm of thermodynamics. In general, higher humps or energy barriers lead to slower transformation rates. Kinetics obviously plays a central role in the study of phase transformations and many examples of kinetic processes will be found throughout this thesis. The microstructural manifestations of the departures from equilibrium achieved by novel processing routes [1,19,96,126-135] can be classified under five headings:

1. *Increased defect concentrations*: These include increased concentrations of vacancies (and to a lesser extent; interstitials), dislocations, stacking faults, twin boundaries and grain boundaries. Another characteristic to be put in this category is an increased level of chemical disorder (that is, reduced order parameter) in ordered solid solutions and compounds, because there is no discontinuity in the metastable phase nucleation in any metastable phase diagram [8].

2. *Microstructural refinement*: This involves finer scale distributions of different phases and of solute due to the shorter diffusion distances. Examples include finer dendrite arm spacings, eutectic spacings and precipitate diameters, forming submicro- and nanostructured metallic structures.

3. *Extended solid solubility*: A stable crystalline phase may be found with solute levels beyond the solubility limit at the ambient temperature, or indeed beyond the equilibrium limit at any temperature, *i.e.*; a supersaturated phase state.

4. *Metastable phases*: A metastable phase may form during processing. It may be one that is not found in equilibrium under any conditions in the system of interest, or it may be so found but appear metastably under different conditions of composition, temperature or pressure. This heading includes crystalline phases, quasicrystalline and intermetallic compounds.

5. *Metallic glasses or amorphous metals*: These are a special case of a metastable phase. When the processing is by cooling a liquid, they form not by a phase transformation but by a *continuous congealing* of that liquid.

For simplicity, we will consider the case of pure liquids. Nucleation during solidification can be defined as the formation of a small crystal from the melt that is capable of continued growth. From a thermodynamic point of view, the establishment of a solid-liquid (S-L) interface is not very easy. Although the solid phase has a lower free energy than the liquid phase below melting temperature, T_m, a small solid particle is not necessarily stable because of the free energy associated with the S-L interface. The change in free energy corresponding to the liquid-solid transition must therefore include not only the change in free energy between the two phases but also the free energy of the S-L interface itself [1,95,102]. From a kinetic point of view, it is possible to arrive at the same result on the basis that the atoms at the surface of a very small crystal have a higher energy than the surface atoms of a larger crystal [5]. Therefore, the equilibrium temperature at which atoms arrive and leave at the same rate is lower for a very small crystal than for a larger one. Consequently, for each temperature below T_m, a solid particle can be in equilibrium with the liquid when its radius of curvature has a particular value, known as the *critical radius*. Because at higher supercooling there is more bulk free energy to compensate for the surface free energy, the critical radius decreases with increasing supercooling [102].

On the other hand, at any supercooling, there exists within the melt a statistical distribution of *atomic clusters* or *embryos* of different sizes having the character of the solid phase (a nucleus is greater than an embryo; an embryo is a cluster of atoms of specific solid phase). The probability of finding an embryo of a given size increases as the temperature decreases. Nucleation occurs when the supercooling is such that there are sufficient embryos with a radius larger than the critical radius [10,102,136].

Finally, it is worthy to note that any metastable structured material has its own validation date and usage like traditional food or any custom product! Do not worry; *the validation date is of order of millions of years*. Though the structure of quench-hardened steel is unstable, yet steels hardened in ancient times retain their structures to our days. Thermodynamically, the equilibration reaction for a metastable phase takes a very huge time to complete if the material is stored, for example, at room temperature. According to Cottrell [6], the speed of this reaction depends on (*i*) the number n of atoms in the metastable position; (*ii*) the frequency v with which an atom in this position vibrates against the potential barrier and attempt to scale it; (*iii*) the statistical probability $p(\Delta Q,T) = \exp(-\Delta Q/kT)$ which is small for solids. Thus, the flux of reacting atoms is $nv \exp(-\Delta Q/kT)$ and rate of reaction is $A_0 \exp(-\Delta Q_m/RT)$, A_0 is a material-dependant constant containing nv, ΔQ_m is the activation energy per mole, $R = N_A k$, N_A is Avogadro's Number, k and R are Boltzmann and gas constants, respectively. The profound effect of temperature on the reaction rate is a result of the aforementioned exponential relations. Let $R \sim 8.4$ J °C^{-1} per mole and $\Delta Q_m = 167500$ J mol^{-1}, then;

1. At room temperature; $T = 300$ °K (27 °C) and $\exp(-\Delta Q_m/RT) \sim 10^{-29}$

2. At $T = 1000$ °K (727 °C); $\exp(-\Delta Q_m/RT) = 10^{-9}$

So, the reaction at 1000 °K has 10^{20} times the speed it has at 300 °K, *i.e.*; if it takes 1 second at 1000 °K it will take 10^{22} seconds or ~ 3 x 10^{12} years at room temperature ! So do not worry. Such behaviour is the basis of the common metal quenching practice that preserves the metastable structure at room or operating temperatures [6]. However, one must be careful when choosing a characteristic material for characteristic operating conditions and environment in order to keep the physical and chemical properties of the material unchanged.

1.4.3.3 Calculation of the Critical Radius and Energy Barrier

The change in the *homogenous free energy* per unit volume, ΔG, to form a solid embryo of spherical shape of radius r from liquid of a pure material involves the variation of the volume free energy and the surface free energy associated with the S-L interface and is given by:

$$\Delta G = \Delta G_v + \Delta G_i = -\frac{4}{3}\pi r^3 \frac{L\Delta T}{T_m} + 4\pi \gamma_{SL} r^2 \ ... \tag{1-9}$$

where ΔG_v is the change in free energy on solidification associated with the volume (= $\Delta G/V_m$, V_m is the molar volume) and ΔG_i is the free energy associated with the interface, γ_{SL} is the S-L interfacial free energy, L is the latent heat per unit volume and ΔT is the supercooling [1,10,102]. The critical radius of an embryo, r^*, occurs when ΔG has a maximum given by the condition, (d(ΔG)/dr = 0), as;

$$r^* = \frac{2\gamma_{SL} T_m}{L\Delta T} \ ... \tag{1-10}$$

Fig. **16**, due to Kurz and Fisher [10], gives a comprehensive picture of the variation of the free energy of an embryo as a function of its radius and ΔT. (a) At temperatures T greater than T_m , both ΔG_v and ΔG_i increase with r. Therefore the sum ΔG increases monotonically with r. (b) at the melting point T_m , ΔG_v = 0 but ΔG_i still increases monotonically. In (c), below the equilibrium temperature, the sign of ΔG_v is negative because the liquid is metastable (hypercooled melt) while the behavior of ΔG_i is the same as in (a) and (b). At large values of r, the cubic dependence of ΔG_v dominates over ΔG_i and ΔG passes through a maximum at the critical radius, r^*. When a thermal fluctuation causes an embryo to become larger than r^*, growth will occur as a result of the decrease in the total free energy. Nucleation in a homogeneous melt is called *homogeneous nucleation* and from Eq. (1-9) the critical energy of activation for an embryo of radius r^* is given by [1,5,8,10,102];

$$\Delta G^* = \frac{16}{3}\pi \frac{\gamma_{SL} T_m^2}{L^2 \Delta T^2} \ ... \tag{1-11}$$

The unlikelihood that statistical fluctuations in the melt can create crystals with a large radius is the reason why nucleation is so difficult at small values of the supercooling. Thus, homogeneous nucleation is only possible for high supercooling (on the order of 0.25 T_m) according to the *Hollomon-Turnbull Theory* [136].

However, small contamination particles in the melt, oxides on the melt surface or contact with the walls of a mould *may catalyze nucleation* at a much smaller supercooling and with fewer atoms required to form the critical nucleus. This is known as *heterogeneous nucleation*. In Fig. **17**, homogeneous and heterogeneous nucleation is compared for a flat catalytic surface and isotropic surface energies. For this simple case, the embryo is a spherical cap that makes a contact angle θ with the substrate given by;

$$\gamma_{cL} - \gamma_{cS} = \gamma_{SL} \cos\theta \ ... \tag{1-12}$$

where γ_{cL} is the *catalyst-liquid interfacial free energy* and γ_{cS} the *catalyst-solid interfacial free energy*. At a supercooling ΔT, the critical radius of the spherical cap is again given by Eq. (1-10), but the number of atoms in the critical nucleus is smaller than that for homogeneous nucleation, as a consequence of the catalytic substrate. Indeed, *the thermodynamic barrier to nucleation, ΔG^**, is reduced by an angular (shape) factor $f(\theta)$ to [137-139] :

$$\Delta G^* = \frac{16}{3}\,\pi\,\frac{\gamma_{SL}^3\,T_m^2}{L^2\,\Delta T^2}\,f(\theta)\;\dots \tag{1-13}$$

where

$$f(\theta) = \frac{(2+\cos\theta)(1-\cos\theta)^2}{4}\;\dots \tag{1-14}$$

and

$$\gamma_{SL} = \alpha_S\,\frac{\Delta S_f}{N_A^{1/3}\,V_m^{2/3}}\,T\;\dots \tag{1-15}$$

Eq. (1-14) is for a planar catalytic site and the nucleus is spherical and $f(\theta)$ is a *catalytic potency factor* in the case of heterogeneous nucleation. ΔS_f is the entropy of fusion, N_A is Avogadro's number, α_S is a factor depending on the structure of the nucleus with numerical values $\alpha_S = 0.71$ for BCC, $\alpha_S = 0.86$ for a FCC or HCP crystal structures, respectively. If nucleation occurs in a scratch or a cavity of the catalytic substrate, the number of atoms in a critical nucleus and the value of ΔG^* can be reduced even more. For a planar catalytic surface, the reduction in the free energy barrier compared to that for homogeneous nucleation depends on the contact angle. Any value of θ between $0°$ and $180°$ corresponds to a stable angle. When $\theta = 180°$, the solid does not interact with the substrate, $f(\theta) = 1$ and the homogeneous nucleation result is obtained. When $\theta = 0°$, the solid *wets* the substrate, $f(\theta) = 0$, and $\Delta G^* = 0$. Thus, solidification can begin immediately when the liquid cools to the freezing point. From the classical heterogeneous nucleation point of view, a good nucleant corresponds to a small contact angle between the nucleating particle and the growing solid [1,5,10,95,102,137]. According to Eq. (1-12), this implies that γ_{cS} must be much lower than γ_{cL}. However, in general the values of γ_{cL} and γ_{cS} are not known and, therefore, it is rather

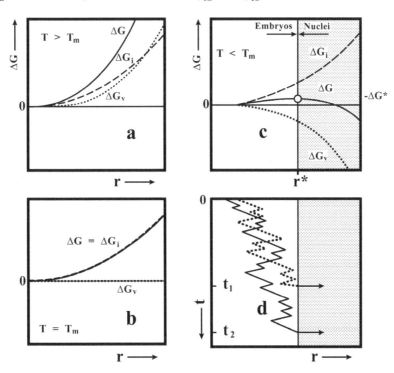

Fig. (16). Volume, surface and total values of the free energy of a crystal cluster of new phase in a supersaturated matrix, as a function of radius, r, at three temperatures: (**a**) $T > T_m$, (**b**) $T = T_m$ and (**c**) $T < T_m$. In (**d**), thermal fluctuations may move the cluster backwards and forwards along the $(\Delta G - r)$ curve due to the effect of random additions or removals of atoms to or from the unstable nucleus [10,102,137].

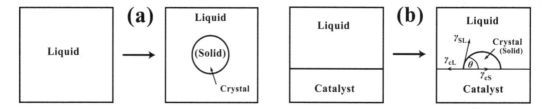

Fig. (17). Schematic comparison of (**a**) homogeneous and (**b**) heterogeneous nucleation of a spherical crystal in a supercooled liquid. The interface energies are assumed to be isotropic and in (**b**) the catalytic surface is assumed to be flat [1,5,102,137].

difficult to predict the potential catalytic effectiveness of a nucleant. Tiller [140] pointed out that there is no clear insight into what determines θ and how it varies with: (*i*) lattice disregistry between substrate and the stable phase, (*ii*) topography of the catalytic substrate surface, (*iii*) chemical nature of the catalytic surface and, (*iv*) absorbed films on the catalytic substrate surface [95].

1.4.3.4 Nucleation Rates

The basis of *the classical nucleation theory* is that when a new phase forms within a parent phase, an interface is formed between the two phases. The formation of the interface creates a local increase of free energy when the first atoms assemble in the new structure. At increasing values of undercooling, r^* is reduced ($r^* \propto \Delta T^{-1}$) and G^* is reduced more rapidly ($\Delta G^* \propto \Delta T^{-2}$). A cluster is often considered to reach the stage of a nucleus capable of continued growth with a decreasing free energy when the size r^* is achieved. But in fact, stable nucleus growth ensues when the cluster size exceeds r^* by an amount corresponding to ($\Delta G^* - kT$), as in Fig. **16** [137,141]. The relationship between cluster size and the number of atoms in a cluster, n^*, is expressed by $(n^* V_a) = (4/9)\pi r^{*3}$; where V_a is the atomic volume.

To relate cluster energetics and fluctuational growth to the rate of nucleation, it is necessary to describe the cluster population distribution. Because the mixture of clusters in an undercooled melt is a dilute solution, the entropy can be described in terms of an ideal solution. The metastable equilibrium concentration of clusters of a given size, $C(n)$, is then given by [10,141]:

$$C_n = C_L \, \exp\left(\frac{-\Delta G_n}{kT}\right) \, \ldots \tag{1-16}$$

where C_L is the number of atoms per cubic meter (or; mol m^{-3}) in the liquid and ΔG_n is given by Eq. (1-9) when r is converted to n, as noted before.

If solid nucleation is regarded as the growth of clusters past the critical size, then the resulting cluster flux or the homogenous nucleation rate I (in m^3 s^{-1}) can be represented kinetically by the product:

$$I = v_{SL} \, S^* \, C_n(n^*) \, \ldots \tag{1-17}$$

where; S^* is the number of atoms surrounding a cluster surface that is roughly ($4\pi r^{*2}/a_o^2$), and $C_n(n^*)$ is the concentration of critical clusters [137]. The full expression for the steady-state nucleation rate is then [142]:

$$I = \left(\frac{D}{a_o^2}\right)\left(\frac{4\pi r^{*2}}{a_o^2}\right) C_L \exp\left(\frac{-\Delta G_n^*}{kT}\right) \, \ldots \tag{1-18}$$

For typical metals, $C_L \sim 10^{28}$ m^{-3}, $D \sim 10^{-9}$ m^2 s^{-1} and $a_o \sim 0.3 \times 10^{-9}$ m, so we get:

$$I \simeq 10^{40} \exp\left(-\frac{16\pi \gamma_{SL} T_m^2 V_m^2}{3kTL^2 \Delta T^2}\right) \, \ldots \tag{1-19}$$

and shows a rather *steep temperature dependence*, as illustrated in Fig. **18**. At high temperatures, the temperature dependence of I is dominated by the driving free energy term, which is contained within the exponential dependence on the activation barrier, and I can vary by about a factor of five per degree Celsius. Eqs. (1-18) or (1-19) indicates that the maximum nucleation rate occurs at $(1/3)T_m$ [8,10,137,142].

In evaluating nucleation rates, accurate values for the activation barrier are important because of the exponential dependence of I on ΔG^*. The evaluation of ΔG_v is based on thermodynamic data or reasonably accurate models [143]. However, separate and independent measurements of solid-liquid interfacial energy γ_{SL} at the nucleation temperature are not available, and calculations based on model-dependent estimates of γ_{SL} are of uncertain accuracy. Similarly, the energetics of heterogeneous nucleation can be described by a modification of Eq. (1-9) to account for the different interfaces and the modified cluster volume involved in nucleus formation. In terms of the cluster formation shown in Fig. **16**, the free energy change during heterogeneous nucleation is expressed by:

$$\Delta G_{het} = V_{SC}\,\Delta G_v \;+\; A_{SL}\,\gamma_{SL} \;+\; A_{cS}\,\gamma_{cS} \;-\; A_{cL}\,\gamma_{cL}\;\cdots \tag{1-20}$$

where V_{SC} is the spherical cap volume and A_{SL}, A_{cS}, and A_{cL}, are the solid-liquid, nucleant-solid, and nucleant-liquid interfacial areas, respectively. For the heterogeneous nucleation, referring to Fig. **17** representing the *spherical cap model*, the shape factor $f(\theta)$ becomes [8,10];

$$f(\theta) = \frac{2 - 3\cos\theta + \cos^3\theta}{4}\;\cdots \tag{1-21}$$

and

$$\Delta G^*_{het} = \Delta G^*_n\, f(\theta)\;\cdots \tag{1-22}$$

In addition, the concentration of critical clusters is represented in terms of the number of surface atoms of the nucleation site per unit volume of liquid, C_a, which is of the order of 10^{20} m^{-3}. Therefore, by a similar way as in homogenous nucleation, the heterogeneous nucleation rate expression I_{het} (m^{-3} s^{-1}) is given by [8,137,142]:

$$I_{het} = f'(\theta)C_a \exp\!\left(-\frac{\Delta G^*_{het}}{kT}\right)\;\cdots \tag{1-23}$$

and

$$f'(\theta) = \left(\frac{2D\pi\,r^{*2}\,(1-\cos\theta)}{a_o^4}\right)\;\cdots \tag{1-24}$$

For typical metals;

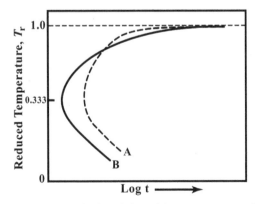

Fig. (18). Comparison between heterogeneous nucleation (A) and homogeneous nucleation (B) in terms of the relative transformation kinetics below the melting point. The reduced temperature $T_r = T/T_m$ and time $t \propto I^{-1}$ [95,109,137,144].

$$I_{het} = 10^{30} \exp\left(-\frac{16\pi \gamma_{SL}^3 T_m^2 V_m^2 f(\theta)}{3kTL^2 \Delta T^2}\right) \dots \tag{1-25}$$

Although only a single transformation curve (the C-shaped curve) is shown in Fig. **18** for heterogeneous nucleation, in reality there will be as many curves as the number of heterogeneous nucleation sites. Each curve for heterogeneous nucleation will be distinguished by a catalytic potency, $f(\theta)$, and a site density [144]. To attain homogeneous nucleation conditions, it is clear that all heterogeneous nucleation sites must be removed or *bypassed* kinetically.

A final treatment in this section is about the supersaturation. Supersaturation is the difference in concentration of solute atoms between the equilibrium phase and the supersaturated phase and drives the diffusional growth processes. This supersaturation (and the related undercooling ΔT) represents the driving force for the diffusion of solute at the dendrite tip in an alloy. In principal, when the supersaturation is equal to zero, the transformation rate $\dot{f}_S$ ($= df_S/dt$, f_S is the fraction solid and $f_S = 1 - f_L$) and nucleation rate will be zero. With increasing supersaturation, the growth rate of the new phase (*i.e.*; the solid phase) will increase [10,102]. At r^*, the growth rate is zero and therefore all of the supersaturation is used to create curvature of the embryo (the Gibbs-Thomson effect) and none remains to drive diffusive processes. For a hemispherical needle-like crystal dendritic tip, the solution of the diffusion equation shows that the dimensionless or relative supersaturation, Ω, is equal to the ratio of the tip radius to the diffusion boundary layer. This dimensionless ratio is known as the *Péclet number* of the diffusion equation through the S-L interface of the growing phase crystal. Also, $\Omega = \Delta T / \Delta T_o$, $k = \Delta C_S(T) / C^*$, ΔT_o is the temperature difference at liquidus-solidus interface ($= T_S - T_L$; solid phase has a different temperature than the liquid phase), k is the phase diagram partition coefficient, $\Delta C_S(T)$ is the supersaturation of the solute at any temperature T and C^* is the solute equilibrium solubility or composition at the S-L interface [10,102].

According to Lacmann [145], if the nucleus is very small, caused by high supersaturation and surface tension, it must be taken into account that the nucleus can only be built of an integer number of atoms. This aggregate should no longer be called a critical nucleus, since it is usually in a labile equilibrium with the mother phase (the supersaturated phase). For a given nucleus size, a corresponding supersaturation should exist. In the case of a finite range of supersaturation, the nucleation is determined by an aggregate of a constant number of n atoms. At higher supersaturation, it is then a smaller aggregate with fewer atoms. Supersaturation can be quite large at not too high temperatures and can grow beyond the solubility limit, resulting in a precipitation (the secondary phase nucleation) or grain coarsening. Then, equilibrium is reached when the supersaturation equals zero with a reduction in free energy, where the driving force per mole of precipitate, $\Delta F_{\alpha\beta}^m$, is proportional to $\ln\Omega$, F is the Helmholtz free energy. In addition, during the coarsening process the supersaturation ΔC_S decreases asymptotically towards zero with a reduction in Gibbs free energy.

The nucleation rate for primary nucleation I is given by [146,147]:

$$I = K \exp\left(-\frac{\Delta G_n^*}{kT\Omega^2}\right) \dots \tag{1-26}$$

and

$$I_{het} = K \exp\left(-\frac{\Delta G_n^*}{kT\Omega^2} f(\theta)\right) \dots \tag{1-27}$$

Therefore, $\ln I$ depends on Ω^{-2}, as in Fig. **19**. Here, there is also a linear dependence between $\ln I_{het}$ and $1/\Omega^2$ (Fig. **19b**). The course of $\ln I$-curve represented in Fig. **19a** and **19b** corresponds with reality, since with low supersaturations heterogeneous nucleation occurs, passing into homogeneous nucleation for higher supersaturations [145]. The reaction path of a supersaturated solid solution can be rather complex, sometimes involving the formation of one or more intermediate non- equilibrium phases prior to reaching the equilibrium two-phase microstructure. Unlike in the classical theories mentioned above, these complications, which are of practical relevance, can be taken into consideration in numerical approaches. Even though they still contain significant shortcomings, numerical simulations and solutions lead to a practical description of the kinetic course of a precipitation reaction that lies closely to reality [148]. For almost all diffusional reactions, the theory is in good agreement, at least qualitatively and semi-qualitatively, with a very large amount of experimental data. A major success of this theory is its ability to account for the observation that nucleation increases from a

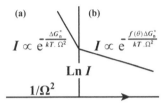

Fig. (19). Rate of nucleation depending on supersaturation [145].

rate that is almost undetectable to rates that are too fast to measure over a narrow range of undercooling ΔT or supersaturation. However, there are many situations of its failure, like in the nucleation of new grains in the recrystallization of a deformed material [149]. Also, the failure of the classical model of boundary nucleation is that it can not explain the stimulated nucleation at grain boundaries by grain boundaries defects like dislocations. In addition, there is a tremendous lack of reliable interfacial-energy information needed to make use of the theory [150] with the lack of experimental verification of heterogeneous nucleation. Moreover, foreign particles are present in every *real* solution or situation, leading to heterogeneous nucleation. According to experiments carried out with foreign particles added to the reactants, the nucleation work is reduced to approximately 10% of the predicted value for homogeneous primary nucleation. In fact, the aforementioned spherical cap model, with that pristine catalytic substrate, isn't realistic at all, because any real substrate has its own tribological topography that presents preferable heterogeneous nucleation sites (due to $f(\theta)$ and the increase in catalytic surface area) with another nucleus shapes rather than the simple spherical one [142,147,151,152]. Although many modifications, including capillarity, had been done to that model [153,154], the consideration of real surface topography and its influence on nucleation energetics has not been carried out in detail [142]. Even the more recent simulations of the deformation-induced solid-state crystallization, nano-crystallization or amorphization can capture some of the experimentally observed behavior [155,156], but due to the complexity of deformation-induced reactions, the simulations are often based upon simplifications such as limited system size, simplified interatomic potentials and two-dimensional arrangements, and so on [157-160]. There are also some basic thermodynamic and kinetic concepts on the initial stages of interfacial reactions that provide general guidance for interpreting the reaction pathway. For a general overview of the broad field of diffusive phase transformations in materials science including heterogeneous nucleation, solid-state induced transformations and discontinuous precipitation not covered in the present study, can be found in the articles by Doherty, Perepezko and others [137,142,161-164].

One of the major assumptions of the classical nucleation theory is that the free energy per unit volume and free energy per unit surface area are independent of the size of the embryo. Since the interface between solid and liquid is usually considered to be *diffuse* on the level of a few atomic dimensions, *embryos that are a few atomic dimensions in radius cannot be described classically*. This leads to a radius (or temperature) dependence of the surface energy as shown by Larson and Garside [165] and Spaepen [166]. Perepezko [95,167] has pointed out that if θ approaches zero for a heterogeneous nucleation process, the thickness of the spherical cap can approach atomic dimensions, even when the cap radius is much larger, a fact that would also necessitate a *non-classical approach* to heterogeneous nucleation using quantum mechanics of atomic clusters; the atomistics of physical metallurgy. This must be conducted with the advanced high speed microscopic analysis tools for the nano-scale investigations (*e.g.*; 4-D Electron Microscopy, Dynamic TEM, High Resolution TEM (HRTEM) and dynamic high-intensity x-rays or neutron diffraction analyses) [102,142,163,168-174] in order to capture the real-time and real-space visualizations of the *fleeting changes* in structure and shape of matter, which is barely 1 nm in size during solidification or any phase transitions.

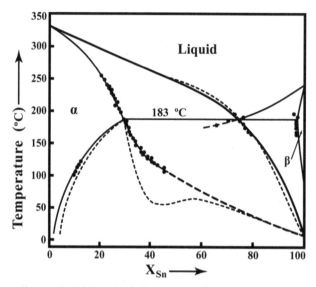

Fig. (20). The stable Pb-Sn phase diagram (solid line), including measured and calculated metastable (α + L) extensions (dashed line) of liquidus and solidus curves, along with experimental data points [176,181].

1.4.3.5 Formation of Metastable Phases by Supercooling

One of the most dramatic effects of large supercoolings prior to solidification is the possibility of forming metastable phases. When one or both of the stable solid phases in an eutectic or peritectic reaction are suppressed, metastable solid phases may develop and yield metastable eutectics and peritectics [15,95,137]. An elegant yet simple example occurs for pure Bi [175]. The bulk free energy change for solidification, ΔG_v in Eq. (1-8), is always largest for the stable phase. However, in the context of the heterogeneous nucleation theory, a metastable phase may make a smaller contact angle with a particular catalytic site than does the stable phase. Thus, the barrier for nucleation of a metastable phase may be smaller than the barrier for the stable phase. Of course, one must always supercool below the melting point of the metastable phase in order for ΔG_v for the metastable phase to be negative. Similarly, metastable phases have been formed in alloy systems. In fact, taking the Pb-Sn system as an example by avoiding the nucleation of the stable Sn phase, the metastable Pb liquidus and solidus curves have been measured more than 80K below the Pb-Sn equilibrium eutectic temperature, as shown in Fig. **20** [176]. When nucleation did occur in this supercooled state, a metastable phase was formed.

1.4.4 Formation of Quasi-Crystals

Also called quasi-periodic crystals, matter formed atomically in a manner somewhere between the amorphous solids of metallic glasses and the precise pattern of crystals. Like crystals, quasicrystals contain an ordered structure, *but the patterns are subtle and do not recur at precisely regular intervals* [11,177-180]. Rather, quasicrystals appear to be formed from two different structures assembled in a non-repeating array, the three-dimensional equivalent of a tile floor made from two shapes of tile and having an orientational order but no repetition, as in Fig. **21**. Although when first discovered such structures surprised the scientific community. It now appears that quasicrystals rank among the most common structures in alloys especially of aluminum with such metals as iron, cobalt, or nickel. While no major commercial applications yet exploit the unique properties of the quasicrystalline state directly, quasicrystals form in compounds are noted for their high strength, hardness, lightweight, optical properties and surface chemistry, suggesting many potential applications in aerospace and other industries.

1.4.4.1 Quasicrystalline Structures

D. Shechtman, a researcher from Technion, a part of the Israel Institute of Technology, and his colleagues at the National Bureau of Standards (now the *National Institute of Standards and Technology*; NIST) in Gaithersburg, USA, discovered quasicrystals in 1984 [177,178,182]. A research program of the U.S. air force sponsored their investigation of the metallurgical properties of aluminum-iron and aluminum-manganese alloys. Shechtman and his coworkers mixed aluminum and manganese in a roughly six-to-one proportion and heated the mixture until it melted. The mixture was then rapidly cooled back into the solid state by dropping the liquid onto a *chill spinning wheel*. When the solidified alloy was examined using an electron microscope, a novel structure was revealed. It exhibited *five-fold symmetry*, which is forbidden in crystals, and long-range order, which is lacking in amorphous solids. Its order, therefore, was neither amorphous nor crystalline. Many other alloys with these same features have subsequently been produced (see Figs. **22-27**). The original electron diffraction pattern of quasicrystalline aluminum-manganese published by Shechtman and his coworkers is shown in Fig. **22**. Rings of 10 bright spots indicate axes of five-fold symmetry, and rings of six bright spots positions. Recalling the earlier result that fivefold symmetry axes are forbidden in crystalline materials, a paradox is presented by quasicrystals. They have long-range order in their atomic positions, *but they must lack spatial periodicity*. The quasicrystalline state is usually obtained either by rapid quenching from the melt, by moderate annealing of an initially amorphous sample, or by solid-state reaction. *Icosahedral alloys* of the Al-Li-Cu system [183] were even obtained by conventional casting, indicating that the icosahedral phase may be an equilibrium phase in a certain temperature, pressure, and composition ranges. *Icosahedral quasicrystals* occur in-between many intermetallic compounds, including aluminum-copper-iron, aluminum-manganese-palladium, aluminum-magnesium-zinc, and aluminum-copper-lithium. Other crystallographically forbidden symmetries have been observed in many materials as well. Local icosahedral order is also postulated and confirmed experimentally in many glasses, and is evident in some complex crystal phases like that of crystalline boron, noble gases and C_{60} (the *buckminsterfullerene*) with atomic cluster scales reaching 147-atom per an icosahedral-cluster near 1 nm size (an i-nanocrystallite) like MacKay and Bergman clusters [184,185]. These include *decagonal* symmetry, which exhibits *ten-fold rotational symmetry* within two-dimensional atomic layers but ordinary translational periodicity perpendicular to these layers. Decagonal symmetry has been found in the alloy systems of aluminum-copper-cobalt and aluminum-nickel-cobalt. The *International Union of Crystallography* (lUC) now defines the term *crystal* as applying to *any solid having an essentially discrete diffraction diagram* (lUC 1992). The large family of solid-state crystals thus defined can then be divided into two classes:

1. Periodic crystals which exhibit a perfectly periodic structure at the atomic scale, *photonic crystals* are the photonic analogues of these crystals.

2. Aperiodic crystals.

Most compounds named thus far contain aluminum. Indeed, it appears that aluminum is unusually prone to quasicrystal formation, but there do exist icosahedral quasicrystals without it. Some, like gallium-magnesium-zinc, simply substitute the chemically similar element gallium for aluminum. Others, like titanium-manganese, appear chemically unrelated to aluminum-based compounds. Furthermore, some quasicrystals such as chromium-nickel-silicon and vanadium-nickel-silicon display *octagonal* and dodecagonal structures with *eight-fold* or *twelve-fold* symmetry, respectively, within layers and translational periodicity perpendicular to the layers. D. Holland-Moritz *et al.* [186] have made investigations on some quasicrystal alloys and proved that an *icosahedral short-range order* (ISRO) prevails in metallic melts, independent on the solid structure (Ni, Co: FCC; Fe, Zr: BCC; $Al_{65}Cu_{25}Co_{10}$: decagonal quasicrystal). This ISRO is observed even at temperatures above the melting temperature and gets more stably pronounced if the temperature is decreased to hypercooling limits. *These results confirmed a more than 50 years old prediction by Frank* [180].

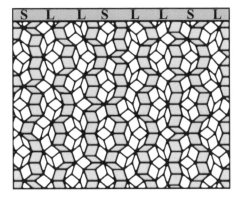

Fig. (21). Penrose pattern, discovered by the English mathematical physicist Roger Penrose, which is an example of a two-dimensional pattern that combines fivefold rotational symmetry, as also in their diffraction pattern, with quasi-periodic translational order. The plane is covered by rhombuses and tiles with parallel edges lie in rows (shaded) separated by large (L) and small (S) intervals in a *quasiperiodic translational order* in the form "LSLLSLLSLLSLLSL" as a Fibonacci sequence [107].

There is also a connection between the annealing of metglasses and *quasicrystallinity*. Bulk metallic glasses (BMGs) offer for the first time the possibility of studying the properties of the glassy and the supercooled liquid states on a temporal and spatial scale previously considered unattainable. A very unexpected result was that: on *devitrification* they produced nanocrystals, paving the way for bulk nanostructured materials. Some of them gave rise to quasicrystals; again, leading to bulk *nanoquasicrystalline* materials and alloys [187,188]. It also indicated a link with icosahedral order for at least some of these glasses. Current interest in this subject is captured in a series of books [189,190], reviews [191-196] and several conference proceedings [197-199]. *Nanoquasicrystallisation* has been observed in Zr-Ti-Ni, Zr-Al-Cu-Ni-Ag, Zr-Al-Cu-Ni-Pd [192,200] and various other hafnium-containing alloys. For Al-base alloys, the quasicrystals so precipitated are metastable, composing of icosahedral particles (= *I-phase* or *i-phase* according to references) and are embedded in a crystalline α-Al matrix [188,201]. This kind of microstructure is produced directly from the liquid state. The first report focusing on the study of the microstructure and its properties could be that from Inoue *et al.* [202] in 1992 on Al-Mn-Ce system. This kind of alloys was developed also in Al-Cr-RE systems and more recently in the Al-Fe-Cr-TM (with TM: Zr, V or Nb) [188,201,203,204]. These alloys have metastable phases, while the alloys from the latter system have a good stability of the microstructure that allows retaining a high strength at temperatures ~ 350 ℃ [203].

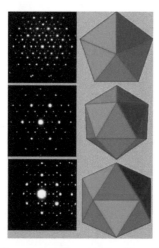

Fig. (22). (Left) Electron diffraction patterns of quasicrystalline Al-Mn alloy. (Top left) View is along the five-fold symmetry axis; (centre left) rotating by 37.38° reveals the three-fold axis; (bottom left) rotating by 58.29° reveals the two-fold axis. (Right) Corresponding views of icosahedrons that quasicrystalline symmetries match those of the icosahedrons [177,178,180].

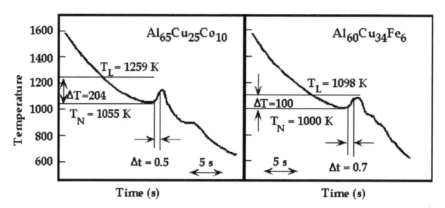

Fig. (23). Temperature-time profiles, obtained from levitation undercooling experiments on quasicrystal forming alloys. Left: $Al_{65}Cu_{25}Co_{10}$, forming a decagonal D-phase. Right: $Al_{60}Cu_{34}Fe_6$ forming an icosahedral i-phase [179].

Amorphous nanogranular Al-based alloys are an intermediate class of nanostructured alloys between nanostructured alloys with amorphous matrix and nanoquasicrystalline alloys. These alloys have a microstructure composed of granules of an amorphous phase with 2 to 5 nm sized embedded in an α-Al matrix [201]. In addition, these alloys are obtained directly from the liquid state, particularly in the Al-Fe-V and Al-Fe-Ti systems [201]. It was suggested that the *amorphous granules* have an icosahedral short-range order [188,201,205], which could be explained as an initial stage of a nanoquasicrystalline alloy. Zr-Ti-Ni is an interesting system to study because it forms nanoquasicrystals over a wide composition range and they are stable up to 4 hours of annealing at the peak temperature of the crystallization reaction [187]. Annealing appears to enhance the stability of the amorphous phase against thermally induced crystallization and same results were found for specimen hot-extruding or hot-rolling. Also, these quasicrystal precipitates were found to enhance the ductility, and then mechanical properties, of the metglass precursor material. Liu *et al* [206] have shown that dislocations are firstly piled up against the I-phase particles with nanosize, but the subsequent dislocations were annihilated at the interface between I-phase and matrix. This *dislocations annihilation effect* at the interface seems to have similar action as that for the passage of dislocations by the precipitation particle, because the dislocations in these situations can not play a role in the further increase in flow stress but flow strain. This mechanism gives a reasonable interpretation of high ductility of alloys with nanosized I-phase. Interfaces between the I-phase and matrix in Mg-Zn-Y alloys are reported to be quite strong [207]. The quasiperiodic property of the I-phase ensures the formation of matching and strong interfaces with the matrix by creating several spacings on a set of planes where the I-phase has tendency to facet on fivefold and twofold planes [208]. This interesting phenomena opened one inspiration that wrought Mg-alloy with high ductility can be obtained through introducing large volume fraction of I-phase with nanosize in the matrix.

1.4.4.2 Origin of Quasicrystallinity

Returning to Eqs. (1-13) and (1-15), the barrier for nucleation, *i.e.*; the *interfacial energy* γ_{SL}, depends on the structure of the nucleus. γ_{SL} is smaller for systems with similar short-range order (*i.e.*; amorphous) of the undercooled melt and the nucleus. Already Frank has pointed out [180] that an icosahedral short-range order should be energetically favoured in undercooled melts of metals and metallic alloys, suppressing any solidification event. Hence, the energy barrier between an undercooled melt and a nucleus of a crystallographic phase with polytetrahedral symmetry should be small in comparison to crystalline phases. The interfacial energy is often determined by measuring the maximum undercooling attainable for a melt [209]. Undercooling experiments on alloys, which form quasicrystalline phases, have been performed using the electromagnetic or aerodynamic levitation techniques [89,186,210,211]. Fig. **23** shows two temperature-time profiles obtained from experiments on an $Al_{60}Cu_{34}Fe_6$ alloy, which forms an icosahedral (I) phase (right), and on an $Al_{65}Cu_{25}Co_{10}$ alloy, which solidifies in a decagonal (D) phase (left). This sequence indicates that an ISRO is present in undercooled metallic melts, which favors the solidification of solid phases having polytetrahedral symmetry elements. The more pronounced the polytetrahedral symmetry, the lower the nucleation barrier and, hence, the maximum undercoolability. The present analysis is based upon the assumption of homogeneous nucleation. Atomization and drop-tube experiments on Al-Mn quasicrystal forming alloys [212,213] indicate that there is a high probability for quasicrystalline phases to be formed by homogeneous nucleation or heterogeneous nucleation of very small catalytic potency [$f(\theta) \approx 1$] in containerlessly-undercooled melts. If, nevertheless, heterogeneous nucleation is tentatively assumed, this would have the following consequences: anticipating the same catalytic potency $f(\theta)$ for all phases and the main conclusions remain unchanged. On the other hand, if the observed change of ΔG^* is exclusively attributed to a change in $f(\theta)$, an anomalous variation has to be postulated, *i.e.*; by a factor of 3 between the two quasicrystalline phases and even higher values between the icosahedral and crystalline phases [210].

For particles interacting *via* a central potential, such as the famous *Lennard-Jones (LJ) potential*, which approximate the potential energy of electrical forces (especially van der Waals forces) between two interacting atoms or spherical molecules [214-216], the fully relaxed energy of an icosahedral cluster is lower than the two close-packed crystalline configurations [180,217]. If a large population of atoms in the liquid phase has a local icosahedral order, this could explain why many liquids can be cooled below their melting temperature without crystallization; a LJ-fluid or glass [218-222]. To form the first crystal clusters, this stable configuration must be replaced by the higher-energy local configuration of the crystalline phase. Atomistically, the possibility of the icosahedral order in the liquid was first suggested for a single component Lennard-Jones liquid, in reality it was found in complex alloy systems in the form of the quasicrystal. This is because the icosahedral order is geometrically frustrated in a single component liquid, and it can be achieved only by introducing atomic size differences (*e.g.*; r(Mn) = 1.32 Å and r(Al) = 1.43 Å [178]), which is loosely equivalent to introducing the curvature in the 4-dimensional space analysis [223]. The phase transition from the liquid to the quasicrystal is so far the first order transition. On the other hand, there still is a possibility open for the second order transition from the liquid to the quasicrystal, through continuous extension of the icosahedral short-range order. Such a phenomenon is likely to be found in a very complex alloy system with a large number of components [224].

According to Egami [224], liquid interacting with a central force potential does not support the icosahedral long-range order, since icosahedral clusters have an *antiferromagnetic* orientational interaction. On the other hand, the long-range icosahedral order was indeed discovered in more complex alloy systems in the form of quasicrystals. Quasicrystals are found in specific composition ranges, typically those that support local icosahedral order. Because describing the structure of liquids and glasses meaningfully is already a major challenge, it is extremely difficult to create a theory that explains various properties of liquids and glasses. Many experimentalists still use *the free-volume theory* [228-230] or at least the free-volume concept to explain the experimental observations. But the validity of the free-volume theory for metallic liquids is questionable, and its atomistic basis is challenged by recent *Molecular Dynamics* (MD) simulations. It is more likely that the apparent success of the free volume theory originates from the fact that the volume is the easiest

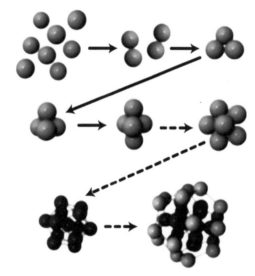

Fig. (24). Development of short-range order in undercooled metallic melts, notice the number of atoms in cluster *h* [89].

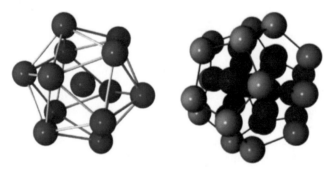

Fig. (25). Quasicrystal atoms packing. In (**a**); an icosahedron consisting of 13 atoms arranged in five-fold symmetry. In (**b**); a dodecahedron, this cluster consists of 33 atoms and is characterized by icosahedral symmetry. It is constructed from a simple icosahedron of 13 atoms (black) by placing further atoms (grey) densely on each of the 20 triangular faces of the icosahedron [89].

Fig. (26). Pentagonal dodecahedral single grain of i(HoMgZn). A US 1cent coin is shown to illustrate the scale of the quasicrystal, which has edges over 2mm long [217,225].

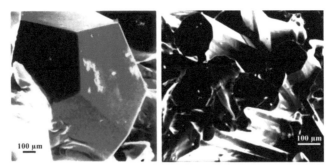

Fig. (27). (**a**) SEM photograph of a single crystal with pentagon-dodecahedral morphology [226]. (**b**) SEM photograph of decagonal Al_{70}-Co_{15}-Ni_{15} with *decaprismatic* (*i.e.*; ten-sided prism-like) growth morphology with sharp glassy edges [227].

property to measure that represents the fictive temperature of the system, as the volume of a glass cube is proportional to the temperature. In other words, the free-volume theory is successful as a phenomenology but not as a microscopic theory. A new theory has to be developed to describe the atomistic movements in glasses and liquids. The *theory of local topological fluctuations*, describing the atomistic movements in glasses and liquids, looks like an alternative and promising one. In this theory, topological fluctuations are represented by the atomic level stresses, and evolution of their distribution with temperature determines various thermal properties [132]. Glass transition, structural relaxation, glass formation, and mechanical deformation have been well described by this theory. Although the details need to be worked out by further studies, this theory promises to be the one that could replace the free-volume theory in elucidating the complex behaviors of metallic glasses [132,231].

1.4.5 Metastable Phases Diagrams

Whether a rapidly solidified microstructure is composed of stable equilibrium phases or metastable phases, it depends on the nucleation and growth kinetics of the competing product phases. A schematic representation of the role of nucleation and growth kinetics for phase selection as a function of processing conditions is shown in Fig. **28**. The thermodynamic relationships for the molar free energy of a pure material as a liquid, a stable phase α, and a metastable phase β are shown. The melting point of the metastable β phase is a well-defined thermodynamic quantity given by T_m^{β} in Fig. **28a**. In Fig. **28b** and **28c**, a possible pair of functions is depicted to illustrate the role of kinetics in phase selection under conditions that favor the nucleation and growth of the metastable β-phase. For the nucleation of α or β from the liquid, the *dominant* product phase is determined to a large extent by the lowest value of the activation energy barrier for nucleation, ΔG^*, of α or β. The value of ΔG^* for each solid phase is a function of the amount of undercooling below the respective melting points, the liquid-crystal surface energy and the potency of any catalytic surfaces present, which act as heterogeneous nucleation sites for α or β. For the nucleation rate relationships presented in Fig. **28b**, nucleation of α dominates at low undercoolings below the stable melting point, whereas nucleation of β can dominate at lower temperatures. Several examples of a nucleation-controlled transition in phase selection with increasing undercooling have been reported [91,92,232]. This behavior is controlled by interfacial energies associated with the nucleus, liquid, and active catalytic sites and occurs even though the bulk thermodynamic driving force is usually greater for the formation of the stable phase.

The growth of a stable phase can also be difficult. Following nucleation of two phases from the melt, one phase may grow so much more quickly that it will dominate the microstructure of an RS alloy (*i.e.*; *phase competition*). A simple example is illustrated in Fig. **28c**, which shows schematically the growth rate of the α and β phases from the liquid as a function of interface temperature in a situation favoring metastable phase formation. At large undercooling, the metastable β phase grows faster than the stable phase α. Such a situation could exist for pure materials when the α-phase grows with a faceted interface and growth is sluggish due to the difficulties of interface attachment. A simple criterion for the development of a *faceted (smooth) growth interface* was developed by Jackson [102,233].

When a dimensionless alpha factor, which is the ratio of the entropy of fusion (latent heat per mole over the melting temperature) to the universal gas constant, is greater than about four, faceted growth is expected for close-packed

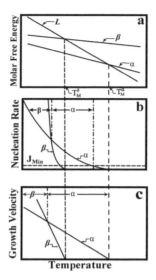

Fig. (28). Schematic representation of the operation of competitive phase selection kinetics which favors the formation of a metastable phase *β* from the liquid L at low temperatures in spite of the thermodynamic stability of α-phase in (**a**). (**b**) shows the temperature range for faster nucleation of *β*-phase while (**c**) shows the temperature range for faster growth of the *β*-phase [95].

interfaces. Then one usually obtains a steep velocity-undercooling curve as shown for the *α*-phase in Fig. **28c**. For alloys, the analysis of growth competition for various phases also requires the incorporation of solute redistribution difficulty in the different growth morphologies present (for example; dendrites, eutectics … etc). The nucleation and growth curves presented in Fig. **28** pertain to relatively low undercoolings. For *thermally activated kinetics*, both rates exhibit a maximum at lower temperatures. For example, such a maximum occurs for eutectic growth due to the temperature dependence of the diffusion coefficient [95,102,234].

While the plots shown in Fig. **28** are schematic, they can be constructed in a quantitative manner for specific systems and metastable phases of interest. The first level of information needed for such analysis is the melting temperature of a metastable phase, which identifies its range of possible formation from the melt. Most often, the melting temperature of a metastable phase is calculated from the analysis of a thermodynamic cycle [234], but there are examples of direct experimental measurement under favorable kinetic conditions [235-240].

Formation temperature is a key element in the competition between alternative constituents and morphologies that results in the final solidification microstructure. Particular attention is given to measurements of nucleation or growth temperature (T_N or T_G due to references) made by the *Bridgman method* in which temperature gradient G_L and growth velocity V can be imposed during solidification from the melt. So, the resulting T_N or T_G can be determined from the distance of the nucleation or growth front from the position of a thermocouple sensor embedded in the sample (for example; determined by longitudinal metallographic sectioning following a quench), where; a cylindrical crucible is moved through at a fixed temperature gradient, G_L, with a constant translation velocity V', with $\dot{T} = G_L V$. It is often assumed that the interface will remain stationary with respect to the furnace during most of the growth period and that the growth velocity of the interface, V, is equal to that of the crucible or the moving furnace (V') [102,240]. In a recent work by Jones [240], the results for Bridgman solidification [241,242], TIG-weld traversing [243], and laser surface melt traversing [244,246] show a good fit (within a factor of three in the number density of equiaxed dendritic grains during solidification, N_V) taking:

$$\overline{N}_V = A\dot{T}^n \ \dots \tag{1-28}$$

with $n = 1$ and the kinetic parameters A for dendrites equals 130 mm^{-3} °C^{-1} s. Applying it to the Al-Si system, the measured formation temperature of polyhedral primary silicon in Bridgman solidified Al-18.3 wt.% Si alloy is governed by heterogeneous nucleation with an effective nucleation contact angle of 30 ± 3°. Corresponding results for Al-30 wt.% Si are also reasonable in terms of heterogeneous nucleation, rather than by growth governed by kinetic undercooling. Also, measured dendritic tip undercoolings or supersaturations of aluminide dendrites in Bridgman solidified hypereutectic aluminium alloys are in good accord with values calculated from theoretical models [10,238,247-251] for the (T_L - T_G) relation (T_L is the alloy liquidus temperature), though the actual level of agreement achieved depends on the values of liquidus temperature, Gibbs-Thomson Γ-parameter and solute diffusivity in the melts that are used [240]. As an example, the addition of third elements can dramatically change the resulting constituents in systems such as Al-Fe, for which several intermetallic phases and their eutectics with α-Al compete for formation during solidification. The introduction of α-Al dendrites into the eutectic on adding 1 wt.% Cu, Mg, Si is notable as is the displacement of Al-Al_6Fe eutectic by Al-αAlFeSi eutectic on addition of 1 wt.% Si [240].

Considering pure metals exhibiting an *allotropic transition* between a low temperature β-phase and a high-temperature α-phase (and *vice versa*, referring to Fig. **28**), the melting of β at T_m^β lies between the melting of α at T_m^α and the allotropic transition at $T_{\alpha/\beta}$ [95]. When specific heat corrections are neglected, the value of T_m^β is given as:

$$T_m^\beta = \frac{\left[\Delta S_{\alpha/\beta}\, T_{\alpha/\beta} + \Delta S_m^\alpha\, T_m^\alpha\right]}{\Delta S_{\alpha/\beta} + \Delta S_m^\alpha} \cdots \tag{1-29}$$

where $\Delta S_{\alpha\beta}$ is the entropy change for the allotropic transition and ΔS_m^α is the entropy of melting for α phase. Therefore, phase transition melting temperature depends not only on the free energy state of this phase but also on its equilibrium or metastable *configurational* state that can be controlled by processing parameters. While allotropic transitions are not observed for many pure components, other crystal structures are stabilized by high pressure or with the addition of alloying elements. In these cases, a thermodynamic analysis of the phase equilibriums permits estimates of the so-called *lattice stability* or the relative stability of different crystal structures of a given component, as compared to the equilibrium crystal lattice structure [95].

Clearly, such lattice stability estimates are sensitive to the accuracy of the alloy phase equilibria or elevated pressure measurements that are used for analysis. For the purpose of comparison, a partial listing of the melting points of the common BCC, FCC, and HCP structures for a number of metals of interest in RS, that are obtained from lattice stability expressions, is provided on the extensive and pioneering work of Kaufman [252], except for the case of Fe [116,253,254]. Lattice stability values may be obtained for other crystal structure types including intermediate alloy phase structures. With this first level assessment, it is possible to identify candidate systems in which alternate metastable crystal structures may be produced during RS. For example, if an undercooling level ($\Delta T/T_m$) of about 0.3, below the stable phase melting point or alloy liquidus, is used as a basis for evaluation, then alloys based on Be, Co, Cu, Fe, Mg, Ni, Ti, Zn, and Zr are likely candidates for the formation of one or more alternate crystal structures. However, Al, Cr, and Nb base alloys will not be good candidates [102].

Beyond the first level of the previous discussions, the elemental lattice stabilities together with thermodynamic models of alloy solution behavior serve as the basis for the computational analysis of binary and higher order phase equilibria, as in Fig. **29**. Thus, there may be a spread in nucleation temperatures even under nominally identical conditions, and consequently the results are best displayed on *microstructure-predominance maps* [102,255,256], as in Figs. **29** and **30**. Such maps have been constructed for binary alloys, with alloy composition and processing parameters (such as; droplet diameter, ejection temperature, wheel speed or material … *etc.*) as coordinates. It is found that: (*i*) microstructure correlates very strongly with processing parameters (which determines the availability of nucleant sites and the cooling rates), (*ii*) the effects of processing conditions (*e.g.*; gas purity in atomization and environmental heat conduction) can be taken into account and (*iii*) correlation with undercooling can be found through comparison with controlled undercooling experiments and growth modeling [257,258]. It is clear, however, that we are very far from being able to predict nucleation undercoolings, the diversity of potential heterogeneous nucleants becomes a key impediment to quantitative modeling of most real situations. In this way, following the early work of Boettinger *et al* [259], a series of solidification microstructure selection maps has been obtained in recent years, which allows a more rational approach to the solidification processing of many technically important alloys [256]. These maps have been used as a tool for analyzing and predicting the microstructures of laser surface treated materials and other metastable processing routes (a specific map for a specific technique). A similar approach, but for undercooled melts with a corresponding maximum velocity criterion, has also been developed [255,256,260,261]. Another valuable example is the *CALPHAD* (CALculation of PHAse Diagrams) methodology, which is one of the most powerful methods and can be applied to study thermodynamic properties of practical alloy systems. It has been developed over nearly 60 years of international cooperation, since Scheil equation [265]. The method is based on the *Bragg-Williams theory* (also known as the *mean field theory* or approximation of crystal lattice ordering and defects) of the entropy and free energy, and hence the mathematical treatment is not difficult [266]. By this effective simulation method, many correct equilibrium and non-equilibrium phase diagrams/maps can be obtained and definitively surveyed experimentally with very great success, as in Fig. **29**. It helped in determining and expecting metastable phases composition accurately that can be checked and seen by modern micro and nanostructures investigation methods, though it was expected a long time ago using the CALPHAD methodology. As an outstanding example for the beneficiations of CALPHAD in the metastability studies, the development of the MS amorphous ribbons/powders of Fe-Cr-Mo-Ni-B alloy system, for gas turbine shafts manufacture [267]. The input of the phase diagram predictions greatly helped in understanding the evolution of microstructure in this alloy. Although these alloys were produced by a highly nonequilibrium route, the calculations also showed that the phases present after extrusion consolidation were the stable phases for the alloy, so design criteria based on equilibrium calculations could therefore be used. A further advantage of the CALPHAD calculation route was that the number of alloys that needed to be examined, in order to achieve the optimum microstructure/property combination for the design criteria of the turbine shaft, could be dramatically reduced [267-269]. Finally, among 70 metallic elements a 70!/3!67! = 54,740 ternary systems and 916,895 quaternary systems are formed. In view of the amount of work involved in measuring even one isothermal section of a relatively simple ternary phase diagram, it is very important to have a means of estimating ternary and higher-order phase diagrams. The most fruitful approach to such predictions is *via* thermodynamic methods. In recent years, great advances have been made in this area by the international Calphad group. Many key papers have been published in the Calphad Journal (the *Computer Coupling of Phase Diagrams and Thermochemistry Journal*). In the Calphad Journal and in the *Journal of Phase Equilibria* there are many articles on the relationships between thermodynamics and phase diagrams. For the measurement of thermodynamic properties, including calculations, properties of solutions and techniques of measuring phase diagrams, the reader is referred to the literature [270-279].

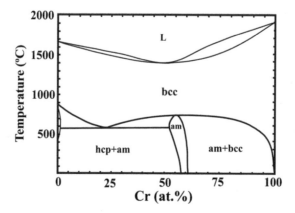

Fig. (29). Metastable phase diagram of the Ti-Cr system as calculated by the **CALPHAD** (**CAL**culation of **PHA**se **D**iagrams) method for modeling phase diagrams and metastability, considering only equilibria between the hcp and bcc solid solutions and the amorphous or the liquid phase [262-264].

1.4.6 Alloy Nucleation and Multiple Nucleations during Continuous Cooling

For a binary alloy, ΔG, in Eq. (1-9), depends not only on the temperature but also on the composition of the liquid and of the solid nuclei. Thus, for a given liquid composition, critical values of nucleus composition as well as size are required to determine ΔG^*. If the surface energy and $f(\theta)$ are constants and independent of cluster composition, the smallest value of r^* (hence easiest nucleation) is obtained if the composition of the critical cluster maximizes ΔG_v. For alloy nucleation, the appropriate expression for ΔG_v is obtained by dividing the expression given in Eq. (1-8) by the molar volume of the solid, V_m. It is apparent from Eq. (1-8) that ΔG_v would be maximized for a composition of the solid, where $\mu_S^A - \mu_L^A = \mu_S^B - \mu_L^B$.

This maximum driving force condition has been proposed [280,281] to find the favoured nucleus composition for a given temperature, pressure and liquid composition. In order to use this condition, one must have a thermodynamic model for the alloy of interest; *i.e.*, the free energy functions for the liquid and solid phases must be known. For simple analysis, regular solution models are often employed for the liquid and solid phases. More precise models, which fit the measured phase diagram and other thermodynamic data, are often available in the literature [267].

In contrast, by including a simple model of the composition dependence of the surface energy, Ishihara and others [282] have shown that the critical nucleus composition can approach the bulk liquid composition at large supercoolings (*i.e.*; partitionless solidification by rapid quenching). Experience with alloy supercooling indicates that the composition dependence of the nucleation temperature, T_N, reflects the composition dependence of the liquidus temperature T_L. For example in the Pb-Sb system, the supercooling results shown in Fig. **31** reveal that T_N follows a similar trend to T_L even for different T_N levels resulting from catalytic sites of different potency, *i.e.*, different surface coatings [102,283]. The maximum ΔG_v condition to determine nucleus composition has been used to successfully predict the composition dependence of measured values of T_N in various alloy systems [281]. While a consideration of competitive phase selection kinetics is necessary, it is incomplete since the thermal conditions immediately following the initial nucleation event have an important influence on the possibility of further nucleation events. Thus, the *thermal history* (T °C vs. time t, $t = 0$ at the moment of ejection of the melt or start of nucleation) reflects the balance between the imposed external cooling and the recalescence due to the latent heat emitted by a rapidly advancing solidification front [95,284-286]. The thermal path following the initial nucleation appears to have the greatest effect under high-cooling-rate condition(s). For example, solidification of undercooled powders during slow cooling invariably yields single-grain structures, but during rapid quenching of undercooled powders multi-grain structures have been reported for a number of alloys. In addition, for melt spinning, the superheating degree with high volumetric flow rates reduces the cooling rates and produces coarser-grained ribbons. The problem quantitatively describing the relationship between the thermal history and solidification behavior is not simple. Several detailed numerical calculations have been developed to treat various aspects of the thermal history after the initial nucleation event [120,175]. However, useful insight into the basic features of this thermal history can be developed from an approximate model. This approach is limited and clearly not rigorous, but it does serve to illustrate the essential features of behavior for different processing methods.

Boettinger and Perepezko [95] have shown that; during continuous cooling at a rate $\dot{T}$ the temperature T in a volume v_s is assumed spatially isothermal, following a single nucleation event at time $t = 0$ at a temperature T_N and can be estimated by;

$$T = T_N - \dot{T}\, t + \frac{4\pi V^3}{3v_s}\left(\frac{L}{C}\right) t^3 \cdots \qquad (1\text{-}30)$$

where V is the radial growth rate of a spherical crystal assumed constant near T_N. For a powder sample, the volume v_s may be thought of as the powder volume. For a sample not limited in size in some dimensions, such as a melt-spun ribbon, the volume v_s may be thought of as the reciprocal of the nucleant density. Clearly, the temperature given by Eq. (1-30) is an average within the volume v_s, and some regions near the growing crystal are above this temperature while regions away from the growing crystal are below this temperature.

From Eq. (1-30), the temperature can be shown to continue to drop for a time, t^*, to a temperature, T_N^*, prior to recalescence, as given by;

$$t^* = \left\{ \frac{v_s\, C\, \dot{T}}{4\pi\, V^3\, L} \right\}^{1/2} \; \dots \tag{1-31}$$

$$\left(T_N - T_N^*\right)^2 = \frac{1}{9\pi} \left(\frac{C}{L}\right) \left(\frac{\dot{T}}{V}\right)^3 v_s \; \dots \tag{1-32}$$

Following time t^*, recalescence causes the temperature to rise, returning to T_N in a time $t^*\sqrt{3}$ and exceeding T_N to some maximum temperature at longer time before final solidification. Clearly, a second nucleation event in v_s will alter the thermal history. The thermal history profile described by Eq. (1-30) is illustrated schematically in Fig. **32**.

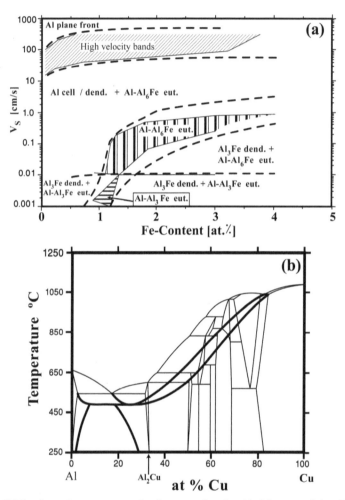

Fig. (30). (a) Calculated solidification microstructure selection map for the Al-rich part of the Al-Fe system (broken lines) superimposed on the experimental results (full lines) [238]. (b) Calculated Al-Cu phase diagram with metastable extensions (thick lines) [255].

Conditions that favor large values of $T_N - T_N^*$ and t^* are of interest. Here, the importance of high values of $\dot{T}$ and v_s is evident, but also the effect of relatively low values of V can be appreciated. Solidification structures with low values of V usually involve growth with significant *solute redistribution* and *segregation*, due to the thermodynamic impossibility of partitionless solidification (because the melt must be highly-supercooled, *i.e.*; hypercooled, firstly to attain the non-equilibrium state causing partitionless solidification) [97,102,113,179]. Maximum velocities in this case are typically estimated to range up to ~ 10 cm s^{-1}. Although Eq. (1-30) is based on a constant V and will tend to overestimate

($T_N - T_N^*$), a *fair upper bound* can be obtained. With $V \simeq 10$ cm s^{-1}, it is possible that ($T_N - T_N^*$) will range from 10 to 100 °C for cooling rates from 10^5 to 10^6 °C s^{-1} with $v_s = 10^6$ cm^{-3} and L/C ~ 300 °C. Much smaller ($T_N - T_N^*$) values occur for *collision-limited growth* [1,8,102]. In all cases it should be noted that Eqs. (1-30)-(1-32) and the model illustrated in Fig. **32** are intended to apply approximately to relatively low values of ($T_N - T_N^*$) (order of 10 °C). Two interesting features can be noted from this simple model. Even with a processing method involving cooling by conduction to an underlying substrate where potent nucleation sites are plentifully present, it is possible to develop bulk melt undercooling levels, which are sufficient to significantly alter the solidification velocity and allow for metastable phase formation. Similarly, due to the strong temperature dependence of the nucleation rate, the additional undercooling ($T_N - T_N^*$) can allow for the development of multiple nucleation events which can yield fine-grain, micro-, nano-crystalline or amorphous structures. Using this approach, the incidence of extra grains has been described by Turnbull [287] and calculations of grain size for pure metals have been performed [113,288-291]. Although such a mechanism can account for the evolution of a microcrystalline structure, there are other possible origins including heterogeneous volume nucleation on a high density of internal sites. Furthermore, severe reduction in V can be realized as T_N^* approaches the glass transition temperature, T_g, due to a reduction in liquid diffusivity, while these conditions imply that transient nucleation effects also become important, a large increase in grain density is expected as well. An example

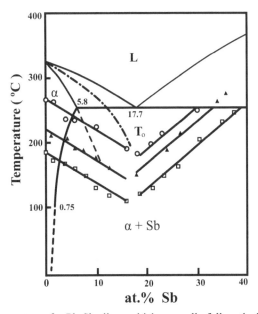

Fig. (31). Summary of nucleation temperatures for Pb-Sb alloys which generally follow the liquidus slope. Supercooling trends at different levels are produced by different droplet surface coating treatments (triangles, squares and circles) [102,283].

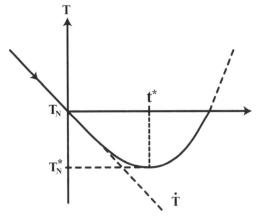

Fig. (32). Approximate thermal history profile following nucleation of an undercooled liquid [95].

for the effect of thermal history is illustrated in Fig. **33** for temperature history in the vicinity of the recalescence peak for a (Fe-24 at.% Ni) alloy [89]. The left peak represents a recalescence event as observed following spontaneous nucleation at 1472 °K (1199 °C) with ΔT = 278 °C. An increase in the temperature up to 1751 °K (1478 °C) is found in good agreement with the equilibrium liquidus temperature of this alloy. The right peak was observed following solidification triggering with the Fe$_{92}$-Mo$_8$ needle/tip at a temperature of 1556 °K (1283 °C) with ΔT = 194 °C. Obviously, the recalescence temperature increase ends at a temperature well below the equilibrium liquidus line, which points to a metastable BCC solidification product. Immediately following the recalescence peak, a *weak hump* is found in the cooling trace, which is due to a solid-state transformation of metastable BCC δ-phase into stable FCC γ-phase. This hump is missing in the temperature-time profile for the spontaneous nucleation. This confirms that during spontaneous crystallization FCC phase is nucleated, whereas triggered solidification leads to nucleation of metastable BCC phase, which however transforms into stable FCC phase during cooling. Furthermore, the cross-sectional TEM micrograph of a melt spun Nb-25 at.% Si alloy ribbon in Fig. **34** provides a striking illustration of the processing effects [292]. *Away from the wheel side of the ribbon*, a mixture of glass and microcrystallites is observed with grain sizes of about 15 nm. *Toward the ribbon center*, grain sizes for a mixture of α-Nb and Nb$_5$Si$_3$ of about 0.1 μm are typical. Clearly, the growth rate of the *two-phase mixture* is slow compared to a *single-phase* material and contributes to the abundant nucleation in this material after the initial crystallization event. Also, the observation of a *glass-crystal transition* region and a *grain size gradient* with distance from the wheel surface supports the importance of both cooling rate (outside and inside the ribbon thickness) and undercooling during processing [95]. A final and related example is in Fig. **35**. A high-resolution image of a selected edge region of a *high-energy* mechanically-alloyed sample reveals the metastability generated by the acquired mechanical energy [293], where a metastable crystalline phase, with an interplanar spacing of *d* = 0.213 nm, coexisting with an amorphous phase.

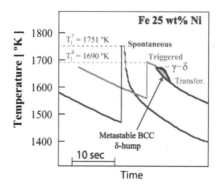

Fig. (33). Two temperature-time thermal history profiles obtained during solidification of undercooled Fe-24at.%Ni melts. Left-hand side shows spontaneous crystallization of a stable FCC γ-phase, right-hand side shows solidification of a metastable BCC δ-phase upon solidification triggering by a Fe-Mo tip [89].

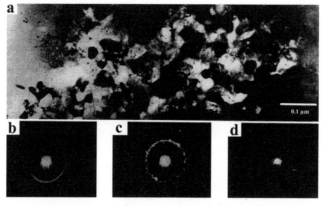

Fig. (34): (**a**) Taper-section thin foil TEM micrograph of a melt-spun Nb-25 at% Si alloy which shows transition from glassy structure (left) to progressively larger grain microcrystalline structure of α-Nb and Nb$_5$Si$_3$ as a function of distance from wheel surface. (**b**), (**c**) and (**d**) are the corresponding *SADP*s for the transition [95,292].

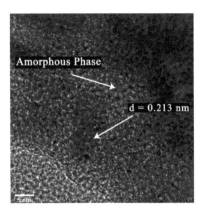

Fig. (35): HRTEM image of the selected boundary region of the Al-10Mg-20Zr (at.%) alloy showing the coexistence of nanocrystallites and amorphous phase [293].

The HR image clearly shows regions of structural disorder having an amorphous structure. In addition, crystalline regions showing clear atomic planes can be seen interspersed in the amorphous regions. This suggests the presence of a nanocomposite structure comprising amorphous and crystalline phases in different volume fractions in this specimen. Finally, the topic of rapid solidification is very large and vast, especially when treating its effects on the alloys and metastable phases formation and needs a stretched real time experiments for compiling more consistent and realistic models.

1.5 SUMMARY AND CONCLUSION

Rapid solidification of metallic melts is an effective means for generating metastable phases and materials that have new and exceptional physical and mechanical properties as a composite material. By rapid extraction of a melt heat, phases nucleation kinetics are changed and growth velocities become very rapid reducing the time needed for equilibrium reactions and diffusion distances. In present study we treated the phenomenon very briefly referring to the current theoretical and empirical models for some common RSP routes, focusing on the effect of cooling rate (for the melt spin casting) and amount of undercooling (especially for the powder atomization) on the nucleation and growth rates. Metastable phases nucleation, including metglass and quasicrystals, have been treated. Theory of RSP until now has presented merely a glance about the metastable nucleation and a reasonable understanding of its kinetics. Clearly, it needs many advanced quantum treatments and modifications based on real time microscopy to catch very quick fleeting micro- and nano-structural changes, for more realistic consistent equations of states, though many modifications and/or numerical methods has been used successfully for model corrections, solutions and controls, but they are still away from reality.

CHAPTER 2

Chill-Block Melt Spin Technique

M. Kamal[a]* and Usama S. Mohammad[b]

[a]Chairman and Professor of Metal Physics, IUPAP-C5 Member, Department of Physics, Faculty of Science, Mansoura University, Al-Mansoura, Egypt and [b]Ph.D. Researcher in Metal Physics and Materials Science, Department of Physics, Faculty of Science in Damietta, Mansoura University, New Damietta City, Egypt

Abstract: Chill-Block Melt Spin (CBMS) technique is the most important rapid solidification process because of its low-costs and flexibility compared to other technologies such as drop tubes and atomization rapid solidification. Mathematical analysis with experiments have revealed that thermal transport is the predominant over momentum transport for ribbon formation and substrate nature and speed control the cooling rates. Microstructural anatomy of the melt-spun ribbons shows a scale-up in the grain size from the wheel side towards the ribbon upper free surface with clear effect of convection-induced instabilities that affect the local cooling rates and microstructures for near Newtonian cooling conditions.

Key Words: Rapid solidification; chill block melt spin; undercooling; metastable phase; cooling rate; thermal momentum transfer modeling.

2.1 INTRODUCTION

From the different rapid solidification process, the melt-spinning process shows big advantages over its competing (industrial) techniques for the following reasons. Melt-spinning can generate the highest cooling rate and scores best in cost-performance ratio and it is a *continuous method* for producing rapidly-quenched *amorphous* or *microcrystalline* materials. Whereas gas atomisation and spray forming are batch processes, the melt-spinning is a continuous process generating billets or compacts with industrial dimensions and fully flexible in length (*e.g.* 7m or more). Also, it should be mentioned that in gas atomisation process the route to the billet is very costly [294].

The cheaper alternative *Cold Isostatic Pressing* (CIP) compacts the powder into 70% density and is *degassed* for several hours before extrusion. As a consequence, degassing is far from optimal, time consuming and may degrade properties during preheating. A more optimal (but more complex alternative) is the *Hot Isostatic Pressing* (HIP) route [18,295]. The powders are canned, degassed (under vacuum) and hot isostatically pressed into a fully dense billet. After decanning, the billet can be used for extrusion. The HIP route will always remain too expensive for mass production, like automotive industry, and will be restricted to niche products. The spray forming route is more economic because directly from the melt a billet is formed. However, also spray forming is a batch process with relatively high costs involved because of extensive handling (each billet needs to be individually handled). Furthermore, it is not only the lower cooling rate of spray forming, but also the larger distribution of cooling rates which create a relative inhomogeneous microstructure compared to melt-spinning and gas atomization. Besides, gas-atomized metal powders usually contain small amounts of the atomizing gas within individual particles, which can cause microporosity and microcracks, especially in the case of argon. For very fast solidification, it is necessary to form a small-dimensioned shape. Also, in gas atomisation and spray forming, a powder or droplet is formed in a gas protected environment. Conventional consolidation of powder to full density through processes such as hot extrusion and hot isostatic pressing normally requires use of high pressures and high temperatures for extended periods of time. Unfortunately, this results in loss of the benefits achieved due to the metastable effects, such as the crystallization of the amorphous phase in the powder, or nanostructures obtained by mechanical alloying (MA). Therefore, novel and expensive innovative methods of consolidating the mechanically alloyed powders are required, as we will see in chapter **3**.

In the melt-spinning process, a continuous ribbon is created in *open air* for most engineering applications. The melt is cast on a rotating copper wheel and instant solidification takes place as soon as the melt hits the wheel. The ribbon is several millimeters wide and around 80 μm thick. In this way not only a very high, but also a very constant (*i.e.*;

*Address correspondence to M. Kamal: Chairman and Professor of Metal Physics, IUPAP-C5 Member, Department of Physics, Faculty of Science, Mansoura University, Al-Mansoura, Egypt; E-mail: kamal422002@yahoo.com

reproducible) cooling rate is obtained. Cooling rate (thus the product quality) is a controllable parameter that can be easily varied by the wheel speed. The effect of process parameters on ribbon formation and geometry has been extensively studied [49,75,135,296-308]. From the ribbon, two more steps are needed to obtain a billet. The ribbon is *chopped* into flakes directly after melt-spinning and then compacted at a *proper* elevated temperature into a billet and the density of the billets reaches a level of 99-100% of the *theoretical density*. The RS billet can then follow the conventional extrusion route; it is sawed to preferred length, preheated in a billet furnace and extruded into a profile.

Processing powders into full density bars encounters big difficulties; one of the major problems is the tenacious nature of aluminium oxide, as an example, on the surface of aluminum-based powders, which inhibits consolidation to high density. The flakes contain significantly less oxide layers compared to powders, which eases a full consolidation (see some of consolidation techniques in the literature [18,309-313] and next chapter). At this moment, RSP technology is capable of processing flakes into rod sections with diameters up to 90 mm. The melt-spinning process is shortened because compaction and extrusion are combined in one step, which means cost reductions and high quality.

2.2 EMPIRICAL TREATMENT FOR CBMS

Several variations of CBMS are described in the literatures with regard to the manufacture of metallic ribbons directly from the melt [19,306]. In the basic process, a molten alloy jet is first formed and then impinged onto a rapidly moving substrate surface, resulting in the formation of a molten alloy puddle from which a ribbon is continuously formed, dragged and chilled. The most common substrate is a rapidly rotating wheel. The structure formed during solidification depends mainly on cooling rate, jet-flow rate and alloy composition [135,306,314]. The cooling rate $\dot{T}$ in this process can be determined by process variables such as wheel properties, surface conditions, material flow rate, ribbon or substrate velocity, melt impingement angle, melt surface tension, the crucible orifice diameter, the melt temperature and composition, substrate properties (*e.g.*; *microfinish* and material) and atmosphere in which casting is conducted.

The unextended length of thermal contact zone (in the downstream part of the melt puddle/pool) is somewhat longer in the CBMS, and by increasing the wheel speed the ribbon thickness decreases [315,316]. The relative importance of fluid flow and solidification rate in determining ribbon thickness is an equally crucial consideration for CBMS and PFC processes.

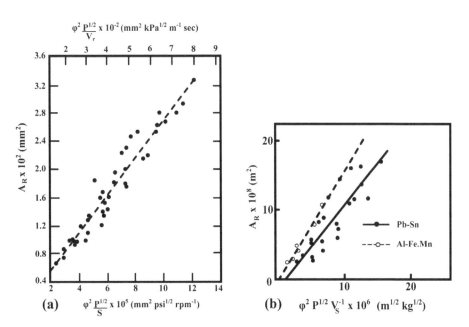

Fig. (1). Ribbon cross-sectional area A_R as a function of $(\varphi^2 P^{1/2} / \omega)$ for melt-spun (**a**) metallic glass (Fe-40Ni-20B, at.%) [314], and (**b**) microcrystalline Pb-Sn and Al-(Fe, Mn) eutectic alloys [317].

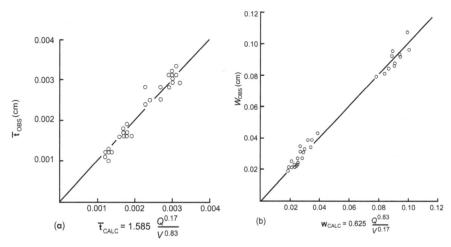

Fig. (2). Observed against calculated values of (**a**) thickness $\bar{t}$ and (**b**) width w for melt-spun glass ribbons of (Fe- 40 at.% Ni-14 at.% P-6 at.% B) alloy [135].

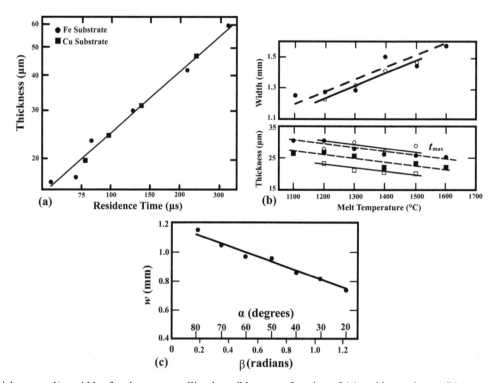

Fig. (3). Thickness and/or width of melt-spun metallic glass ribbon as a function of (**a**) residence time t, (**b**) temperature of the melt and (**c**) jet impingement angle β [19,305,321,325,326].

Measurements for CBMS of metallic glass [314] and of two microcrystalline alloys [317] showed that ribbon cross-sectional area increases linearly with $(\varphi^2 P^{1/2} / \omega)$ as in Fig. **1**, in accordance with Bernoulli and continuity equations of fluid dynamics [compare with Eq. (1-6) in section (1.4.1) of RSP empirical treatment], where φ is the *melt orifice diameter*, P is the *expulsion or ejection pressure*, ω is the chill-block *rotational speed*. Such and further measurements [19,135,305,316,318-325] indicate that ribbon average thickness $\bar{t}$ (calculated by dividing the ribbon mass by the product of length, width w and density ρ) and width w are proportional to (Q^n / v_r^m) and (Q^m / v_r^n) respectively. Q is the volumetric flow rate $(= v_r w \bar{t})$, v_r is the chill-block *surface speed*, n and m are typically ~ 0.2 and 0.8 respectively (see Figs. **2** and **3**). Good agreement with measured values of $\bar{t}$ as a function of v_r for metallic ribbons has been obtained (Fig. **4**) both for momentum control (by controlling of v_r and P) alone and its combination with thermal control (by controlling of the heat sink/substrate material and heat conduction to the surroundings) [19].

Measurements (as in Fig. **3**) also show that the average thickness $\bar{t}$ is proportional to t^p , where t is the *residence time* defined as melt pool length l divided by v_r, and p is $\sim$ 0.5 [318-320] or $\sim$ 0.7 [321-324]. Therefore, $\bar{t}$ is decreased and w is increased by increased superheat [325], and that w is decreased with increasing jet impingement angle (between the axis of the melt orifice and the tangent at the impingement point on the wheel surface) [305], as in Fig. **5**. Liebermann [305] have showed that;

$$\bar{t} = \frac{Q v_p}{\left[\varphi v_p + \left(Q - Q_{min} \right) \left(1 - \beta/2 \right) \right]} \frac{1}{v_r} \; \dots \tag{2-1}$$

$$w = \varphi + \frac{Q - Q_{min}}{v_p} \left[1 - \frac{\beta}{2} \right] \dots \tag{2-2}$$

where; Q_{min} is the minimum material flow rate,

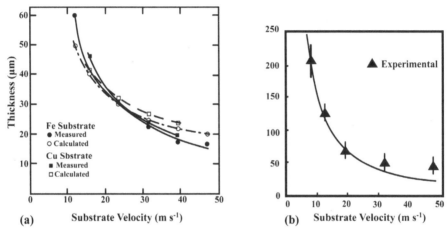

Fig. (4). Observed and predicted thickness as a function of chill-block surface speed for melt-spun ribbon of **(a)** [Fe-13 at.% P-7 at.% C] (soft magnetic) metallic glass [321-324] and **(b)** Microcrystalline (Al- 4.5 wt.% Cu). In each case there is a limiting approached non-zero value [*e.g.*; 70 μm in **(a)** and 260 μm in **(b)**] That is specific for the technique, operating conditions and alloy composition [74,327].

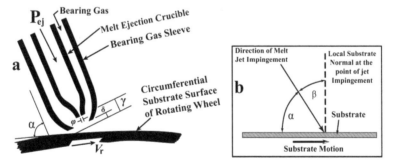

Fig. (5). **(a)** The co-axial jet melt-spin technique showing the composite nozzle used to modify the melt puddle perturbation (instability), and the processing parameters are denoted. **(b)** The angles of impingement ; α and β [305, 306].

$$Q_{min} = \frac{\pi \phi^2}{4} \left(\frac{8 \gamma_1}{\rho_1 \phi} \right)^{1/2} \dots \tag{2-3}$$

φ is the crucible orifice diameter, γ_1 and ρ_1 are the surface tension and density of the melt (assuming that frictional forces are neglected). β is the melt jet impingement angle, in radians, v_p is a constant of the same units as Q and known as the *dynamic melt puddle viscosity*. Q_{min} is produced by minimum driving pressure P_{min} required to overcome the surface tension forces, $P_{min} = 4\gamma_l/\varphi$. Lieberman got very good agreement between Eqs. (2-1) and (2-2) and experiment, as in Fig. **6a** for the ($Fe_{40}B_{20}Ni_{40}$) alloy at 1400 °C, and the results was alike for different atmospheres, as in Fig. **6b**. The slop of each data set yields a value for v_p and it is appearing that v_p is not sensitive to the volumetric flow rates, alloy composition or ambient atmosphere too. In addition, from conducted experiments, v_p is not sensitive to either melt-superheat or surface roughness of the chill substrate. This insensitivity of v_p to certain process variables can be rationalized on the basis that: a change in the melt viscosity has an insignificant effect on the average value of the dynamic melt puddle viscosity v_p, $v_p \approx 900$–1100 mm^2 s^{-1}, that is in consistence with Eqs. (2-1) and (2-2) in this section and with the work of others with various glass-forming alloy systems and CBMS conditions [305,314,318,328].

2.3 THEORETICAL TREATMENT FOR CBMS

Here, we will try to make a mathematical kinetic model for the CBMS tech by a general equation or by a set of equations. A model is a way to describe and usually simplify some processes or phenomena that one tries to understand. An equation(s) may have its analytical solution. In many cases, however, the model is too complex to solve all the equations on a piece of paper, and we have to resort to computers to solve the equations by numerical methods and simulations, as we will see later.

Of all the process parameters of CBMS, the melt fluid properties, the melt jet volumetric flow rate and substrate properties are all the key controllers of the properties of the CBMS products. Therefore, transport phenomena thermodynamics accompanied with nucleation and growth (or; interface) dynamics must be used in order to investigate and account for the process. Nagashio and Kuribayashi [308,329,330] have proved, with no doubt, that the ribbon solidifies just at the underside of the melt pool and the thermal transport control (in which the ribbon solidifies in the melt puddle) is the more dominant than the momentum transport (in which the melt film is dragged out of the melt pool by the rotating wheel and solidifies further downstream) for all ranges of rotation speed. They used a high-speed IR camera for imaging the melt spinning of molten silicon on a chill rotating block, which is a thick silicon wafer. Kavesh [135] was the first trying to model the melt-spin technique; he first pointed out the importance of the formation of a *stable* melt puddle on the moving chill substrate to produce geometrically uniform ribbon. He also proposed the two limiting transport models realizing that thermal transport is the more dominant and faster than the momentum one (caused by viscosity) because of the high heat transfer coefficient at the interface between the melt puddle and the heat sink (for typical metals like copper and iron).

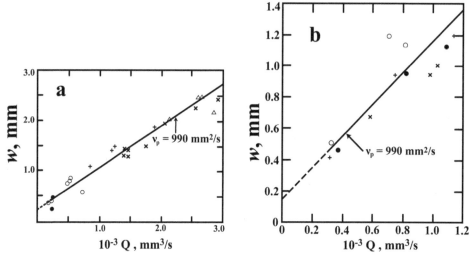

Fig. (6). A volumetric flow-ribbon width plots for; (**a**) Fe-Ni$_{40}$-B$_{20}$ alloy at 1400 °C melt-spun in air with substrate velocities 26.2 m s^{-1} $\leq v_r \leq$ 45.8 m s^{-1} and orifice diameter 0.25 mm $\leq \varphi \leq$ 1.25 mm; (**b**) Fe-B$_{14}$-Si$_4$ alloy at 1500 °C melt-spun in various ambient gaseous atmospheres at 100 kPa and in vacuum with a substrate velocity $v_r = 39.9$ m s^{-1}, orifice diameter $\varphi = 0.25$ mm, and 0.51 mm, and a melt jet impingement angle $\alpha = 90°$. (•) helium cast; (O) N$_2$ cast; (+) air cast; and (x) vacuum cast [305].

2.3.1 The Momentum Transfer of the CBMS Technique

It is known also as the *boundary layer problem* for the CBMS process. Kavesh [135] first pointed out the importance of the formation of a stable *puddle* on the moving chill substrate in order to produce a geometrically uniform ribbon. He applied the concept of thermal and momentum transport modes for the analysis of the engineering parameters of ribbon formation out of the puddle. In the case of the thermal transport-controlled model, ribbon formation occurs by solidification of liquid due to the heat flow into the chill block. The solidified ribbon is extracted from the puddle; hence, the thickness of a ribbon corresponds to that of the solidified layer. On the other hand, in the case of the momentum transport controlled model, the liquid film is dragged out from the puddle to solidify farther downstream. It was thought, in early stages of the study of chill-block casting, that the thermal transport mechanism dominates the ribbon formation process. This was substantiated by Kavesh's calculations based on the *Prandtl number* of the melt, which indicated that thermal propagation was roughly three to nine times faster than momentum propagation. However, much experimental evidence of single roller experiments revealed the existence of unsolidified liquid downstream of the puddle, generally supporting the dominance of momentum *boundary layer* propagation [307,331]. Recently, it has been proved by experiment and calculations [329,330] that the thermal transport control is the most dominant, as we will see afterwards.

When molten alloy is supplied onto a rapidly moving substrate, it will be repelled by the substrate as liquid globules, unless wetting between the melt and the substrate takes place. Such globulization of the melt tends to occur when the velocity of the substrate is insufficient. Increased substrate velocity induces molecular contact between the melt and substrate to enhance wetting. When sufficient wetting is attained, a puddle is formed from which a film of liquid is drawn. The liquid film is formed within the liquid puddle as a momentum (or velocity) boundary layer. This boundary layer is usually defined as the contour of the position within the liquid wherein the local velocity has achieved a certain given fraction of the velocity of the moving solid substrate. Thus, the 1% boundary layer means that the fluid flow velocity is less than 1% of the substrate velocity outside this layer. The analysis of the velocity boundary layer on the rotating wheel may be treated most simply as a modification of the Blasius problem introduced by Schlichting [332]. According to Shingu and Ishihara, with referring to Fig. 7 for the notations of coordinates and velocity directions, the basic equations for the flow of liquid on the moving substrate are [307,333];

$$u \frac{\partial u}{\partial x} + v \frac{\partial u}{\partial y} = v \frac{\partial^2 u}{\partial y^2} \dots \tag{2-4}$$

$$\frac{\partial u}{\partial x} + \frac{\partial v}{\partial y} = 0 \dots \tag{2-5}$$

The boundary conditions are;

$$u = v_r, v = 0 \text{ at } y = 0 \dots \tag{2-6}$$

$$u = 0 \text{ at } y = \infty \dots \tag{2-7}$$

where v is the *kinematic viscosity* of the liquid, u and v are the x- and y-components of the vector velocity **u** of the boundary layer element in two dimensions. The simultaneous partial differential equations and boundary conditions (2-4)-(2-7) can be transformed to an ordinary differential equation and boundary conditions by the use of a similarity variable $\eta = y / \sqrt{v x / v_r}$ and the flow function $\phi = \sqrt{v v_r x} \, f(\eta)$ [4,333], giving;

$$2 f''' + f'' f = 0 \dots \tag{2-4}'$$

with boundary conditions ;

$$f = 0, f = 1 \text{ at } \eta = 0 \dots \tag{2-6}'$$

$f' = 0$ at $\eta = \infty$ $\qquad\qquad\qquad\qquad\qquad\qquad\qquad\qquad\qquad\qquad$ (2-7)'

This two-point boundary value problem was solved numerically using the physical properties of the Fe-Ni-P-B alloy, which is a typical transition metal-metalloid type glass-forming alloy [298].

Fig. **8** shows the calculated result of 1% and 10% boundary layer thicknesses. In this treatment, the length of the puddle was obtained from experimental results, since the analytical estimation of puddle length is quite difficult. Difficulties exist in the mathematical treatment of the free-liquid surface. Recent developments in the numerical methods of calculating transient fluid flow with free boundaries may be applied for this problem. Fig. **9** [334] shows examples of the results of such calculations, indicating the decrease in puddle size with the increase in substrate velocity. This tendency is in qualitative agreement with the observed results. When the thermal contact between substrate and molten alloy in the puddle becomes extremely good (reaching infinity), the assumption of momentum transport domination cannot be asserted. This thermal contact may well be expressed in terms of $\bar{h}$; the interfacial heat transfer coefficient between substrate and melt. For good thermal contact (large $\bar{h}$), sufficient reduction of temperature in the puddle takes place to cause solidification of the liquid. In the case of glass-forming melts, the effect of temperature reduction of liquid in the puddle is associated with an increase in viscosity. In this case, the partial differential equations to be solved are;

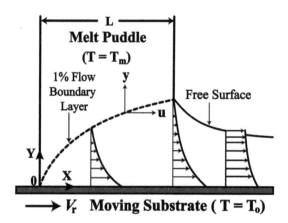

Fig. (7). Schematic drawing showing the formation of flow boundary layer on the moving substrate [307, 333].

$$\frac{\partial u}{\partial t} + u\frac{\partial u}{\partial x} + v\frac{\partial u}{\partial y} = \frac{\partial}{\partial y}\left(v\frac{\partial u}{\partial y}\right)\ldots \qquad\qquad (2\text{-}8)$$

$$\frac{\partial u}{\partial x} + \frac{\partial v}{\partial y} = 0 \ldots \qquad\qquad\qquad\qquad\qquad (2\text{-}9)$$

$$\frac{\partial T}{\partial t} + u\frac{\partial T}{\partial x} + v\frac{\partial T}{\partial y} = \alpha\frac{\partial^2 T}{\partial y^2}\ldots \qquad\qquad (2\text{-}10)$$

Eqs. (2-8)-(2-10) are the momentum, the continuity and the energy equations, respectively, for a *non-steady state incompressible fluid*, where T and α are the temperature and thermal diffusivity, respectively [333,335]. Initial and boundary conditions are as follow;

$$u = 0, \frac{\partial v}{\partial x} = 0, \frac{\partial T}{\partial x} = 0, \text{at } x = 0 \ldots \qquad\qquad (2\text{-}11)$$

At $y = 0$:

$$u = v_r, \ v = 0, \ k\frac{\partial T}{\partial y} = \overline{h}\left(T - T_o\right) \cdots \tag{2-12}$$

$$u = 0, \ T = T_m \text{ at } y = \infty \ldots \tag{2-13}$$

$$u = 0, \ v = 0, \ T = T_m, \text{ at } t = 0 \ldots \tag{2-14}$$

where k is the thermal conductivity. T_o and T_m, respectively, refer to the roller surface and the initial melt temperatures. The change in kinematic viscosity of undercooled liquid may be given by the *Vogel–Fulcher–Tammann ansatz* (approach) relation [336,337];

$$v = \frac{A}{\rho} \exp\left(\frac{B}{T-C}\right) \cdots \tag{2-15}$$

$$\delta = 4.92\sqrt{vx/v_r} = 4.92\frac{x}{\sqrt{Re_x}} \cdots \tag{2-16}$$

where A, B, and C are numerical constants and ρ is the density of the melt. Re_x is the *Reynolds number*; $Re_x = v_r x / v$ and it specifies the relative influences of inertial and viscous forces in a fluid problem.

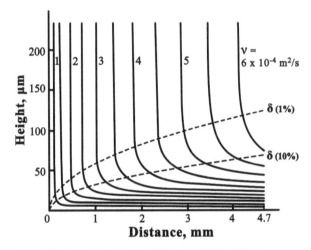

Fig. (8). Fluid flow lines within the puddle. Boundary layers (1%) and (10%) indicating, respectively, the positions of 1% and 10% of the substrate velocity and are shown by dashed lines. δ is the thickness of the boundary layer; $y = \delta = f(v_r, \rho, v, x)$, v is the kinematic viscosity and ρ is the fluid (melt) density [307].

The subscript on Re_x (x in this case) tells what length or co-ordinate it is based upon [333]. The result of calculations indicates that, for $\overline{h}$ value of as high as 4.2×10^5 W m^{-2} °C^{-1}, the process is still dominated by momentum transport [307]. After a liquid boundary layer is dragged out of the puddle, the final thickness of a ribbon will be reached when the flow velocity of ribbon free surface reaches to that of the ribbon-roller surface. Fig. **10a** [318] shows the calculated final ribbon thickness, assuming a *purely momentum transport mechanism*.

The $\left(\overline{t} \propto v_r^{-0.8}\right)$ relation was obtained experimentally by Hillmann and Hilzinger [318] and is excellently matched. The good agreement may be accidental, considering the many simplifications made for calculation. Nonetheless, this result supports the importance of momentum transport in ribbon formation of amorphous materials. In the case of a crystalline material, the melt viscosity becomes almost infinitely large when crystallization takes place. Owing to the crystal growth shape and the direction of dendrite growth, the direction of solidification may be deduced in many cases [326]. Such observations revealed that the crystallization mode varies, depending on the alloy species and composition. In most cases, ribbons show the development of dendrites growing from the substrate contact to the free surface [338] and this indicates that the solidification proceeds in the same sense. Some experimental studies [304,307,339] of the ribbon thickness dependence on roller surface velocity resulted in the relation $\left(\bar{t} \ \alpha \ v_r^{-0.6} \right)$ as shown in Fig. **10b**. This exponent is slightly smaller than that for the case of amorphous alloys. Actual temperature measurement of crystalline materials [340], as shown in Fig. **11**, clearly showed that solidification (crystallization) takes place downstream from the melt puddle for relatively thick ribbons.

2.3.2 Thermal Transport Control during CBMS Process

Heat flow in a thin ribbon in contact with a cold substrate is mainly perpendicular to the contact surface and can be described by the *transient heat diffusion/conduction equation* in three dimensions;

$$c\rho \frac{\partial T}{\partial t} = \nabla \bullet \left(k \nabla T \right) + \dot{q} \ \ldots \tag{2-17}$$

where; c is the specific heat, ρ is the density, T is the absolute temperature, q is the heat flux or the rate of heat transfer, $\dot{q}$ is the time derivative of q (q has the units of W m^{-2} or J m^{-2} s^{-1}) and ∇ is the vector *differential operator del*,

$$q = -k \nabla T \ \ W \, m^{-2} \ldots \tag{2-18}$$

where k is the thermal conductivity [333,335]. The term *transient* is used when the temperature in a heat conduction process depends upon both time and spatial coordinates. Eq. (2-17) is equivalent to a *Newtonian-* or *near Newtonian-cooling* conditions and no convection exists, the left-hand side is the total incoming thermal energy, the first right-hand side term is heat consumed in conduction from the melt to the chill substrate and the second right-hand side term is the heat consumed in solidification [49]. To simplify the analysis, we consider the melt as an *ideal incompressible fluid*, and with no convection existing [341] because the thickness/thinness of the forming melt puddle zone suppresses it. Besides, we have neglected the radiation and conduction to air (after the ribbon separation away from the wheel) and environmental heat losses for simplicity purpose, though they are present [342,343]. Here, it is known that the cooling conditions in melt-spinning are controlled by the melt-substrate interface (*i.e.*; Newtonian cooling) due to the thinness of the ribbon, even though the solidification rate is very high. Therefore;

$$q = \frac{Q}{A} = \bar{h} \left(T_2 - T_1 \right) \ \ldots \tag{2-19}$$

Q, here, is the transferred heat energy and A is the contact area. So, the heat flux from the impinging melt becomes;

$$q = \bar{h} \left(T_s - T_o \right) \ \ldots \tag{2-20}$$

and the heat consumed in solidification process is [344-346];

$$q = \bar{h} \left(T_m - T_o \right) = \frac{R \Delta H_f}{A} \ \ldots \tag{2-21}$$

where; $\bar{h}$ is the heat transfer coefficient over the melt puddle/substrate interface, without the bar ($^-$) it becomes local at any surface element on the interface. T_S, T_o and T_m are the superheating, substrate and melting temperatures, respectively.

R (= dm/dt) is the growth rate; mass per unit time, normal to the substrate, ΔH_f is the fusion enthalpy, its sign is negative for solidification (= mL, L is the latent heat), A is the area of the growing solid and kept constant in our calculations for fixed and optimized CBMS process conditions [49, 344]. Strip thicknesses are typically 20-50 μm, with measured cooling rates ($\dot{T}$) of 10^4-10^6 °C s^{-1} [49,347] and solidification takes place at a few ms, corresponding to near-Newtonian conditions with heat transfer coefficients ($\bar{h}$) of 10^4-10^5 W m^{-2} °C^{-1} or somewhat higher, with little or no thermal gradient in the strip [285,348,349]. Substituting to Eq. (2-17) from Eqs. (2-18)-(2-21) we get;

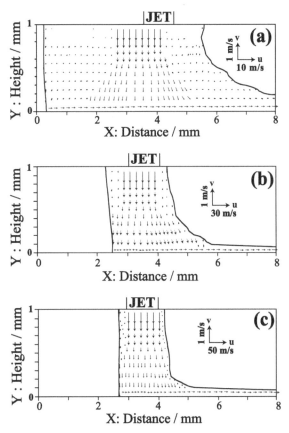

Fig. (9). Calculated fluid flow velocities in the puddle for three different substrate velocities. (**a**) $v_r = 10$ m s^{-1}, (**b**) $v_r = 30$ m s^{-1}, (**c**) $v_r = 50$ m s^{-1}. The orifice length is 1 mm for all three cases [307,334].

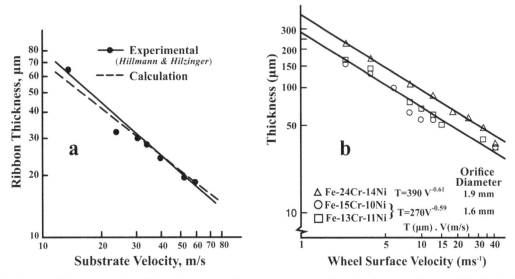

Fig. (10). (**a**) Relationship between the ribbon thickness and substrate velocity for glass-forming alloys and for Fe-Cr-Ni alloy system in (**b**). Ribbon thickness is roughly proportional to the (-0.8) power of substrate velocity [49,307,318]. See also Fig. 4.

Process Time: 0.2 s

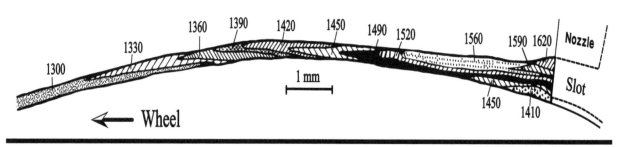

Process Time: 3 s

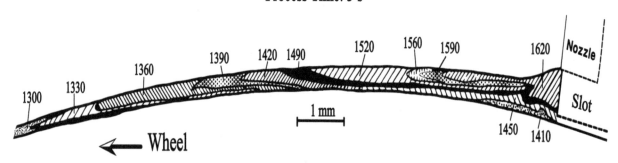

Fig. (11). Infrared photograph of ribbon temperature distribution for a high-speed steel ribbon during melt spinning [340] and solidification takes place far downstream from the melt puddle with the development of dendrites growing from the substrate contact to the free surface in most cases [307,338]. See also Figs. **17, 30** and **31**.

$$R\rho\left(L + C_L\Delta T\right) = \overline{h}\left(T_m - T_o\right) \ \dots \tag{2-22}$$

ρ is the density (solid density is assumed to be the same as that of the liquid density), C_L is the specific heat of the liquid [= dU/dt, $C_p = C_v$ for an incompressible system, U = ρcV($T_2 - T_1$)], ΔT is the superheating degree [8,49,308]. Eq. (2-22) is *a simple form* representing the direct effect of heat transfer coefficient on the solidification speed or rate that is explicit in thinner ribbons and implicit in thicker ones (and can be solved by an *explicit finite difference method*). In fact, the present situation is more complicated because R, $\dot{T}$, $\overline{h}$ and even ρ are all time-, spatial- and surface roughness dependent. The substrate surface wettability and tribology plays a vital role in heat transfer for this process. Besides, Eq. (2-22) represents the solidification heat extraction (thermal control of solidification process), and can be derived under the assumption that: the growth rate continues to increase as the cooling rate increases, and this assumption is only valid, in this study, for one-component systems like pure metals and single-phase alloys, for simplicity. The multi-component systems are discussed elsewhere, including the study of multi-phase solidification kinetics and multi-component transport phenomena [95,102,121,125,283,301].

If the growth kinetics is very slow (*e.g.*, glass forming metals and oxides), especially when using low heat conductivity substrates, solidification may occur during cooling but the remaining melt will cool further away from the cooling wheel, which results in the mixture state of crystalline phase and glass phase (*i.e.*; solid and frozen/congealed melt) [95,335,347]. In this case, the solid forming during cooling is embedded in the glass. This microstructure has been observed in metallic glass materials [175,350,351], resulting in a ribbon formation process controlled by momentum transport [135,308]. Therefore, slower growth kinetics will lead to a momentum transfer control.

Using a high-speed IR-photography technique combined with the melt-spin apparatus, Nagashio and Kuribayashi [308,329] showed that solidification started at the time of the next frame after impingement. The thickness of the solidified layer must be known in order to calculate the heat flux per unit time to the chill-block substrate, where;

$$h = \frac{k}{\delta'} \quad , \quad \alpha = \frac{k}{\rho c} \ldots \tag{2-23}$$

δ' is the *effective* ribbon or metal-film thickness for heat transfer, k is the thermal conductivity of the melt, α is the heat diffusivity [5,344]. The solidification in the puddle is shown in the photos taken in Nagashio's experiment (as in Figs. **12** and **13**) that is in great agreement with the pattern given by Katgerman *et al* [304,297,352]. Nagashio used a silicon rotating substrate with heat conductivity near that of copper and the melt was B-doped silicon.

Returning to Eq. (2-22); if we made the right-hand side constant, one can find that the growth rate is inversely proportional to the latent heat of the melt metal/alloy. Despite of the latent heat of the silicon is barely three times larger than that of typical metals ($L_{Si} = 51$ kJ/mol, $L_{Fe} = 15$ kJ/mol, *i.e.*; the solidified volume in unit time for Si is smaller than that for typical metals), the direct IR high speed camera show the solidification at the underside of the melt puddle directly following the melt impingement on the chill substrate. This indicates explicitly that thermal transport is more dominant than the momentum transport, though the later controls the geometry of the emerged ribbon from the melt puddle. Davies [299], within the framework of the combined model describing simultaneous propagation of thermal and momentum boundary layers within the melt pool, showed that the mechanism of ribbon formation depends on the value of the heat transfer coefficient at the melt-ribbon interface. Earlier, Ruhl [44], on the basis of a rather simple mathematical model, showed that the rate of cooling of a thin melt layer in contact with a massive heat sink is strongly affected by $\bar{h}$ and proved that, for ideal cooling, $\bar{t} \propto \left(\dot{T} \right)^{-1/2}$. Not to forget also that the melt pool viscosity and solid structure growth are all temperature dependent. Therefore, the value of the interfacial heat transfer coefficient appeared to be the most important parameter of the melt-spinning process, because it governs the mechanism of ribbon formation and its thickness as well as the rate of heat extraction [44,308,329,342,343,353].

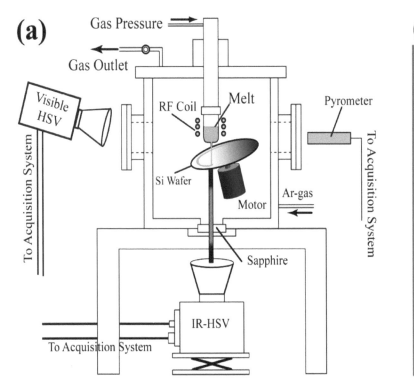
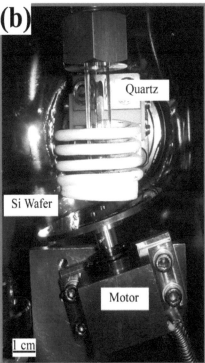

Fig. (12). (**a**) Schematic of the experimental apparatus used by Nagashio and Kuribayashi. (**b**) A photograph of the main setup of the transverse melt-spinning system. The single crystal silicon wafer has a high thermal conductivity, of the same order of magnitude as that of copper, and is transparent for wavelengths longer than 1.1 μm. The tilt angle of the 3 mm thick Si-wafer was kept at 15° for moderate solid-liquid (melt) area ratios in order to obtain rationalized photos. Increasing the flat wafer tilt angle increases the solid-liquid area ratio to become unity at 30° [308,330].

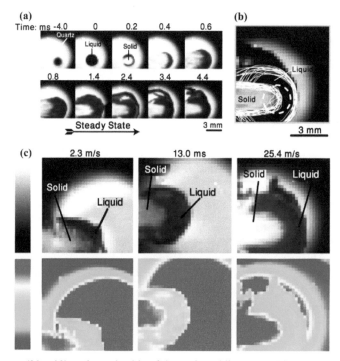

Fig. (13). (a) Successive images (32 x 32) at the underside of the melt puddle captured by high-speed IR imaging at 5 kHz. The time unit is milliseconds (ms). In these images, the solid appears brighter than the liquid due to the higher emissivity of the solid. At the wheel speed of 25.4 m/s, a single rotation of the Si wafer takes 7.8 ms. **(b)** Representative image in which the positions of the solid-liquid and liquid-gas interfaces were traced at $t > 1.4$ ms in **(a)** and superimposed. The white dotted circle is the initial position of the ejected melt just before impact [*i.e.*, $t = 0$ in **(a)**]. **(c)** Temperature uniformity within the solid region as a function of rotation speed. The tilt angle of the Si-wafer was constant at 15°. Upper images are shown in gray scale, while lower images are shown in color scale [308,330].

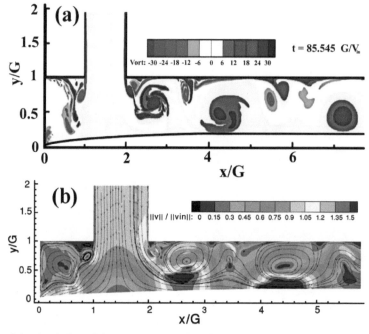

Fig. (14). (a) A snapshot of the simulation of the z-component of vorticities within the melt puddle versus time, $t = 85.545$ G/V_{in} time units, G is the nozzle-wheel gap (= 1 mm) and V_{in} is the vertical fluid velocity component at the center of the inlet. The recirculation regions (*i.e.*; the turbulent eddy flows) restricts the mass flow and are colored according to the positive or negative values of the local vorticity; x/G and y/G are the scaled coordinates. **(b)** The streamlines corresponding to the vorticity snapshot in **(a)** [369].

2.3.3 CBMS Process Computer Simulation

Simulation means experimentation with models and *models* means mathematical analysis and algorithms. For example, classical physics predicts continuous changes of quantities such as position, velocity, or voltage with continuous time. By simulation, we can relate model-system states to earlier states, benefiting the speed of the computer system to give us results and predictions on the screen in a variety of graphical forms in 2-D or 3-D forms, saving time and costs. As mentioned before, for many cases, the model is too complex to solve all the equations on a piece of paper, and we have to resort to computers to solve the equations. Microstructures are typically complex structures and it is always *impossible* to solve the equations analytically when the goal is to obtain local information, rather than bulk statistics. Therefore, *Microdynamic Simulations* employ a whole variety of numerical methods with the use of computers to solve the equations. These methods are not specific to microdynamic simulation, but are employed in all fields of science. In order to practice scientific simulation, one should be familiar with partial differential equations, Lagrangian mechanics, numerical analysis, tensor and asymptotic analyses, and a programming language; basically FORTRAN or C languages or any other logical programming language like BASIC or JAVA, for creating simulation algorithms and codes. Thus, we can use the common numerical and molecular dynamics simulation methods, such as Monte Carlo Methods, Phase Field, Finite Difference or Finite Element methods, for creating two- and tree-dimensional models of the process, like that in Fig. **14**, that help us to expect values of parameters, process controls and creating microstructure selection maps [354-361]. By these models we can determine the upper and lower limits of the present process and device, the so-called *operation window*.

Another complicated and specific models (including microstructure development) can be found in literature with numerical analysis and simulation methods (though some of them are about polymer fibers melt spinning but have good relevant ideas) for both CBMS and PFC-MS [362-373]. As a general, these models may be successful in many sides but fail in others, especially those don't include perturbations effects that cause discrepancies, and that is successful for an alloy system may not be thus for another. It's early to make the microstructure-property diagrams or tables for this technique.

2.4 CBMS RIBBON FORMATION AND ANATOMY

2.4.1 Formation

The ribbon formation process is qualitatively summarized using Fig. **15** and Table **1**. A solid boundary layer forms, at the wheel chill surface within the melt puddle, and propagates into the melt puddle to form a ribbon [338,135,308,365]. For simplicity we will neglect the surroundings and irradiation cooling, though they are of significant effect. The shape of the melt puddle is illustrated schematically based on a previous work [374].

First, a low rotation speed of ~ 1-5 m s^{-1} is considered (Fig. **15a**). In this case, the typical ribbon thickness is ~ 300 μm. The current cooling rate just at impingement is high ($\dot{T} \sim 10^5$ °C s^{-1}) since the melt is in ideal contact with the substrate in terms of the impingement rate and the contact area. The heat transfer coefficient is expressed as h', and the region where the melt first solidifies is defined as stage I. The heat extracted in stage I is expressed as $C_P^L \, dT + L$, where C_P^L is the specific heat for the melt and L is the latent heat. Once the solid is formed on the substrate, the heat resistance increases, since the heat is extracted into the substrate through the solid formed on the substrate and the solid/substrate interface. Moreover, the contact area between the ribbon and substrate is considerably decreased due to the volume change during solidification [338,350], as shown in Fig. **16a**. During solidification typical metals and alloys shrink; however, if thermal extraction from the ribbon to the substrate is discussed, the important point is the change in contact area during solidification. All metals and alloys exhibit similar behaviors; that is, a decreasing contact area. This region is defined as stage II and the heat transfer coefficient at the solid/substrate is expressed as h'', which is smaller than h'.

The heat extracted in stage II is also $C_P^L \, dT + L$. However, the rate of heat extraction is much lower than that in stage I due to the lower value of h'', leading to an expected decrease in cooling rate between stages II and III, as detected by temperature analysis at the top surface [350,374-376]. Therefore, the grain formed in stage II is large, as shown in Fig. **17a**. Although many researchers measured the cooling rate from the top surface of the ribbons, it is difficult to distinguish stages II and I. Finally, all the melt solidifies (stage III). The heat extracted in stage III is $C_P^S \, dT$ and the heat transfer coefficient is again h'' [308].

Next, a high rotation speed of ~ 30 m/s is considered (Fig. **15b**), with $\dot{T} \sim 10^6 \, ^\circ\text{C s}^{-1}$. A characteristic of this second case is the thinness of the ribbon, ~ 20 μm. At the first contact between the melt and the substrate, the melt solidifies rapidly due to the high h' in stage I ($h'|_{30 \, m/s} \gg h'|_{5 \, m/s}$ in reality). Then, the ribbon formed in stage I is dragged out from the melt puddle through the short stage II, because the rotation speed of the substrate is faster than the solidification rate in stage II. The ribbon in stage III is cooled slowly because of low h''. That is; in the case of high rotation speed, there is almost no stage II, which results in better physical properties because the microstructure (which is formed only in stage I) is fine throughout the thickness. Although h' in the case of high rotation speed (Fig. **15b**) is, of course, larger than h' in the case of Fig. **15a** for low rotation speed, the difference was neglected here for simplicity.

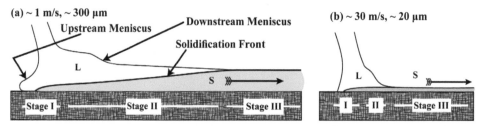

Fig. (15). Qualitative ribbon formation processes. (**a**) Low rotation speed (~ 1 m s⁻¹) and thick ribbon (~ 300 μm) and the melt puddle looks like a long foot and the upstream meniscus being its *heel* that is featured with fluid recirculation. (**b**) High rotation speed (~ 30 m s⁻¹) and thin ribbon (~ 20 μm) [304,308,369].

Table 1. Heat transfer coefficients and extracted heat for different stages [308].

Stage	I	II	III
Heat Transfer Coefficient	h'	h''	h''
Extracted Heat	$C_P^S \, dT + L$	$C_P^S \, dT + L$	$C_P^S \, dT$

2.4.2 Melt Puddle and Solidification Front Stabilities

Puddle serves as a local reservoir from which ribbon is continuously formed and chilled. For a puddle of smooth and unchanging shape during the process, ribbons with well-defined edges and surfaces will emerge. Such geometric uniformity is desirable because it results in a uniform local quench rate through the ribbon thickness. As mentioned before, the ribbon cross-section depends markedly on the material flow rate [305] and on the angle of impingement [318,377], while the uniformity in the ribbon shape and geometry are dependent primarily on the melt puddle stability [306,341]. This stability is crucially affected by the gas boundary layer accompanying the rotating wheel (air or inert gas in most cases) causing *gas pockets*, which, in turn, degrade the thermal contact, substrate *surface roughness*, vibration and *eccentricity ratio* of the rim of the wheel itself under the melt crucible nozzle [304,338]. As the metal is solidified, it is held between the hot nozzle face and cold chill wheel by surface tension. Commercially available melt-spinners are typically unable to deliver ribbon thicknesses reproducible to better than 20%, *i.e.*; any melt spinning technique or device is unique [58]. The greatest technical difficulty in the experiments is the precise control and measurement of the nozzle/wheel gap (in PFC melt spinning; the gap is less than 1 mm). When the effective cooling rate is close to the critical cooling rate (*i.e.*, the minimum rate required for forming an amorphous alloy), pressure fluctuations from a poorly ballasted crucible pressure system (±10%) and small variations in the orifice opening (±10%) can result in a mass flow rate change of 30%, yielding irreproducible results [327]. To understand the results of the experiments it is useful to distinguish steady-state process behaviour from process stability. Only under certain conditions, a uniform ribbon is observed to form. For example, if the overpressure is sufficiently high, the upstream molten metal meniscus will blow out, spraying metal droplets to cool in the ambient air, or separates into *fingers* of metal, for

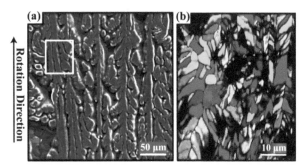

Fig. (16). (**a**) Magnified scanning electron microscopy image of the bottom surface in the ribbon obtained at 25.4 m s^{-1}. (**b**) EBSP (Electron Backscattering Diffraction Pattern) orientation map of the area in the white box in (**a**) [308].

example. Thus, operation data can be used to refine the critical process characteristics, which may serve as a guide in improving run-to-run (and system to system) reproducibility and provide a more physical basis for adjusting the quench rate to achieve the desired materials properties through a controlled microstructure. Charting (or predicting) these operating limits (*i.e.*; stability boundaries) in the parameter space is of vital importance to commercialization [58,59,327,378]. The melt-pool shape or perturbation arises from the balance between the viscous shear drag imposed by the wheel, the feeding properties of the incoming stream, the Reynolds number of wheel gas boundary layer (critical or a hyper-critical, as in Fig. **18**), and the surface forces at the melt-gas interface, without any other physical constraints. Several improvements were made to the melt spinning system, including the improvement of crucible pressure control, the reduction of crucible vibration, and the implementation of a precision nozzle design. Both run-to-run and single-run variability in ribbon thickness were decreased by these steps [76,306,371,379]. Clearly, the *blow-out instability* will depend on the particular metal or alloy under consideration through its surface tension. On the other hand, when the surface tension is sufficient to prevent the blow-out instability (and other instabilities are avoided) the thickness of the resulting ribbon may not be expected to depend on surface tension. Also included is the *air* or *gas entrainment instability* that violates the contact-line integrity and tends to occur for large dynamic contact angles (the angle the liquid metal makes) and for high wheel speeds. Thus, it is clear that the operating limits will generally involve more parameters than the steady-state behaviour [58,59,326]. Many of the cited instabilities may alternately be attenuated by making the proximity between the melt ejection crucible and jet very close, *i.e.*; short enough to ensure a stable melt-stream but long enough to avoid any physical interaction between the crucible and the melt pool, or using the coaxial jet casting technique, as in Fig. **5a** [306]. In this later way, the flowing melt is entrained or stabilized and made resilient against perturbations by the coaxial gas stream that bears down on the molten alloy puddle, attenuating the accompanying gas boundary layer on the rapidly moving substrate surface. However, the planar flow casting (PFC) approach requires accurate control of the crucible-substrate gap. In any real melt-spinning system, the substrate expands during processing and makes the gap smaller, which can constrict melt flow entirely. On the other hand, the '*free jet*' CBMS has a substantial advantage in that: the crucible maybe situated so sufficiently far from the substrate that no alteration or disruption of melt flow occurs [338].

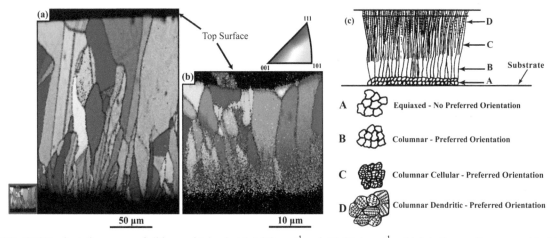

Fig. (17). EBSP orientation maps of ribbons obtained at (**a**) 2.3 m s^{-1} and (**b**) 25.4 m s^{-1}; which is the small square aside (**a**) for dimensional comparison. The direction from the bottom to the top surface for each grain is colored using a reference triangle [308]. (**c**) A representation of solidification microstructure, which is found in CBMS ribbons [338].

All previous process instabilities affect the overall thermodynamic stability of the solidification front. The major effect is that: apart from the interface stability, there is an *absolute stability*, indicating that the decomposition in the solid/liquid interface can no longer occur above a certain solidification velocity. This condition is shown in Fig. **19** (for a Fe-C-Si alloy) and Fig. **20** (for a Pb-Sn-Sb alloy) at highest solidification rates. In Fig. **19a**, there is a supersaturated solid solution that still shows some carbon segregation, the formation of carbide, however, is avoided. At lower solidification velocities, in Fig. **19b**, a dendritic morphology is found. Variations in the quenching rate with constant (G_L/V_S) ratio result in the same morphology but with a different size of the relevant microstructural dimensions (for example, a different size of secondary dendrite arm spacing or lamellar spacing) [380]. During growth of the solid in the melt pool, the shape of the solid-liquid (S/L) interface controls the development of microstructural features. The nature and the stability of the S/L interface are mostly determined by the thermal and constitutional conditions (*e.g.*; constitutional supercooling that depends upon alloy composition) that exist in the immediate vicinity of the interface. Depending on these conditions, the interface growth may occur by planar, cellular, or dendritic growth [8,10,381].

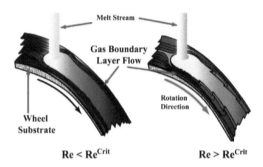

Fig. (18). Schematic diagram showing the influence of gas boundary layer Reynolds number on FJMS ribbon edge definition. Notice the melt puddle serration in case of hyper-critical Reynolds number; $R > R^{crit}$, resulting in serrated and non-uniform metallic ribbons [58,338,382].

Derivations of the most important characteristics of the diagram in Fig. **21**, *i.e.*, the *interface destabilization*, which corresponds to the transition between planar (a) and cellular (b) interface, the periodicity of the cellular structure (b), the characteristics of the dendrite field (c), and the transition toward rapid solidification (d), are given in this figure. Another case is included, when a eutectic structure is obtained with two different solid phases growing simultaneously from a single liquid. In practical applications, three regimes are used in order to grow materials.; (*i*) *planar interface growth* is used to get high-quality single crystals for electronics, optics, detectors, and so on, (*ii*) *dendritic growth* is the most common because it is used universally for all metallurgical processes: steel, cast iron, Al- and Cu-based alloys, and others. The reason is that the dendritic regime leads to tiny microstructures, furthermore accompanied by eutectic areas, and this gives excellent mechanical properties to these alloys. (*iii*) Rapid solidification is used in order to obtain very small *cellular* micro-/nano-structures and amorphous materials; it is also a way to get metastable phases that cannot be solidified at lower growth rates [383].

According to the *morphological stability theory* developed by Sekerka, Coriell and Mullin [384-387]; during directional solidification (as in the melt-spinning and chill-casting techniques), the interface becomes unconditionally stable and the absolute stability solidification velocity V_{Sa} becomes;

$$V_{Sa} = \frac{m_L D_L (1-k) C_o}{\Gamma k} \; \ldots \qquad\qquad (2\text{-}24)$$

where m_L is the liquidus slope (for the alloy system used). Fig. **22** shows a typical, simplified phase diagram of a binary alloy. For a given temperature, T_2, the composition of the solid is related to the composition of the liquid by the segregation or partition coefficient k, $k = C_S/C_L$ at certain temperature and composition. D_L is the diffusion coefficient at the S/L interface, C_o is the initial composition of the growing solid at steady-state solidification and Γ

is the Gibbs coefficient (close to 10^{-7} °C m for metals using the Gibbs's classical treatment of nucleation theory, $\Gamma = \gamma_{SL}/\Delta S_M$, ΔS_M is the solute atom migration/diffusion entropy). V_{Sa} is independent of the thermal gradient G_L because its value becomes negligible compared with the solute gradient in front of the interface. With classical material parameters, the absolute velocity is in the range 0.1–1 m s^{-1} [383]. At very high growth rates, the solute boundary layer in the liquid decreases so much that it becomes of the order of the interface thickness and the S/L interface velocity becomes so high that the atoms or molecules do not have enough time to rearrange at the interface. Then, the liquid is solidified without any segregation or distribution; the solute segregation coefficient *k* tends toward unity and, according to Aziz [102,390], this favors the *absolute stability* of the front in the case of directional solidification. Then, for the growth from the undercooled melt, dendrites become purely thermal (giving equiaxed grains/cells) once *diffusionless* or partitionless solidification conditions are reached at the tip of the dendrite, because the growth direction and the heat flow direction are the same in equiaxed growth and the melt becomes its own heat sink [10,383]. Of course, the chemical potentials of the solid and the liquid are no longer equal, as they are not in equilibrium because of the thermal and chemical gradients at the solidification front. In any case, the interface velocity can not be higher than the velocity of sound in the liquid, which is of the order of 10^3 m s^{-1} for metals.

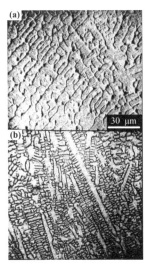

Fig. (19). Microstructure of a rapidly solidified Fe-Si-C alloy [380].

Morphological stability theory remains an active area of research. On a larger scale, the first phenomenon which becomes apparent is the morphological instability of solid-liquid interfaces, which occurs on a scale of the order of micrometres and even nanometers for hypercooled melts and droplets, and this is only a transient stage in the development of the steady-state growth forms of cells, dendrites or eutectics. Equiaxed grains of pure or alloyed materials are inherently unstable when their diameter exceeds a critical value (of the order of some micrometres).

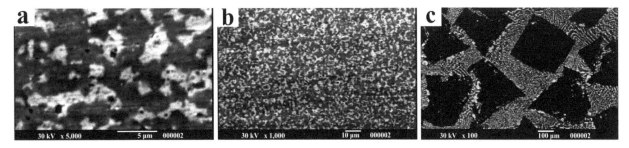

Fig. (20). SEM micrographs for the $Pb_{40}Sn_{48}Sb_{12}$ alloy. **(a)** and **(b)** show the MS ribbon samples of cellular (mottled) ternary and fined-eutectic structure. **(c)** shows the bulk annealed (*homogenized*) equivalent with dark SbSn cuboids and triangles, embedded in the ternary lamellar eutectic. Dark phase is the tin rich and the lighter one is the lead rich. Notice the large black cuboids and triangles of the SbSn phase in the bulk micrograph which are absent in the MS-ones due to the grain growth suppression imposed by the process ($v_r \sim 30$ m s^{-1}), though SbSn is still existing in the MS samples in crystallite forms within the dark and light phases, confirmed by x-rays analyses, so increasing the hardness of the ribbon samples by many degrees. Nital etchant was used [14,388].

On the other hand, columnar grains are always stable (plane front growth) in pure samples and can be stable in the case of alloys if the temperature gradient is sufficiently high. At very high growth rates, the onset of so-called absolute stability will lead to the occurrence of solidification involving a structureless solid-liquid interface [10,102]. The main implication of such confinement is the change in the system total energy and hence the overall thermodynamic stability. The emphasis is on the use of instability behavior as a means of understanding those processes that ultimately determine the micro- and nano-structure of a crystalline solid. Experimentally, Kramer *et al* [327,391] had performed experiments for determining the melt pool area oscillations (that affect the final ribbon thickness, average area and instantaneous cooling rates) with time, using a high-speed (26000 frame s⁻¹) CMOS camera, as in Fig. **23**, and measured the melt pool surface temperatures. With appropriate care, the temperatures at the surface of the melt pool can be determined to an absolute accuracy estimated at 10 °C and a relative accuracy of ± 2 °C (Fig. **24**), over a length of approximately 5 mm along the central axis of the ribbon.

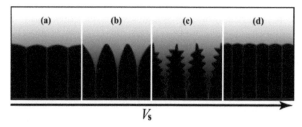

Fig. (21). Evolution of the solid-liquid interface morphology with increasing velocity. (**a**) Planar interface generally observed in single-crystal growth. (**b**) Cellular structures. (**c**) Dendrites are observed in classical metallurgical processes. (**d**) Flat interface obtained in rapid solidification. The interaction between transport and capillary factors controls the shape of the S/L interface [8,10,383].

Beyond 5 mm from the pool, the surface temperature was both too low to measure and had dropped out of the focal plane of the camera. The resulting temperature profiles, shown in Fig. **24**, clearly reveal the effect of wheel speed on the ribbon surface temperature profiles (in accordance with Figs. **4, 10** and **11**). They have shown that the wheel surface quenching rates are largely independent of the ambient pressure or temperature until the ribbon separation from the surface (the so-called free flight cooling stage), and the resulting amorphous ribbons leave the wheel surface with high temperatures (red-hot temperatures; ~ 800 °C) but with no change in structure or dimensions due to the low cooling rates of the ambient atmosphere (even with low wheel speeds but not lower than 7.5 m s⁻¹). Here, based on the experimental results of Fig. **24** and observed melt pool shapes, the distance that the ribbon remains in contact with the wheel at 20 m s⁻¹ is about 12 mm. This is consistent with the rapid drop in quench rate observed for larger contact distances. Since below 800 °C the contribution from radiative cooling diminishes rapidly, the chamber gas and pressure will influence the overall quench behavior [392]. Of course, the residence time on the quench wheel can be increased by increasing the quench wheel diameter and this is a common procedure for minimizing the wheel curvature and concentricity. Also, for initial experimental results, they found a large degree of scatter in the thickness measurements, indicating a high degree of both run-to-run variability as well as variability within any given run, indicating an oscillatory behaviour of the melt pool, as in Fig. **23**. The frequency of the variation in ribbons width was found to be 314 s⁻¹, as in Fig. **25**, independent of wheel speed [391], indicating that the melt pool oscillations cannot be attributed to any asperities in the wheel surface relief. However, a closer examination of the time-resolved image sequence coupled with the ribbon width measurements and the measured ejection pressure, suggests that the reduction in area is apparent, due to harmonic oscillations. These free body oscillations or perturbations can only be minimized by lowering the chamber pressure or constraining the free boundary conditions of the melt stream and pool, as in PFC. A comparison between the oscillation time scale and the melt pool residence time reveals that: while the oscillations are present even at very high spinning rates, they have a significant effect on melt pool dynamics only in the low velocity regime and may be a key contributor to the unsteady behavior observed below 10 m/s. Indeed, this is supported by the *in situ* imaging showing that the transition to unsteady melt pool behavior occurs at some wheel speeds that are consistent with the quarter-wavelength oscillation in the melt pool, determined from the ribbon-width periodicity [327,391]. Eventually, it is evident from previous discussion that the melt puddle stability and geometry has a significant control for the phase nucleation kinetics and competition and suggests upper and lower limits for phase nucleation or selection (*i.e.*; a *phase solidification window*) for the present system.

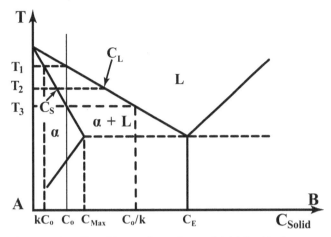

Fig. (22). Solidus-liquidus relationships for dilute binary alloys. For an initial liquid alloy composition of C_0, the equilibrium solid composition is kC_0. For a solid of composition C_0, the equilibrium liquid composition is C_0/k [8,10,102,383,389].

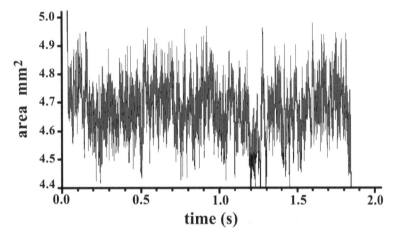

Fig. (23). Measure of the time varying area of the melt pool obtained using a CMOS camera with 256 x 256 pixels taking an image every 0.74 μs. The frequency of these oscillations is in accordance with the measured variability/periodicity in ribbon thickness and width [327,373,391].

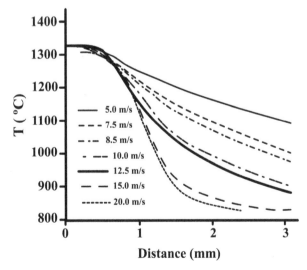

Fig. (24). Temperature profile from a 2-D thermal image that is averaged along the ribbon width, measured at the melt pool surface over a length of approximately 5 mm along the central axis of the dragged-away ribbon [327].

2.4.3 Ribbon Anatomy and Geometry

Ribbon geometry control and uniformity is of great importance in the fabrication of a product having uniform surface roughness and it is process-parameters dependant. Imperfections in CBMS ribbon geometry arise from a variety of sources which include the nature of the molten alloy stream, the substrate, and the atmosphere in which melt-spinning is conducted. Therefore, factors such as melt ejection nozzle geometric imperfections and vibrations imposed upon a free liquid jet [393] can cause premature disruption of the jet into droplets [338]. Also, melt puddle disruption caused by substrate surface roughness elements can result in molten alloy droplet spraying and in irregularities in ribbon width and thickness. In addition, the presence of foreign contaminant films on the substrate surface during CBMS can result in complete non-wetting of the melt on the substrate [394]. Fig. 26 shows the effect of melt-spinning on amorphous alloy ribbons in various gaseous atmospheres. In addition to the ribbon edge serrations, melt-spinning in a gaseous environment may cause increased ribbon surface roughness [395] and penetration of gas beneath the melt puddle to form small pockets in the underside of the ribbon [320,396], as shown in Fig. **27a**. The situation is met in any melt spinning device. Due to the accompanying process instabilities, there are some common surface textures of melt-spun ribbons other than the smoother ones. According to Steen and Carpenter [58], the ribbon surface not in contact with the wheel usually contains non-uniformities even under normal casting conditions. These textures appear to be a result of poor contact between the liquid metal and the chill wheel, as has been observed [320]. The first texture will be called *dimple pattern*. This imperfection is made up of relatively large depressions found on the meniscus side of the ribbon, which correlate with smaller depressions at the same location on the metal-wheel contact surface of the metal strip (Fig. **27a**).

This texture is found on all experimental runs and is by far the most common non-uniformity. Depressions on the wheel side are caused by the problematic air/gas entrainment or pockets when the liquid metal first contacts the wheel, leading to non-uniformities in heat transfer. The correlation with the dimple pattern on the upper surface has several possible explanations. At such locations in the puddle, the metal is not solidifying as quickly and, therefore, the ribbon may be expected to be thinner. Alternatively, such points in the melt puddle will be locally hotter and may cause a decrease in the surface tension on the upper surface. A *Marangoni instability* [397,398], which is a convection-induced instability, may then ensue and ultimately be frozen into the ribbon (inhomogeneity may exist due to the presence of capillary forces on the solid-liquid interface, developing *Marangoni convective flow*, which leads to morphological instability of the interface and inhomogeneous distribution of phases during solidification reaction) and these locations is characterized by coarser grains at the wheel side [58,338], as in Fig. **28** [327,399].

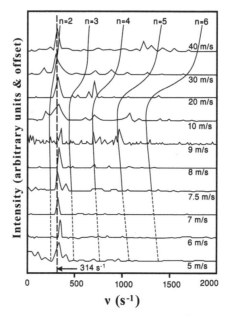

Fig. (25). Fourier transform of ribbon width oscillation showing that the characteristic frequency is approximately 314 s^{-1} for all wheel speeds, for the MS setup by Napolitano and Meco, falling between predicted n = 2 and n = 3 oscillation modes [391].

For the *streak pattern* (Fig. **27b**); this type of imperfection appears as long thin grooves in the direction of casting on the upper surface of the ribbon. The cause of the streak pattern may be similar to the dimple pattern. On the wheel side of the ribbon, thin lines of poor contact between the metal and the wheel are found, with the possibility of enhancement of the streaks by a *painting type instability*, like that found in coating of thin films.

The last non-uniformity is the *herringbone pattern* found on the wheel contact side of the ribbon (Fig. **27c**). This imperfection is caused by small nonuniformities in the fluid flow. The herringbone texture would be found, typically, when a small piece of metal protrudes slightly into the nozzle slot flow path. Some of the liquid metal would flow around the protrusion causing the liquid metal to contact the wheel in a nonuniform manner and to produce a characteristic pattern in the solid ribbon. This pattern may be desirable in some cases because it suppresses other instabilities such as the dimple pattern. Although the herringbone pattern on the meniscus side of the ribbon is produced by a similar mechanism to the dimple pattern, the upper surface of the ribbon is smoother overall. In most industrial applications, these ribbon patterns are not important because the MS ribbons are mainly pulverized and powdered for further consolidation routes, so producing bulk amorphous, nano-, or fine microstructural billets or sheets for the majority of industrial purposes. However, many applications and products demand regular ribbons such as metal fibers, machine blades and composites (*e.g.*; metal-metal or metal-ceramic composites). But to ensure less gas entrainment and thus guarantee a more consistent cooling rates and long range fine structure, one should perform the CBMS in a chamber of *rough vacuum* (104-103 Pa) in He-gas or He-4H (at%) mixture for excellent heat transfer with lower coasts [89,382,392]. Surface profilometry of ribbon that has been cast on a wheel with a regular pattern of tooling marks parallel to the wheel rotation shows that the wheel surface of the ribbon tends to mimic the quench wheel morphology (Fig. **29**). The degree of replication of the wheel surface is highly dependent on the gas pressure within the melt-spinning chamber [400], with poor replication at 1 At. and very good replication occurring at low chamber pressure, clearly indicating that trapped gas pockets are present at higher chamber pressures [392]. For more smoother ribbon surface, free of tool marks, one should make the substrate/wheel surface smoother by grinding and polishing it with a fine grit paper and finer polishing powder just before the run onset [304].

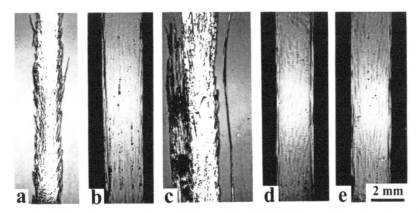

Fig. (26). Micrographs of MS ribbons' upper surfaces. (**a**) Conventional CBMS in air (dimpled and serrated or saw-like). (**b**) Coaxial jet casting under proper conditions. (**c**) Coaxial jet casting under improper conditions. (**d**) Vacuum-cast CBMS geometric equivalent of sample (**b**). (**e**) Helium-cast CBMS equivalent of sample (**b**) [306].

To analyze the microstructure, the cross sections of process ribbons shall be polished firstly by a 0.1 µm diamond or corundum grit powder and then etched, if the test needs, and with using a high resolution scanning electron microscope we can gain a clear fine structure images. Fig. **17a** and **17b** show the *Electron Backscattering Diffraction Pattern* (EBSP) orientation maps of ribbons obtained at 2.3 and 25.4 m s^{-1}, respectively. Typical thicknesses are ~ 200-600 µm and ~ 20-50 µm for 2.3 and 25.4 m s^{-1}, respectively [308]. The direction from bottom to top surface for each grain was colored using a reference triangle. In the case of a low rotation speed (Fig. **17a**), grains were fine near the bottom surface and then became larger (cellular-to-dendritic transition). Similar variations have already been reported [338,376] for thick ribbons. In contrast, fine grains were found throughout the thickness in the case of a high rotation speed (Fig. **17b**). Although the grain size appears to be non-uniform, it can be understood that all the grains in Fig. **17b** are reasonably fine when the magnification is adjusted with that of Fig. **17a**.

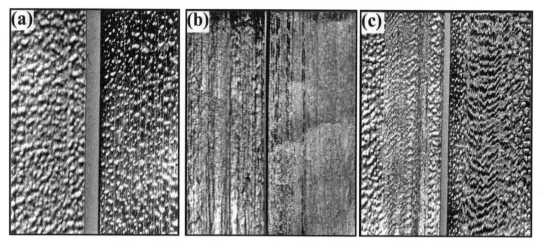

Fig. (27). MS ribbons surface geometries. (**a**) The dimple; (**b**) streak; and (**c**) the herringbone patterns [58].

The small EBSP image, at the left of Fig. **17a**, is the same image as Fig. **17b** but shown at the same magnification as Fig. **17a** [308]. Moreover, the grain boundary in Fig. **17b** is indistinct, in contrast to that in Fig. **17a**, which indicates that the solidification rate was more rapid [304,308,326]. Fig. **16a** shows a magnified scanning electron microscopy image at the bottom surface of the ribbon obtained at 25.4 m s⁻¹ and an arrow shows the rotation direction. It seems that the melt (B-doped silicon in this figure) solidified dendritically parallel to the rotating substrate. The EBSP orientation map of the area in the white box in Fig. **16a** is shown in Fig. **16b**. The direction from the bottom to the top surface for each grain was also colored using the reference triangle shown in Fig. **17**. The random colors indicate that no dendritic solidification occurred but rather *successive nucleation events* occurred at the melt/substrate interface.

When considering the solidification kinetics, the solidified morphology is the manifestation of solid-liquid interface stability (with the planar interface being the most stable mode), and is generally controllable by adjusting processing parameters according to stability guidelines. At ordinary solidification rates, where the *constitutional undercooling principle* applies (which determines that; when the speed of solute equals that of the S-L interface, partitionless solidification can occurs, depending on the constitution of the alloy system) [89,135,296,326], the interface increases its stability with an increase in the ratio between the liquid thermal gradient and solidification velocity, G_L/V_S, where;

$$R = -\rho A_R \, Y \frac{df}{dt} = \rho A_R \left(-Y \frac{df}{dt}\right) = \rho A_R \, V_S \, \dots \tag{2-25}$$

$$G_L = \frac{dT}{dy} = \frac{1}{\alpha}\left(T_m - T_{SL}\right) V_S \, \dots \tag{2-26}$$

and;

$$\dot{T} = V_S \, G_L = \beta \, Y^{-2} \, \dots \tag{2-27}$$

Y is the growing thickness ($Y = \overline{t} = Y$ (t), t is time), f is the *fraction liquid, fraction solid* is $f_s = 1 - f$. V_S is the solidification rate, *i.e.*; extraction of heat into the substrate drives the solid-liquid interface across the alloy layer with a velocity V_S [49,326]. α is the *thermal diffusivity* of the melt, T_m is the melt ejection temperature, T_{SL} is the solid-liquid interface (*i.e.*; solidification front) temperature (in °K) and β is the proportionality constant. At high solidification rates, where the *absolute stability theory* prevails [296,297], the interfacial stability increases with the rate of solidification, V_S. Note that the solidification velocity has reverse effects on the interface stability in the two solidification regimes. In addition, in the rapid solidification regime, the solidification velocity becomes the sole process parameter controlling the interface stability and morphology (at a fixed alloy composition). An adequate knowledge of the rapid solidification kinetics is therefore critical for an understanding of the rapid solidification microstructure.

The microstructure of a ribbon section parallel to the casting direction of a vacuum melt-spun Ni-superalloy (the Haynes™ 230 superalloy; 53Ni-14Co-17Cr-8W-1.75Al-3.5Ti-0.65Nb-2Mo-0.02C-0.01B, all in wt.%) is shown in the micrograph in Fig. **30a** by Huang *et al.* [326]. Columnar cells in the ribbon normal direction can be seen in regions close to the ribbon bottom surface. Above approximately the center of the ribbon thickness, the cells developed *side branches*, which became more pronounced near the ribbon free surface. The columnar dendrites are noted to lean toward the leading end of the ribbon at an angle of about 15 degrees, as also in Fig. **34**. The dendrite branches maintain an orthogonal relationship to the inclined main stems.

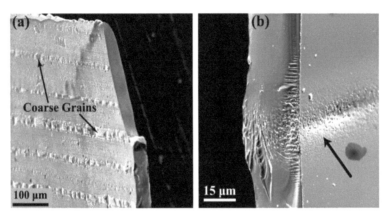

Fig. (28). SEM image showing the effect of regions of poor quench due to gas pockets between the melt-pool and the quench wheel for a magnetic alloy $Y_2Fe_{14}B$ melt spun at 12 m/s. The wheel side (**a**) shows numerous pockets aligned in the direction of the wheel rotation. These are regions of poor quench which propagate up to the free surface resulting in coarsely crystallized grains as seen also in (**b**) and indicated by the arrows [327].

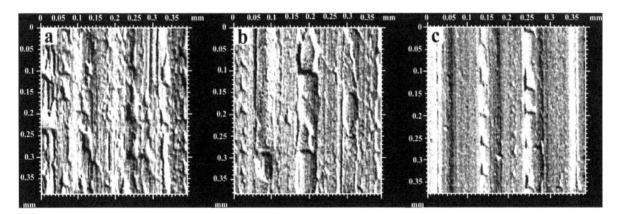

Fig. (29). Reconstructed profilometer images of the ribbon underside (*i.e.*; wheel surface side) for an alloy melt spun at 25 m/s. Image (**a**) was taken from a ribbon quenched using a chamber pressure of 750 torr Ar, image (**b**) was in a chamber pressure of 750 torr He and image (**c**) was in a chamber pressure of 250 torr He. Note how well the tool marks on the wheel surface are replicated in image (**c**) with the low chamber pressure [327].

The orthogonal side branches can be seen more clearly in Fig. **30b**, which shows the extensive dendritic columnar structure developed near the top surface of an 80-μm thick ribbon. The microstructural dimensions and morphologies at various positions in the 35-μm ribbon are further compared in the transmission electron micrographs of Fig. **31**. The TEM specimens were prepared by electro-polishing from either one or both of the surfaces. Fig. **31a** shows that the chill crystals at the wheel-contact surface of the ribbon are polygonal, straight-edged and of sizes ranging from 0.1 to 0.5 μm. Every crystal represents a single nucleation event. The nucleation density thus approaches 10^8 mm^{-2}. In Fig. **31b**, the cross-sections of the columnar grains at the middle of the ribbon are seen. The grains show the cellular morphology, having mostly *high-angle boundaries*, since they were developed from separate nucleation sites. Near the ribbon top surface, Fig. **31c**, the grains show the morphology of branched dendrites which seem to be in directions near the **[100]**-directions, judging from the branch symmetry, indicating a lower thermal conductivity at the free side of the ribbon by the run atmosphere [326].

Returning to Fig. **30**, microstructures of Fig. **30a** and **30b** show a deflection in the solidification orientation. The observed growth orientation adjustment is attributed to the convection effect of the liquid metal flowing transversely in front of the solid-liquid interface. The deflected solidification orientation due to the convection effect would have been suppressed; there were a layer of stagnant liquid metal separating the growing solid from the convected liquid. Based on this observation, the present superalloy ribbon-formation process is believed to be controlled by solidification heat extraction (thermal control). The result of Eq. (2-25) specifically describes the solidification rate occurring at the non-contact surface of a ribbon which has a thickness $\bar{\tau}$. For example, the solidification rate at the top surface of a 35-μm ribbon is ~ 100 mm per second (see Fig. **32**).

As implied by the continuity of the ribbon formation mechanics, Eq. (2-25) should also describe the local solidification rate at various positions through the ribbon thickness. This is valid, particularly when the boundary effect of radiation heat loss at the downstream surface of the melt puddle is negligible. For the latter interpretation of Eq. (2-25), $\bar{\tau}$ represents the distance away from the wheel contact surface. Eq. (2-25) thus shows that the local solidification rate within the ribbon increases rapidly toward the substrate contact surface (see also Figs. **32** and **33**). The present improved interfacial heat transfer in Figs. **30** and **31** is attributed to melt spinning in vacuum. However, it is inappropriate to assign a single value to the cooling rate for any chill block casting since the cooling rate can vary dramatically through the thickness of the ribbon and through the run itself, but rather it is better to give a range or a mean value for it [401]. This is particularly important when trying to produce amorphous ribbons in systems where the effective cooling rate is close to the critical cooling rate.

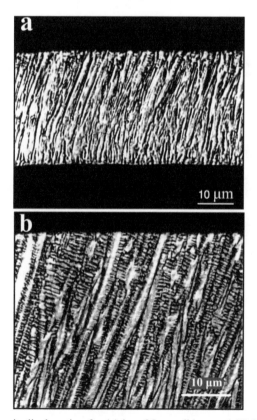

Fig. (30). Optical micrograph of a longitudinal section for (**a**) in a 35-μm ribbon and (**b**) in an 80-μm ribbon. The *non-contact surfaces* of the ribbons are shown at top. The right-hand end of the ribbons was cast first [326]. Note the well-developed dendritic secondary arms near the top after the middle of the ribbon.

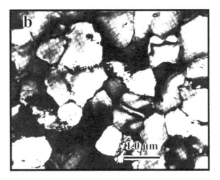

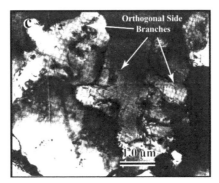

Fig. (31). Transmission electron microstructure in a 35-μm ribbon: (**a**) at the wheel contact surface, (**b**) at the middle and (**c**) at the non-contact surface [326]. Notice the developed secondary arms in upper left-hand corner and in the center of picture (**c**) for the free ribbon surface.

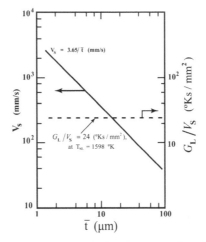

Fig. (32). Relationships of solidification rate, V_S, and constitutional supercooling characteristic ratio, G_L/V_S, as a function of ribbon thickness or distance, $\bar{t}$, from the ribbon bottom surface for the used Haynes™ 230 superalloy [326].

Such a situation often results in a ribbon that is amorphous on the wheel side and crystalline on the free side, as determined using X-ray diffraction methods [327]. Direct evidence of the improved ribbon-wheel contact *in vacuum* is the *uniform contact pattern* [378] and the high nucleation density ($\sim 10^8$ mm^{-2}) observed (Fig. **31a**). For comparison, the contact surface of air-cast ribbons is marked with air pocket [402,403], as in Figs. **17** and **28**, and the nucleation density there is less than about 10^6 mm^{-2} [388,402]. Finally, Fig. **30** also shows a deflection in growth direction from essentially along the ribbon normal in the lower one-third of the ribbon to $\sim$ 15 degrees inclination toward the wheel rotation direction in the upper two-thirds of the ribbon. The deflected growth [296,405] can be attributed to the effect of melt *counter convection*, which arises as the solidifying layer travels with wheel rotation relative to the stationary melt puddle. The melt convection is present throughout the entire ribbon formation process, but Fig. **30a** suggests that the convection influence is effective only in the upper regions of the ribbon thickness. This *crossover* in mechanism is also shown to result from the local solidification rate variations within the ribbon (Fig. **32**). The convection-affected ribbon microstructure is schematically illustrated in Fig. **34a** [326]. Note in the figure the orthogonal relationship between the dendrite arms and the inclined dendrite main stems, which is consistent with the ribbon microstructure (Fig. **30**). By comparison, previous observation [405-409] showed that a transverse convection (*i.e.*; advection) and convection-driven instability, that are related strongly to the melt surface tension and composition, lead to a growth direction deflection but not a crystallographic reorientation (as in Fig. **34b**) [326,383,408]. Here, the dendrite side branching direction is not affected by the melt convection; despite the dendrite main stem inclination. The discrepancy in results, albeit not fully understood, seems to be associated with the differences in the regimes of solidification, local cooling rates and melt flow rates.

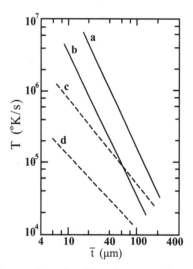

Fig. (33). Correlation of cooling rate vs. distance from the ribbon bottom surface characterized for the vacuum melt spinning process for the same alloy: (**a**) by the dendrite arm spacing measurement technique [8] and (**b**) by the melt puddle length measurement technique (for the same specimen). For the air melt spinning process: (**c**) by the dendrite arm spacing technique, and (**d**) by the melt puddle photo calorimetry technique [374]. All suggests that $\bar{t} \propto \dot{T}^{-1/2}$ [326,405].

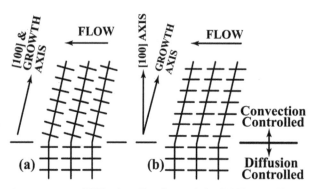

Fig. (34). Relationship between the apparent solidification direction and the [100]-crystallographic orientation observed in: (**a**) the present Haynes™ 230 alloy melt-spun ribbons, and (**b**) the transverse forced convection experiment. Dendrite branches, which were suppressed in the diffusion-controlled region in Fig. **30a**, are added to (**a**) for clarity [326,406,407].

2.5 SUMMARY AND CONCLUSION

In the present chapter we have reviewed the controlling parameters of the CBMS technique. All transport phenomena of the melt flow and solidification kinetics confirmed that the melt-substrate heat transfer coefficient is the key factor in this process that is dominated by the heat transfer not by the momentum transfer, leading to the formation of solid amorphous, nano- or micro-structured metallic ribbons. All experimental studies show that heat transfer coefficient is directly proportional to the wheel speed, and ribbon thickness is in inverse proportion with it. In addition, the thickness is inversely proportional to the square root of the *average* cooling rate. Melt puddle residence time can be increased by increasing wheel diameter, resulting in long range of higher local cooling rates with more finer microstructures. Instability of the melt puddle is strongly affecting the ribbons formation and fine microstructure and can even stop ribbon formation process. Running the process in low pressure chamber (in *rough vacuum*, between 10^4-10^3 Pa for better cost saving) reduces the gas entrainment instability and improves ribbons fine microstructure. Though it is still an open field of research, melt puddle instability must be controlled for regular ribbons and cooling rates with no model discrepancies.

CHAPTER 3

Physical Properties and Applications of Rapidly Solidified Metals and Alloys by CBMS Technique

M. Kamal[a]* and Usama S. Mohammad[b]

[a]Chairman and Professor of Metal Physics, IUPAP-C5 Member, Department of Physics, Faculty of Science, Mansoura University, Al-Mansoura, Egypt and [b]Ph.D. Researcher in Metal Physics and Materials Science, Department of Physics, Faculty of Science in Damietta, Mansoura University, New Damietta City, Egypt

Abstract: Rapid solidification processing by CBMS tech affects the transport phenomena controlling the phase formation and growth kinetics which lead to the emergence and trapping of new phases in the solid state. These new amorphous, nanostructures and/or microstructures make the new bulk material to have new physical properties differing from those prepared by conventional manners. Many new CBMS materials found their way to the new and advanced applications and replaced also the oldest ones.

Key Words: Rapid solidification properties; melt spin; metastable amorphous powder consolidation; particle size; light alloys; quasicrystal application.

3.1 INTRODUCTION

The actual product of rapid solidification is micro- or nano-crystalline, amorphous, quasicrystalline, or a mixture of the previous phases (stable or, frequently, metastable) and it is sensitive to both alloy composition and solidification conditions. Early attempts to represent this behaviour [410] involved mapping fields of occurrence of different products as a function of composition and cooling rate. While composition can be determined with some confidence, cooling rate on such plots was normally estimated from thickness or *dendrite arm spacing*, and is thus subject to uncertainty.

The general practice is to map constitution as a function of composition alone [19] for just one condition of quenching from the melt (Fig. **1**) [411], yielding *compositional limits* for formation of various products for the particular set of processing conditions involved. Cooling in this context is simply a direct means of achieving the supercooling or solidification front velocity required to achieve the desired effect. For example, the production of direct amorphous ribbons can be achieved by applying a liquid helium stream on the lower part of the wheel/rotor in the CBMS device, for low temperature physics studies. Corresponding nucleation-and-growth analyses for competition between alternative crystalline phases and morphologies at high solidification rates have to be developed yet. Complications arise from the difficulty of realistically specifying the nucleation and growth kinetics of so many possible alternatives, when quite small differences can evidently determine the outcome. So, more use of *continuous-cooling-transformation (CCT) diagrams* and *freezing diagrams* should be well-considered.

Much use has been made of power relationships between dendrite arm spacing λ or eutectic spacing Λ and cooling rate $\dot{T}$ or solidification front velocity V_S for estimating locally operative $\dot{T}$ and V_S from measured λ or Λ, where [96];

$$\lambda = B\dot{T}^{-n}, \quad \Lambda = BV_S^{-n} \; \dots \tag{3-1}$$

Eqs. (3-1) have some theoretical basis [8,10,96,102,412] when the secondary dendrite arm coarsening controls the

---

Address correspondence to M. Kamal: Chairman and Professor of Metal Physics, IUPAP-C5 Member, Department of Physics, Faculty of Science, Mansoura University, Al-Mansoura, Egypt; E-mail: kamal422002@yahoo.com

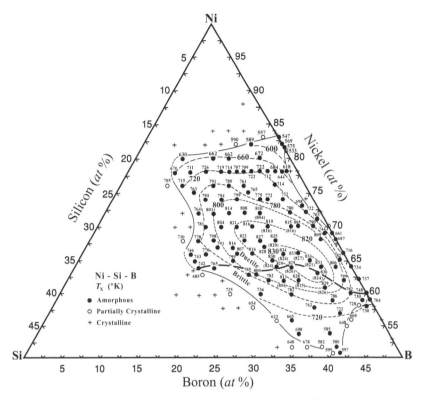

Fig. (1). Ternary composition field glass formation map for the Ni-Si-B Melt-Spun ribbons of thickness 17 ± 3 μm for just one condition of quenching from the melt. On the graph, lines represent compositional limits for formation of various products for the particular set of processing conditions involved [19,411].

final spacing (for λ, n = 1/3) or when the eutectic is regular (for Λ, n = 1/2), B in Eqs. (3-1) depends on the alloy system (see Fig. **2**). The outcome of predictions of the dependence of *grain size* on cooling rate is especially sensitive to values chosen for physical parameters governing rates of nucleation and growth, and there is an almost a large absence of *exact* experimental data on dependence of grain size on cooling rate during solidification that could be compared with predictions; a shortage in solidification theory that may be resolved in future by advanced experimental techniques [142,163,415]. However, it has proved that these relations apply on some alloy systems and deviate from the predicted optimum values for many ones. This is attributable to: (*i*) the approximation made in the microstructure coarsening models, (*ii*) contributions from primary spacing predicted [10,413,414] to have a more complex dependence on V_S and G_L, and (*iii*) increasing suppression of secondary arms at high cooling rate to stabilize a cellular structure.

3.2 MECHANICAL PROPERTIES OF RSP PRODUCTS

The recent progress in rapid solidification processing (RSP) of crystalline alloys has allowed a degree of microstructural control which has never been achieved before by any conventional ingot metallurgy methods. The microstructural refinements include extended solid solubility, increased chemical homogeneity, refined grain size, and formation of metastable phases. These attributes of RS have been utilized in many alloy systems in designing alloys which exhibit principally improved mechanical properties [415]. Particular examples of alloy systems, which have received a wide attention, are; high-strength corrosion-resistant magnesium alloys [416-419], magnesium-lithium alloys [420-422] (the lightest high strength alloys ever made), magnesium-zinc alloys [421,423], reduced-density aluminum lithium alloys [424-427], high-temperature aluminum alloys [188,193,428-433], high-temperature magnesium alloys [434-441], and titanium alloys [442-446]. Other alloy systems, such as copper- [447-451], nickel-[452-457], iron- [458-461], niobium- [462-465], tungsten- [466,467] zirconium- [468- 471] based alloys and intermetallic compounds [371,472-474] which show potential benefits of RS are also of interest [415]. Aerospace industry is the common customer for new RS materials (alloys and composites) that must be of thermally stable, high-strength and light-weight properties [85,475-478]. One of the greatest achievements of RS-CBMS technology is the development of *light alloys*, where the thermally-stable high-strength aluminium alloys are a design engineer's dream and cheaper than titanium. Rapid solidification is the key technology, especially by the universally favoured CBMS technique, for making them by grain refinement and extended solubility that make the phase crystal constants to increase and thus producing a lighter alloy composite structure. Bulk fine-grained light alloys (mainly; RS Al-base, Mg-base and many Ti-Al alloys) are featured with amorphous, nanostructure and/or quasicrystalline phases that give the final product its exceptionally improved high mechanical properties, high thermal stability (more than 400 ºC), wear and corrosion resistances so as to manufacturing aircraft parts with more strong, complex and thinner sections [188,193,475,479-486]. The same is said for other hi-tech industries like modern automobiles, robotics, MEMS (Micro-Electro-Mechanical Systems) and NEMS (Nano-Electro-Mechanical Systems), especially amorphous soft magnetic components, with suitable modifications. Even the new armored vehicles use thick plates of mixed nano-amorphous Al-base composites to increase operation ranges and speeds, instead of the common heavy steel counterparts. For chemistries based on Al-Fe, Al-Li, Mg-Li and Mg-Al-Zn-Re (*Re* is the Rare Earth element in the alloy system), the high solidification rates provided by substrate quenching have provided rapidly solidified alloys with fine-scaled *uniform microstructures* and outstanding sets of balanced properties [415].

Another benefit of CBMS technology is the manufacture of high strength, superplastic and high thermal creep resistance intermetallic alloys that are used world-widely in aircraft turbine engine parts. Normal intermetallics are brittle, difficult/impossible to work and show stoichiometry-dependant mechanical anisotropic behaviour, for both single crystal or polycrystalline forms [487], despite of other valuable features like high melting points, low density, good oxidation resistance (for aluminides and silicides), high strength and strain-hardening rates at elevated temperatures. By *CBMSing*, a fine-grained intermetallic bulk material is produced with improved mechanical and thermal resistance properties, enabling the manufacture of many complex sectors for turbine engines parts. An outstanding example is the development of γ-TiAl alloys that led to the invention of the dominant lightweight Ti-48Al-2Nb-2Cr (Ti-48-2-2; at.%) alloy composition for turbine engine blades and compressor case. γ-TiAl based materials are inherent *non-burning chemistry* (where titanium alloy use is limited by the risk of titanium fires), and pursued mainly because of the desire to raise the thrust-to-weight ratio of high performance supersonic and hypersonic aircraft engines. Niobium acts for reducing any thermal creep rates so extending the thermal stability and for *β-stabilization*, while chromium for better low temperature ductility, excellent high-temperature strength retention, high corrosion and oxidation resistances and more stiffness [488-490]. Melting of this alloy is performed by the *skull vacuum arc melting* technique [371] (the melting arc may be replaced by an electron or laser beam) that is modified to pour the melt from the water-cooled crucible on the rotating wheel in order for producing the MS ribbons for later P/M consolidation. Turbine rotor and blades, made by this alloy, were run for more than 1600 rejected takeoff cycles without any observed damage [446]. The database established for advanced γ-TiAl based alloys indicates that most of the properties (when adjusted to density) appear to be comparable to, or better than, those of many nickel-based superalloys, for which they may be substituted [490-492].

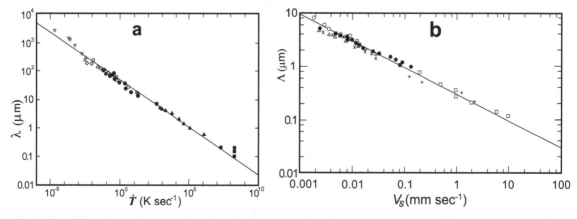

Fig. (2). (a) Dendritic arm spacing λ as a function of cooling rate for Al-4 to 5 wt.% Cu (open points) and Al-7 to 11 wt.% Si (filled points). **(b)** Eutectic interlamellar spacing Λ as a function of growth velocity for Al-Al₂Cu eutectic [19,96].

In general, commercial application of a new alloy is possible only if an important material property is significantly improved without concurrently sacrificing important secondary properties. For example, for structural alloys the primary mechanical properties such as strength, ductility, density, and stiffness should be improved while at least maintaining the secondary properties such as fatigue, creep, fracture toughness, corrosion, and oxidation resistances. There are numerous examples of cases where attempts to improve one property results in a degradation of other properties. For example, improvement in strength is generally associated with a decrease in toughness. Although the design of new alloys to take advantage of RS will vary depending on the particular property requirements, there is one basic requirement for the development of high-performance structural alloys by RS, *i.e.*, retention of beneficial modifications of the as-cast microstructures during subsequent *powder metallurgy* (P/M) processing, as we will see later. In general, the as-cast microstructure is difficult to preserve if one uses conventional P/M consolidation techniques. Thus, in order to take advantage of RS for the development of high-performance alloys, *thoughtful alloy design is necessary* such that the desirable microstructure, that is present in the rapidly solidified alloys, is not lost during high-temperature consolidation [18,19,313,415].

It is relatively easy to generate increased levels of strength through formation of dispersoids, but it is difficult to produce a fine uniform dispersion which will yield, simultaneously, a high strength, a high toughness, and corrosion resistance, in comparison to the conventional ingot coarse-grained ones, especially bulk amorphous alloys containing chromium, *e.g.*; the amorphous Ni-Ta alloy is completely immune against the boiling concentrated HNO_3 [19,415,493-495]. Similar attempts are under way to produce other high performance alloys; these efforts show promise, but much remains to be done. Classic work established the effect of increased cooling rate in increasing tensile strength and ductility of casting alloys through its effect in reducing dendrite spacing [8,496]. Its effects in multiplying strength through dispersion/age hardening [111,497,498] and in increasing fatigue and stress rupture

Fig. (3). (**a**) Forged rapidly-solidified aluminium tow bar made by CBMS for automobiles; high stiffness and light weigh. (**b**) Different products by CBMS and atomization rapid solidification processing technologies for automotives, robotics and aerospace industries [294].

lives through refinement of metallic impurity inclusions [127,499] have been demonstrated, because fine-grained bulk metals and alloys have more grain boundary than coarse-grained ones, so that more hindering exists against the atomic dislocation and slip motions. Such comparisons are essential if the precise contribution of rapid solidification to observed property changes is to be assessed in any given circumstances. These properties actually resulting from rapid solidification evidently depend on the structural changes produced in each particular case. For example, metallic glasses are characterized by high mechanical strength with ductile behaviour in bending, shear and compression; high reactivity or high corrosion resistance depending on composition; superconductivity or high resistivity depending on temperature; and outstanding soft magnetic behaviour for appropriate compositions [96]. High mechanical strength is evidently also attainable in micro- or nano-crystalline materials as a result of refined microstructure combined with increased alloying (due to the extended solubility accompanying the RSP) [500]. This combination can at the same time give rise to *superplastic behaviour* at elevated temperatures ($T \geq 0.4T_m$), giving elongations of 500% (in many cases the elongation reaches over 3000%) of the original specimen length without rupture [11,19,501]. Improvements in such properties as thermal stability and elastic stiffness can result from being able to make alloy additions prohibited to ingot processing [502]. So, for materials reliability, there is along two related paths. The straightforward first path is simply *to optimize the strength-toughness durability through materials development*. Some approaches to this goal are: (*i*) control of grain structure and second-phase morphology, usually by increasing uniformity and decreasing grain and precipitate size; (*ii*) design the material as a composite, making use of the properties of matrix and fiber to optimize overall results; (*iii*) attention to embrittlement factors such as external chemistry, temperature, ... etc [503]. The appreciable capability of achieving more consistent properties, through consolidated sections independent of their size, is itself a notable effect of rapid solidification compared to ingot solidification, leading to more predictable and controllable behaviour in processing and in service (as in Figs. **3**, **8**, **11** and **17-19**).

3.2.1 Elastic Properties

Among early observations of the unusual mechanical behavior of materials possessing a fine-grained structure was the reduction of the elastic constants, in comparison with the corresponding coarse-grained materials; this reduction was often of the order of 20-30%. The first measurements of the elastic constants and in particular of the Young's modulus E of nanocrystalline materials were performed with samples prepared by the inert gas condensation method. One of the first proposed explanations [504] was that elastic moduli of the grain boundary regions are much smaller than the corresponding moduli of the grain bulk region. In terms of the linear theory of elasticity established for polycrystalline materials, the elastic constants are then expected to be reduced by a fraction determined by the volume fraction of the grain boundaries:

$$E = f_{sc} E_{sc} + f_{gb} E_{gb} \ldots \tag{3-2}$$

where f_{sc} and f_{gb} are the bulk (single crystal) and grain boundary volume fractions, E_{sc} and E_{gb} are the corresponding elastic moduli (see the coming subsection of *theoretical density* for more information). This last equation represents the rule of mixtures for the elastic modulus directly, and since $E_{gb} \ll E_{sc}$, so as the grain size decreases so the total modulus E, and E_{gb} is the modulus lower limit. Of course, nanostructured bulk materials are of lesser Young's modulus than their polycrystalline counterparts [505]. Presence of defects, especially pores that make cracks grow out of them, reduces the modulus clearly, so the consolidation technique must be chosen carefully. For example, nanocrystalline copper samples, processed by *equal channel angular pressing* (ECAP), have values of elastic moduli lower by 10-15% as compared to coarse-grained copper [506]. Besides the effect of the sample geometry, there is a size effect, due to f_{gb}, when the grain boundary thickness becomes comparable to the specimen thickness. This length-scale will exist in any polycrystalline material and is not specific to thin films. Therefore, the nanostructured thin ribbons/films have much lower E-values than the micro- or sub-microstructured ones [507-509].

3.2.2 Strength and Hardness

The strength of an alloy can be improved by increasing the strength of the matrix (solid solution strengthening), by introducing second-phase dislocation barriers (precipitation and dispersion strengthening), by increasing the boundary area (boundary strengthening) or by all aforementioned mechanisms. Decreasing the grain size increases the yield strength, toughness and hardness, and decreases ductility and elasticity [18,19,56,127,128,313-512], as in Fig. **4a**. For melt spinning, this means that as the wheel speed increases, the hardness and strength of the final product increases as the ribbons grain size decreases. As an example, the remarkably high hardness of martensitic iron can be largely, or even wholly, attributed to *grain-size strengthening*. A study by Mawella *et al* [513] of a melt-spun ultra-strong steel indicated that the considerable strength increase generated by melt-spinning was primarily due to refinement of the martensite structure. It has also been established that high-carbon cast iron (2.4-3 wt% C) following RSP by centrifugal atomization consists largely of retained austenite. Subsequent annealing produces a microcrystalline (α+carbide) structure. Even at 650 °C this microcrystalline structure is so stable that the material behaves superplastically [514]. The conversion of brittle cast iron into a superplastic material is a particularly striking indicator of the changes in metallurgical characteristics obtainable by RSP [127]. Even the malleable bearing alloys benefit more strength and mechanical energy dissipation by CBMSing [2,388,515-517].

Such an increase in strength can be attributed to the sole effect of grain refinement through the well-known *Hall-Petch relationship*, which is one of the best-known and effective relationships between microstructural geometry and properties. It is relating the interaction of dislocations with grain boundaries and predicting a *linear dependence* of the yield stress of polycrystalline materials on the square root of reciprocal grain size [520-524]. Zener first suggested that there should be an inverse square root of grain size dependence for the polycrystal yield stress based on analogy of a slip band and a shear crack [525]. His theory conforms qualitatively to the experimental observation that the dislocation density increases with decreasing in grain size where grain boundaries acts as a barrier to dislocation motion. The *minimum* yield strength of the material follows a Hall-Petch (H-P) law by:

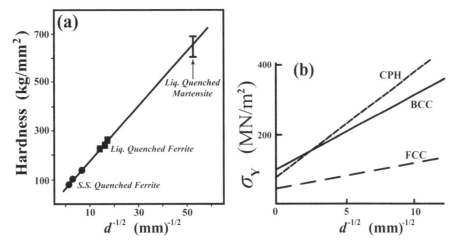

Fig. (4). (**a**) Microhardness vs grain size for pure iron [127,518]. (**b**) Grain size-dependence of the yield stress for crystals of different crystal structure [519]. Note that most FCC metals (Cu, Al, Au, Ni) have small K_Y, while most BCC metals (Fe, Mo, W, Ta) have high K_Y.

$$\sigma_Y = \sigma_o + K_Y d^{-1/2} \dots \tag{3-3}$$

or (substituting with the Vickers hardness);

$$H_V = H_{Vo} + K_Y' d^{-1/2} \dots \tag{3-4}$$

and;

$$H_V = H_{Vo} + K_Y'' \dot{T}^{1/2} \dots \tag{3-5}$$

where σ_o is *the friction or threshold lattice stress* (exerted by the grain boundary against the *unlocked* dislocations along the slip band/plane and represents the overall resistance of the grain crystal lattice by all strengthening mechanisms put together against dislocation movement) and K_Y is the *locking* or *hardening coefficient*; $K_Y = 0.16$ $\sqrt{b}$, b is the Burger's vector and d is the grain size (The Y-subscript is for the *Yielding* process) [519,522,524]. H_V is the Vickers hardness, H_{Vo} , K_Y' and K_Y'' are constants analogous to those in Eq. (3-3). K_Y depends on many factors including the crystal system geometry, number of slip planes within the local crystal system, the nature of impurity/solute atoms that block or lock the dislocations and the density of dislocations pile-ups [526]. Strong locking implies larger K_Y. A multiplicity of slip systems enhances the possibility for plastic deformation and so implies a small K_Y. A limited number of slip systems available would imply a large value of K_Y, as in Fig. **4b**. The slope of the plot is higher for ordered than disordered materials, as it is expected since cross slip is more difficult in the ordered condition, due to grain boundaries [487].

While the H-P relationship is a very general one and can be applied for any grain boundary type, it must be used with some caution. The approach breaks down or deviates in two limits. First, at very large grain sizes; where the observed yield stress does not necessarily approximate that of the single crystal same. Second, the H-P dependence of yield strength on grain size appears to be inaccurate for very small grain sizes. This is sometimes known as H-P breakdown or deviations from H-P behavior. Indeed, at very small grain sizes, it has been claimed that the yield strength of materials decreases with grain size, sometimes called the inverse H-P effect [528]. Approximately, most references don't agree about the fine grain size limit for the H-P relation and most of them put the limit near the borders of the nanosize (about 200 nm), but experimental results are not always fully convincing. Matters are greatly complicated in nanoscale materials, as differences in specific defect distribution or even the complete lack of pre-existing dislocations can greatly alter the point at which the material yields. As a result, nanoscale material volumes can sustain stresses significantly higher than those sustained by their bulk analogues. More recent investigations, using hardness measurements with the *nanoindentation* techniques [529-533], show that the Hall-Petch law seems to break down below a critical grain size of the order of 20 nm (because of the increasing fraction of atoms that are contained in grain boundaries) [534-539] that should be the lower limit for the *classical linear* H-P law. However, at the smallest nanocrystal grain sizes (< 20 nm), different experiments have evidenced different behaviors for the dependence of hardness on grain size. All the following behaviors were found by different researchers [540-542] (*i*) positive slope (*normal* Hall-Petch behavior), often with a decreasing gradient at some critical size, (*ii*) essentially no dependence (almost zero slope) and (*iii*) in some cases, a negative slope (sometimes below a critical size) [508]. Many modifications and models have been done for the H-P approach giving many expressions and values for K_Y, but experiments didn't confirm them clearly [512,526,543-548]. This is mainly due to the difficulty in manufacturing a material with very small grain size and the confusion in data relating to the commercial and industrial routes of the preparation of test samples. For instance, different elaboration processes (sintering, large deformation, … *etc.*) generate different types of grain boundaries (low angle sub-boundaries, various types of high angle boundaries, twins, … *etc.*) which oppose different resistances to dislocation movements and hence give rise to different values of K_Y [549] and thus grain size changes significantly after the consolidation technique.

To sum up, according to Ramesh and others after reviewing many of literatures about this law, there is no so much an inverse Hall-Petch effect (*i.e.*; a decrease of the yield strength as the grain size decreases beyond the critical size), but rather a *reduction in the degree or rate of strengthening* that is obtained as the grain size is decreased, as in Fig. **5** [509,550]. So, the H-P relation is still valid until the grain boundary width or thickness becomes comparable to the bulk or single crystal grain size at the aforementioned nanosize limit where the linear relation turns into an increasing curved one, with positive or negative slopes. Finally, may the rapid evolution of consolidation techniques make this controversy to be resolved in the near future! But be careful. Whatever the material, grain size effects are very sensitive to temperature and of course to strain rate.

In addition, grain size has opposite effects on the flow stress (*i.e.*; the stress to cause plastic flow) at high and low temperatures; the finer the grain size, the higher the flow stress at low temperature and the lower the flow stress at high temperature. Therefore, a cross-over temperature can be defined where the effect is reversed [551]. To increase the creep resistance at elevated temperatures for fine-grained materials, the alloy system must contain some amounts of low diffusion elements with traces of other elements that make intermetallics with the base metal. These trace intermetallic-making elements result in the formation of nanosized intermetallic phases that must be dispersed uniformly and homogeneously within the alloy, and, therefore, surrounding the grain boundaries of the fine structure, hindering any diffusion mechanism (including the grain boundary diffusional switching) and saving the bulk structure unchanged by recrystallization. Common examples are the RS superalloys. Other fascinating examples are the thermally stable high-strength aluminium alloys. These alloys contain high concentrations of slowly diffusing transition metals (TMs) such as chromium, iron, nickel, titanium, zirconium or manganese. Some of these metals (Fe, Ni, Mn, Cr) are already employed to increase the thermal stability of aluminium alloys, especially iron, as they suppress any induced recrystallisation. Various RS Al-TMs alloys have been studied so far. Definitively, they are Al-Fe, Al-Cr and Al-Ni alloys with ternary and other additives like; vanadium, silicon, scandium, yttrium, zirconium, neodymium, samarium and titanium are among the most important [203,552-555]. These systems have been regarded as possible competitors of more expensive Ti-based alloys or heavier steels in some applications as they achieve strength levels above 1000 MPa in the RS state (Al-Ni) [553] and acceptable strength up to 400 °C (Al-Fe) and more [433,436,486,555-559].

A similar size-dependence is found for *strain-hardening* of bulk fine grained structures. Since more slip systems are usually operative near the grain boundary than in the grain center, the hardness usually will be higher near the boundary than in the center, due to the dislocations pile-up head stress at the boundary [519,522]. As the grain diameter is reduced, more of the effects of grain boundaries will be felt at the grain center. Thus, strain hardening of fine-grained metallics will be greater than in a coarse-grained polycrystallines. The exception is in the case of nanostructured metallics. As the grain size decreases the dislocation density/storage decreases, so nanostructured bulk samples strain hardening is small, very small or zero, especially for grain sizes < 30 nm, even though quantitatively the additional strengthening provided, due to the dislocation density, may be the same as in the conventional grain sized version of the material [309,508,509]. As a general, the stress-strain curves of nanostructured specimens exhibit essentially elastic-perfectly plastic behavior; meaning no measurable strain hardening was observed. Strain hardening is not observed and thus conventional dislocation mechanisms are not operating. The lack of dislocations is the result of the image forces which act on dislocations near grain boundaries. Besides, shear banding is also the deformation mode observed in amorphous metallic alloys and in amorphous polymers [508]. This suggests an evident similarity between deformation mechanisms in nanocrystalline materials and in amorphous materials, even though not all tensile data on nanocrystalline materials exhibit a lack of strain hardening. In general, amorphous materials exhibit many of the phenomenological characteristics of deformation in nanomaterials; that is, shear banding, asymmetry between tensile and compressive behavior, and perfectly plastic behavior. Unfortunately, deformation mechanisms are not completely understood in amorphous materials either [508]. This explains why most nanocrystalline materials are often extremely hard and brittle, for instance; with a low tensile ductility at room temperature with a very small elongation to failure, as a result of the absence of strain hardening. To improve high strength, satisfying ductility and incorporating a strain-hardening mechanism, bulk nanostructured and metglass materials should be carefully annealed to increase the density of submicron-sized grains (that have good dislocation storage ability) within the bulk structure which becomes a heterogeneous nanostructure with a small percent of micro- or sub-microstructures [509]. In addition to their dominant wear resistance, these composite amorphous, fine-grained or mixed structures are the main candidates for the common practice of MS bulk products as wear-resistant cutting tools and automotive spring components in many engineering fields, even for shaving razor blades and metallic penetrators [510].

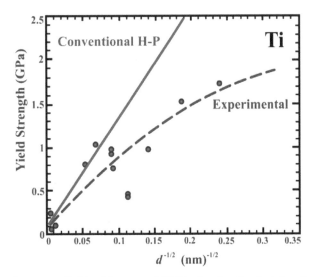

Fig. (5). Hall-Petch plots for titanium, compiled by Meyers et al [550], showing the deviation from classical Hall-Petch at small grain sizes. Similar behavior was observed for other metals like copper, iron and nickel, with some different degrees of data scattering and line curvature. Note the strength appears to plateau but not decrease with decreasing grain size [509].

3.2.3 Superplastic Behavior

Interface diffusion, in combination with a high fraction of interfaces, gives rise to modified physical properties of fine- and ultrafine-grained solids. When studying the creep properties of a material (with a grain size d) by applying a constant load M at a certain temperature T, one gets the relation:

$$\sigma = C\left(d\varepsilon/dt\right)^{m} = C\,\dot{\varepsilon}^{m}\,\big|_{T,M,d}\,\cdots \qquad\qquad (3\text{-}6)$$

where *C* is the *strength coefficient* (not the Young's modulus of elasticity confined to the isotropic elastic deformation); it is a temperature- and microstructure-dependent parameter, $\dot{\varepsilon}$ is the true strain rate and *m* is the *strain-rate sensitivity exponent* and can vary from 0 (perfectly brittle) to 1 (perfectly ductile) and its reciprocal value, the *stress-sensitivity*, *n*, is also commonly reported. Most metals and alloys normally exhibit m < 0.2 whereas superplastic alloys typically have values of m > 0.33. The magnitude of *C* at room temperature typically ranges from as low as approximately 10 MPa for pure aluminum to about 1 GPa for titanium; it decreases with increasing temperature. At high homologous temperatures and low stresses where dislocation creep rates become very small, inelastic deformation is dominated by diffusional flow involving shape changes of polycrystalline samples caused purely by *diffusional material transport*. This deformation is of particular importance in metals, intermetallics and ceramics in which very high lattice resistance to dislocation motion extends to very high homologous temperatures. It relies on material transport by *volume diffusion* and is dominated at lower temperatures, especially in fine grained material, by *grain-boundary sliding and diffusion fluxes* rather than volume fluxes. Many very fine-grained materials exhibit, at high temperatures and relatively low stresses in the diffusional flow response regime, a remarkable stable extensibility and avoidance of ductile or creep fracture resulting in extensionally large strains to failure often by rupture. This behavior, quite characteristic, has been called *superplasticity*; a manifestation of an exceptionally enhanced strain to failure. It offers the possibility of forming complicated shapes at high temperatures, accompanied by increasing their room temperature strength, ductility and ability to be machined as well. Originally discovered in 1912 by Bengough in (α+β) brass, superplasticity is now employed industrially in metal shape forming operations around the world, with primary applications in the aerospace and automotive fields [522,560-564].

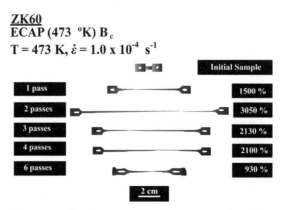

Fig. (6). Appearance of specimens of the ZK60 alloy (Mg-5.5Zn-0.5Zr, wt.%) tested to failure at 200 °C with an initial strain rate of 1.0 x 10^{-4} s^{-1} after processing by *ECAP* (as we will see later) for different numbers of passes. The upper specimen is untested [569].

We can define superplasticity as the capability of polycrystalline materials, particularly two-phase eutectic or eutectoid alloys and peritectically-formed intermetallics, to exhibit very large tensile deformations without showing fracture or necking. Homogenous elongations in the range 100-3000% and more (at strain rates on the order of 10^{-4} to 10^{-2} s^{-1}, as in Fig. **6**, depending on certain temperature ranges) are considered to be the typical defining feature of this phenomenon, with respect to conventional behavior that has elongations of the order of some percent [508,519,522,565,566]. As an example, electrodeposited pure copper with a 20 nm grain size has been elongated by cold-rolling at room temperature reaching about 5000% elongation till rupture [567,568]. In spite of the fact that it is quite slow in providing these large strains, it is, nevertheless, an industrially attractive means of forming complex parts of materials, using *vacuum* or *blow forming* to produce intricate shapes with high draw ratios, in comparison to the limited deformability in the more conventional means of hot forming. The general requirements for superplastic behavior is a fine and stable grain size in the submicron range (between 1 μm and 100 nm or lesser), a high, *homologous* and *critical* temperature (not less than 0.4 T_m) and deformation under a stress with a small mean normal component. As the grain size decreases and strain increases, the plastic deformation becomes more homogeneous [522]. Under these conditions, the overall stress exponent, *n*, of the tensile strain rate is found to be close to but somewhat larger than unity. This is conducive to continued *stable* extensional deformation without localization and attendant rupture; both on the macroscopic scale as well as on the microscopic scale of cavities and pores which tend to harmlessly stretch out rather than expanding in volume [570,571]. Examination of the microstructure of superplastically deforming fine-grained polycrystalline solids shows a number of remarkable characteristics. In spite of the very large accumulating strains, *grains remain equiaxed*, and these materials show also the superplastic behavior when having a conventional (> 1 µm) or coarse-grained structure at the same critical temperature [508]. When the grain size is decreased, it is found that the superplastic *threshold temperature* can be lowered, at the same time, and an increase in the strain rate for the occurrence of superplasticity is observed. When these behaviors were first reported, it was believed that creep rates could be enhanced by many orders of magnitude by simply reducing the grain size below 100 nm, and that superplastic behavior could be observed also in nanocrystalline materials at temperatures much lower than $0.5\ T_m$. Subsequent creep experiments have not fully confirmed this prediction, and instead the observed creep rates are comparable to or lower than those measured in coarse-grained samples of the same material at temperatures $< 0.5\ T_m$. In any case, it must be observed that tensile superplasticity is *limited* to materials that already exhibit superplasticity for their coarser-grain-sized same (1-10 µm). Since creep is primarily

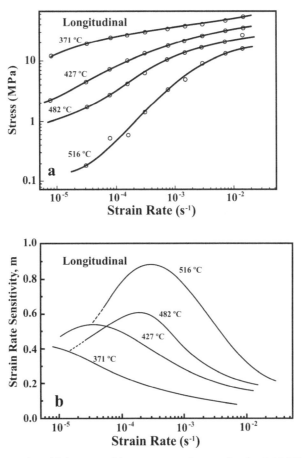

Fig. (7). (**a**) Stress versus strain rate sigmoidal curves; (**b**) *m* versus strain rate, for the AA7475 Al-alloy with a maxima at m ~ 0.85 [572].

controlled by diffusion, it is expected to occur quickly in nanophase materials because of the large volume fraction of grain boundaries and the short diffusion distances. Diffusion along grain interfaces is very rapid, and mechanisms similar to those that drive sintering and neck formation at room temperature in nanophase metals may enhance also the creep rates in these materials. Creep rates may also be influenced by the level of porosity in nanophase samples, since free surfaces tend to increase diffusion rates relative to grain boundary rates. However, individual marked grains are found to show substantial rotations and move apart with other grains undergoing lateral displacements to fill the gaps with the whole assembly of grains showing a mode of deformation. They resemble a *collection of sand grains embedded in a viscous medium*. Clearly, since such a medium does not exist, the individual grains must undergo not only complex shape changes in squeezing past each other, but also large amounts of the grain boundary sliding.

Here, a strong and nearly inverse cube dependence of the strain rate of grain size and net activation energy of the deformation rate, close to that of grain-boundary diffusion, indicate that diffusional flow plays a prominent role in the deformation process. Careful examination, however, shows also considerable dislocation activity that manifests itself in the stress exponent of the strain rate being somewhat larger than unity. Approximately all melt spinning products show superplasticity and even a larger ductility in room temperature. This is attributed to the presence of dislocations, particularly in the submicrometer-sized phases, and the present metastability. On the other hand, the dispersion of intermetallics, together with the grain boundaries of the metastable phase, effectively hinder dislocation motion, thus are causing the observed strength increase. Besides, these melt-spun alloys exhibit a high strain rate sensitivity exponent (m = 0.6) and as well as very large elongations, revealing that they can be deformed by high strain rate superplasticity. This makes them also attractive for shaping operations at moderate to elevated temperatures [193,565,573-576], as in Fig. **8**. In contrast with oxides and alloys, single-phase submicro- or nano-structured metals, such as melt-spun ribbons and their products, usually exhibit high strength and hardness but brittle behavior at low temperatures ($< 0.5T_m$). Experimental results on the ductility of single-phase nanocrystalline materials indicate a little ductility in tension for grain sizes less than about 25 nm, and this is true both for materials that are ductile when being coarse-grained (*e.g.*; elemental metals) and materials that are brittle when being coarse-grained.

Tensile creep measurements on nanophase metals have been performed at room temperature for Cu, Ni, and Pd, which exhibited a logarithmic stress-strain behavior as a function of time, apparently resulting from dislocation activity (knowing as the *viscous glide of dislocations*) or grain boundary sliding. The plastic strain rate increases with decreasing grain size [309,508,522,524]. This suggests that, in nanophase metals, no damage accumulation is happening; a characteristic of superplasticity. The results of studies on two-phase nanostructured materials suggest that the combination of high hardness/strength and toughness/ductility may be possible in multiphase nanostructured materials. Many of Al-, Mg-, Fe-, Ti- and Ni-based alloys, with a nanoscale structures, consist of nanocrystallites in an amorphous matrix and show a superplastic behavior characterized by high m-values (≥ 0.5), since this leads to increased stability against necking in a tensile test with a *sigmoidal relationship* between logarithms of stress and strain rate $\dot{\varepsilon}$, as in Fig. **7** [565,570,572-585]. Some Al-rich alloys containing nanoscale particles are surrounded by crystalline face-centered cubic Al. These multiphase nanostructured alloys possess extremely high strength coupled with some ductility. Ductility is high in compression, but uniform elongation in tension is limited. This behavior is analogous to that exhibited by ductile amorphous alloys, as noted before. These results again suggest the possibility of the development of nanostructured multiphase composites that combine extremely high hardness and strength with toughness and ductility. Such materials could in principle be useful for many applications where they may be employed as unique structural materials in supersonic airplanes and racing automotives [508,565,586]. Another class of superplasticity, rather than the previous *Microstructural Superplasticity*, is the *Internal-Stress Superplasticity* that does not require a fine microstructure, but rather it depends upon a renewable source of internal stresses within the material. Internal stresses are generated by, *e.g.* thermal expansion mismatch between phases or anisotropic grains, or by a polymorphic phase transformation. When an external stress is applied concurrently, the internal mismatch strains are biased to develop preferentially in the direction of the applied external stress, leading to a macroscopic plastic deformation. In general, both an internal and a superimposed external stress are required for this deformation mechanism to be active, and the internal stresses must be renewed many times to accumulate large deformations. *Thermal cycling* is a common means of renewing internal stresses, especially in systems with a thermal expansion mismatch or a thermally-driven phase transformation. Upon each thermal cycle, an increment of plastic strain is developed, and many cycles are required to achieve superplastic strains. Examples are conventionally pure Ti and some γ-TiAl alloys. Unlike microstructural superplasticity, which can exhibit a range of strain-rate sensitivities, internal-stress plasticity occurs under ideal *Newtonian conditions* for the viscous creep flow, with m = 1 such as hot glass, where a *steady-state linear dependence* of the applied stress on the applied strain rate occurs. This is resulting from localized slip processes, which result in the permanent displacement of basic structural units of the material with respect to each other. Unfortunately, this mechanism operates only at low strain rates, in the range 10^{-7} to 10^{-4} s^{-1}, but because it does not require a fine microstructure, it can be applied to hard-to-form

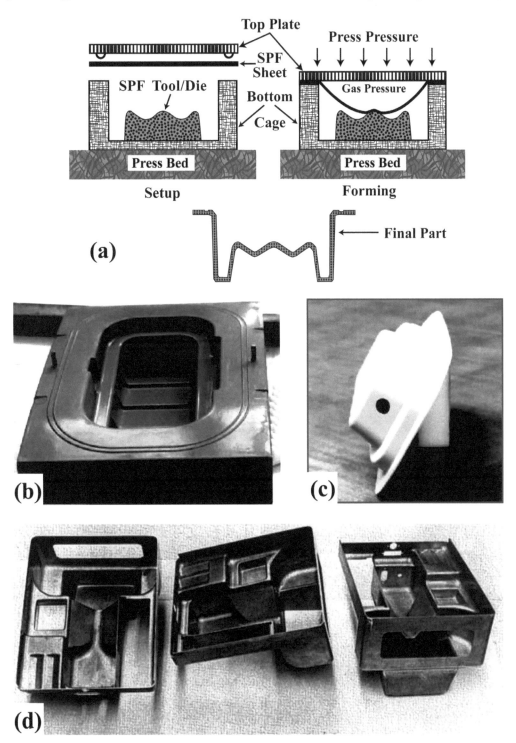

Fig. (8). (**a**) Schematic drawing of superplastic forming of a titanium sheet with the optimum and homogeneous thickness distribution. (**b**) One piece die and (**c**) is the resulting sheet metal part. (**d**) Complex one-piece part made from a sheet of a Zn–22%Al alloy where the very high strains in the walls require superplasticity [446,524,577].

engineering materials such as metal-matrix composites or alloys with coarse microstructures. Internal stress superplasticity has primarily been studied during thermal cycling for the case of thermal expansion mismatch in aluminium matrix composites, as well as during allotropic phase transformation (transformation superplasticity) in iron and titanium alloys [561,587-598].

While grain boundary sliding can contribute to the overall deformation by relaxing the independent mechanisms of slip, it can not give rise to large elongations without bulk flow of material (*e.g.* grain boundary migration). In polycrystalline solids, triple-junctions grain boundaries obstruct the sliding process and give rise to a low *m*-value. Thus, to increase the rate sensitivity of the boundary shear it is necessary to lower the resistance to sliding from barriers, relative to the viscous drag of the boundary. This can be achieved by grain boundary migration. Indeed, it is assured that superplasticity is controlled by grain boundary sliding and diffusion. Sliding continues until it becomes obstructed by a protrusion or a *flange* in a grain boundary when the local stress generates dislocations which, in turn, slip across the blocked grain and pile up at the opposite boundary until the back stress prevents further generation of dislocations, and thus further sliding. At the temperature of the test, dislocations at the head of their *pile-up sites* can climb into and move along grain boundaries to annihilation sites. The continual replacement of these dislocations would permit grain boundary sliding at a rate governed by the rate of dislocation climb, which in turn is governed by grain boundary diffusion. It is important that any dislocations created by local stresses are able to traverse yielded grains and this is possible only if the dislocation cell size (*i.e.*; the new cellular grain formed under the stimulating strain rate) is larger than, or at least of the order of, the grain size, *i.e.* a few microns. At high strain-rates and low temperatures the dislocations begin to tangle and form new cell structures and superplasticity then ceases [519].

In industry, *superplastic formation* (*SPF*) is accompanied by the *Diffusion Bonding* (*DP*) for making a homogenous and isotropic bonding and thicknesses between different metal parts for high precision and closer tolerances products (*i.e.*; the SPF/DB process), mainly in aerospace manufacturing to make large one-piece, consistent unitized structures. Diffusion bonding is a solid state joining process that relies on the simultaneous application of heat and pressure to facilitate a bond that can be as strong as the parent metal. It is only one of many solid-state joining processes wherein joining is accomplished without the need for a liquid interface (brazing) or the creation of a cast product *via* melting and resolidification (welding). In its most narrow definition (which is used to differentiate it from other joining processes such as deformation bonding or transient liquid phase joining) diffusion bonding (DB) is a process that produces solid-state coalescence between two materials [599]. DB, as shown in Fig. **9**, occurs in four steps: (1) the development of intimate physical contact through the deformation of surface roughness by creep at high temperature and low pressure; (2) formation of a metallic bond; (3) diffusion across the faying surfaces; and (4) grain growth across the original interface. Clean and smooth faying surfaces are necessary to create high quality bonds. To prevent the two sheets from sticking together prior to forming of the DB with the lower sheet (as in Fig. **10**), stop-off agents such as boron nitride or yttria suspended in an acrylic binder can be used, making a diffusion-barrier mask by a silk-screen method, or a slight positive pressure can be used initially to keep the two sheets separate. One attractive advantage of titanium is that at temperatures exceeding 622 °C (~ 1150 °F) it dissolves its own surface oxide (TiO_2) and exhibits extensive *plastic flow* at diffusion bonding temperatures, leaving a surface that is very amenable to DB. Another advantage is that many titanium alloys display superplasticity at the bonding temperatures (900-960 °C) which greatly facilitates intimate contact. Typical DB parameters for Ti-6Al-4V (*or*; the Ti-6-4 alloy) are 1650-1740 °F at 200-2000 psi (~ 1.38-13.8 MPa) for 1-6 h [600-605]. The β-rich ($\alpha + \beta$) alloy SP700 (Ti-4.5Al-3V-2Fe-2Mo) shows a better superplasticity than Ti-6-4 alloy and was designed to be more β-stable than Ti-6-4 by addition of molybdenum and iron.

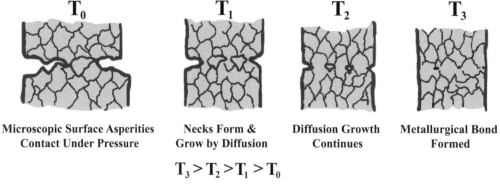

$$T_3 > T_2 > T_1 > T_0$$

Fig. (9). Stages in Diffusion Bonding [599,600].

The SP700 alloy has a fine grain structure and good superplastic formability at low temperatures [606,607]. Although the processing temperatures are lower for solid state diffusion bonding, they are often still high enough to cause significant reinforcement degradation. In addition, the pressures are almost always higher for the solid state processes. Diffusion bonding can be done using either hot pressing or hot isostatic pressing. The perfect joint is indistinguishable from the material surrounding it. Although some processes, such as diffusion bonding, can achieve results that are very close to this ideal, they are either expensive or restricted to use with just a few materials. There is no universal process that performs adequately on all materials in all geometries. The diffusion bonding of aluminum was simplified by using a silver interlayer, whereas titanium required smooth, clean surfaces and protection from atmospheric contaminants.

There have been a number of models proposed to explain the mechanism of superplasticity. Of these the one that captures the basic essence of the complex processes, which is the *Ashby and Verrall model* (or; the Ashby-Verrall power-law of superplastic creep) [608,609]. It recognizes that the essential unit process in the flow of relatively equiaxed grains past each other is a mesoscopic boundary sliding or *a grain switching exchange*, depicted in Fig. **12**. Here, the grains taking part in switching are required to undergo a recurring set of pulsating shape changes, which are primarily accomplished by diffusional flow in the boundary regions, with apparently a small amount of additional dislocation plasticity, and considerable amounts of *trivial* grain-boundary sliding, known as *Lifshitz-type* grain boundary sliding [610]. This can lead to a change in the external shape, but the grains remain equiaxed, maintain their size and orientation and slide over one another by diffusional accommodation at the interfaces. The reason why this mechanism should be activated when the grain size is decreased in a material is that sliding would occur because when a crack is opened at an interface, then diffusion can contribute to fill it. In small grains, the number of atomic rearrangements required for healing is possible in real time during the sliding, while it would take long times in conventional grain-sized materials. Thus, the overall superplastic tensile strain rate $\dot{\varepsilon}_{SP}$ is given mainly by the volume averaged transformation strain associated with the grain switch (from a specific grain conformation to another one) divided by the characteristic time constant of the required shape changes accomplished by local diffusional material transport along grain boundaries. This strain rate is [570,608,609,611,612]:

$$\dot{\varepsilon}_{SP} = A\frac{\Omega D_V}{kTd^2}\left(\sigma - \alpha\frac{\chi_B}{d}\right)\left[1+\beta\left(\frac{\delta}{d}\right)\frac{D_B}{D_V}\right] \ldots \tag{3-7}$$

where σ is the tensile stress, χ_B is the grain-boundary energy, d is the grain size, δ is grain boundary thickness, Ω is the atomic volume, D_V is the volume diffusion constant and D_B is the boundary diffusion constant, where;

$$D_V = D_0\exp\left(-Q_V/kT\right), \ D_B = D_0\exp\left(-Q_B/kT\right) \ldots \tag{3-8}$$

A, α and β are numerical constants with magnitudes of 100, 0.72 and 3.3, respectively. Q_B and Q_V are the activation energy for boundary diffusion and volume diffusion, respectively. All other quantities have their usual meaning. This model does not include the consideration of the dislocation creep component in superplasticity. These effects and much additional phenomenology associated with discrepancies (especially when applying for metals and alloys), fracture avoidance and industrial applications of superplasticity have been discussed with considerable detail in literature [524,562,613]. Karch *et al* [614] also attributed ductility to enhanced diffusional creep providing the plasticity, where conventional grain-size materials would fail in the elastic regime. Their model relates the strain rate (or creep rate) to grain boundary diffusion by:

$$\dot{\varepsilon}_{SP} = \frac{B\sigma\Omega\delta D_B}{d^3 kT} \ldots \tag{3-9}$$

where σ is the applied stress and B is a constant. However, their results obtained for nanocrystalline CaF_2 at 80 °C and TiO_2 at 180 °C have not been reproduced, and it is believed that the porous nature of these samples was responsible for the apparent ductile behavior.

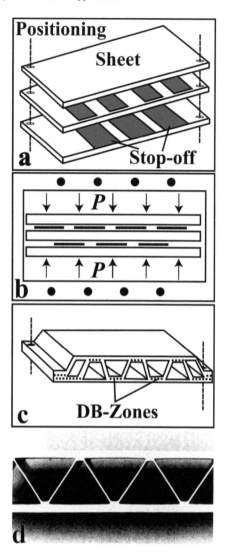

Fig. (10). (a), (b) and (c) are a schematic of SPF/DB process to make the *honeycomb* structure. Three sheets were diffusion-bonded at a few locations and then internal pressurization of the unbonded channels (stop-off sites) caused the middle sheet to stretch. Superplastic behavior with a slower strain rate (~ 10⁻³ s⁻¹ or slower) is required because of the very high uniform extension required in the middle sheet. (**d**) A titanium-alloy aircraft panel made by diffusion bonding and superplastic deformation. Same design may be found in the core structure of a hollow turbine fan blade made of a heat-resistant Ti-alloy [464,524].

Besides, creep measurements of nanocrystalline Cu, Pd, and Al-Zr at low temperatures found creep rates comparable to or lower than the corresponding coarse-grained material rates. One possible explanation is that the observed low creep rates are caused by the high fraction of low- energy grain boundaries in conjunction with the limitation in dislocation activity due to the small grain sizes [508,615]. Another model, provided by J. Bird *et al* [616,617], shows the steady-state strain rate in the superplastic regime, including the stress-activated dislocation effect, and $\dot{\varepsilon}_{SP}$ can be given by the constitutive relation:

$$\dot{\varepsilon}_{SP} = A \frac{DGb}{kT} \left(\frac{b}{d}\right)^p \left(\frac{\sigma}{G}\right)^n \ldots \tag{3-10}$$

D is the appropriate diffusivity (volume or grain boundary), G is the shear modulus, b is the *Burger's* vector, k is the Boltzmann constant, T is the test absolute temperature, d is the grain size, p is the grain size exponent or sensitivity (usually 2 for lattice diffusion controlled flow and 3 for grain boundary diffusion controlled flow), σ is the applied

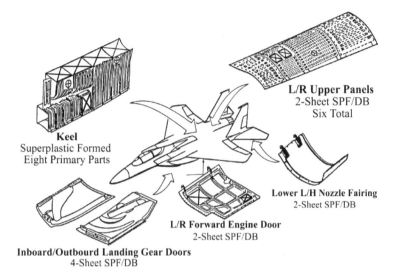

Fig. (11). SPF and SPF/DB applications on a Boeing F-15E fighter [600].

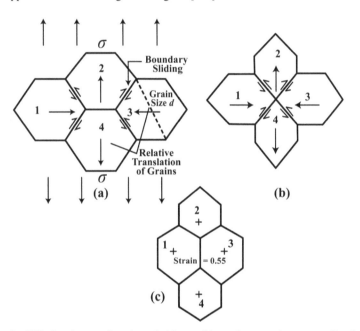

Fig. (12). The unit process of a diffusional-control grain switching and boundary rotation among four hexagonal grains, observed in two dimensional row of an oil emulsion, as in a superplastic medium. (**a**) Initial state. (**b**) Intermediate state. (**c**) Final state [570,611]. Metallic grains don not slide over each other like sand grains, but rather change their configuration by diffusional boundary plastic (yield) deformation.

stress and n is the stress exponent; $n = m^{-1}$, $n \geq 2$ for common superplastic metals and alloys). Eq. (3-10) has been widely verified [309,562,565,571,613,617,618]. Consequently, a reduction in grain size can lead to a reduction in the optimum superplastic temperature at constant strain rate, or an increase in the optimum superplastic strain rate at constant temperature. These factors are very important from a technological viewpoint.

Also, Padmanabhan *et al.* have provided another successful models for the phenomena relating the internal stresses with the grain boundary sliding and the threshold stress (*i.e.*; the applied stress at which the plastic creep begins just after the elastic deformation region and prior to the superplastic behaviour. It is a function of temperature and decreasing in magnitude as temperature increases.), and a steady state equation for superplastic strain rate has been deduced as a function of stress, grain size and temperature. They showed that only grain boundary sliding-related deformation is responsible for high strain rate superplasticity and, on this microscopic scale, the obstacles to sliding are localized, non-periodic and can be overcome by thermal activation. These models is predicting that when the grain size is in the lower range of the nanometer scale, boundary migration needed to cause mesoscopic sliding will take place entirely by diffusion. At coarser grain sizes, a combination of dislocation emission and diffusion will give rise to boundary migration, with true and successful experimental verification *for some metallic alloys*, as in literature [619-624].

To sum up; the total elongation increases as *m* increases and with increasing microstructural fineness of the material (grain-size or lamella spacing), so the tendency for superplastic behaviour is increased. The complete and general explanation of superplasticity is still being developed and can not explain many experimental results and model-discrepancies due to the lack of quantum electronic considerations, especially for the related descriptions of the process activation and grain boundary energies. For metals and their alloys; the phenomenon is nearly confined to pure metals, eutectic, eutectoid alloys and peritectically-formed intermetallics; it is a phase diagram related feature. During deformation, individual grains or groups of grains, with suitably aligned boundaries, will tend to slide. Exceptional ductility, due to diffusional creep in nanocrystalline brittle ceramics or intermetallics at temperatures significantly less than 0.5 T_m, has not been realized, while effective enhanced ductility has been obtained at somewhat higher temperatures, and an improvement of creep properties can be seen in terms of *a lower activation temperature* for superplasticity and a higher strain-rate regime. Experimental results and interpretations are still in contrast and not very clear, so that it can be said that the real effect of the nanostructure on these properties has not been fully understood. Again, it must be underlined that grain boundaries in nanostructured materials are metastable, and even in a material with a fixed grain size a wide range of properties can be obtained and measured, depending on the sample preparation and on the measurement procedure. Also, the presence of metastable phases may be important for the comprehension of the ductile and plastic properties of nanomaterials and must be in mind [508,519,625].

3.2.4 INTERNAL FRICTION

The thermoelastic effect is one of many mechanisms by which vibrational energy is dissipated internally by the material. Other mechanisms which produce anelastic effects in crystalline solids are stress-induced ordering of interstitial and substitutional solute atoms, grain-boundary sliding, motion of dislocations (or; lattice friction), and

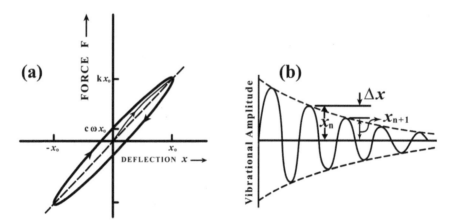

Fig. (13). (**a**) Mechanical *elastic* hysteresis loop (or; Bauschinger effect) caused by internal friction, for a vibrating spring or rod with a viscous damper in parallel, or for any material under cyclic elastic stress; $x = x_0\sin(\omega t)$ and $F = kx_0\sin(\omega t) + c\omega x_0\cos(\omega t)$. (**b**) Typical elastic vibration decay curve with a high decay rate and large damping [522,524,627,628].

intercrystalline and transcrystalline thermal currents arising from the elastic anisotropy of crystals and all are time-dependent processes. *Thermoelasticity* occurs when an elastic stress is applied to a specimen and it is too fast for the specimen to exchange heat with its surroundings and so cools slightly. As the sample warms back to the surrounding temperature, it expands thermally, and hence the dilatation strain continues to increase after the stress has become constant. The diffusion of atoms can also give rise to anelastic effects in an analogous way to the diffusion of thermal energy giving thermoelastic effects. A particular example is the stress-induced diffusion of carbon or nitrogen atoms in iron. The maximum energy is dissipated when the time per cycle is of the same order as the time required for the diffusional jump of the carbon atom. These various energy dissipating effects can be grouped under the generic heading of *internal friction* [522,626]. Internal energy dissipation at larger stresses and strains generally is called damping. In elastic situation, this energy is called the *elastic hysteresis (energy) loss* which is proportional to the forcing frequency [522] (as in Fig. **13**). Many other forms of internal friction exist in metals arising from different relaxation processes other than those mentioned above, and hence occurring in different frequency and temperature regions. One important source of internal friction is that due to stress relaxation across grain boundaries. The occurrence of a strong internal friction peak due to *grain boundary relaxation* was first demonstrated on polycrystalline aluminium at 300 °C by Kê [519,629-632], using the widely-preferred *dynamic resonance method* [633-635] and/or *low-frequency forced torsion pendulum* [519,632,636], and has since been found in numerous other metals. It indicates that grain boundaries behave in a somewhat viscous manner at elevated temperatures and grain boundary sliding can be detected at very low stresses by internal friction studies. Also, movement of low-energy twin boundaries in crystals, domain boundaries in ferromagnetic materials and dislocation bowing and unpinning all give rise to internal friction and damping [519]. Precision measurements of the energy dissipation under very small strains (in the range of elastic deformation) are one of the most sensitive tools for detecting changes in solid-state structure such as precipitation, diffusion, and impurity concentration. Internal friction studies generally are conducted at low stress levels (50 to 500 kPa) and small strains. High damping capacity is important in minimizing machine noises and suppressing vibrations in high-speed machinery (*e.g.*; vehicles suspension systems, shock absorbers and bearings).

A high damping capacity is of practical engineering importance in limiting the amplitude of vibration at resonance conditions and thereby reducing the likelihood of fatigue or creep failures. While damping capacity is not very dependent on frequency of vibration, it depends on stress level (or strain amplitude) and it varies significantly from material to material [522]. For example, the exceptionally high damping capacity of cast iron arises from the presence of graphite flakes which do not readily transmit elastic waves and renowned for their excellent damping capacity, their absorption of vibrations and deadening of noise, making them an ideal choice for the beds of machine tools. Since in electrical circuit theory the reciprocal of the damping capacity value is called the quality factor Q of the circuit, the symbol Q^{-1} has been adopted as a measure of internal friction where;

$$Q^{-1} = \frac{f_2 - f_1}{f_o} \quad \cdots \tag{3-11}$$

The value $(f_2 - f_1)$ is taken at the half-width of the frequency spectrum of the elastic vibrator or at the $(1/\sqrt{2})$ of the maximum amplitude (Fig. **14**). When using the frequency amplitude curve obtained from the dynamic resonance measurements, Eq. (3-11) becomes;

$$Q^{-1} = 0.5773 \frac{f_2 - f_1}{f_o} \quad \cdots \tag{3-12}$$

The dynamic resonance method, which originated from the early measurements by Ide [633] and modified by Kamal *et al* [634] (see Fig. **15**), is based on a standing wave phenomenon. The technique is compatible with the ASTM Standards E 1875 (2002) and C 885-87 (2002) for testing methods of dynamic Young's modulus by sonic resonance [635]. It consists of a variable-frequency audio oscillator (*e.g.*; a small radio speaker), used to generate a sinusoidal voltage and suitable transducer to convert the frequency-controlled sinusoidal electrical signal to a mechanical driving vibration. A frequency meter (preferably digital) monitors the audio oscillator output to provide an accurate frequency determination. A suitable suspension-coupling system supports the test specimen that attaches the specimen (rod or ribbon with well-defined rectangular dimensions) from one tip and the other tip is left free. The incident laser beam is reflected from the free tip on the counterparting screen drawing the amplitude of the specimen vibration, where it converts the horizontal vibration of the ribbon into a vertical measurable one on the screen. When a specimen of thickness a and density ρ is undergoing longitudinal or torsional vibration (*i.e.*; standing wave), the length of the specimen, L, contains an integral number n of half-wave lengths, where;

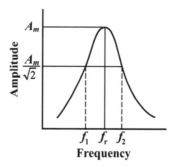

Fig. (14). Resonance curve of a standing wave in a solid rod/ribbon showing the half width amplitude [522].

$$L = \frac{n\lambda}{2} \dots \tag{3-13}$$

$$v = \lambda f_o = \frac{2Lf}{n} \dots \tag{3-14}$$

v is the velocity of the wave, f_o is the resonant (*i.e.*; critical or peak) frequency for a fixed vibration mode. The melt-spun ribbon is forced to oscillate at various frequencies with constant input. When the impressed frequency is equal to a critical resonance frequency f_o of the specimen, the amplitude of vibration becomes a maximum. This amplitude decreases as the impressed frequency deviates to either side of f_o. The resulted data were fitted *via* a Peak-Fit computer program to obtain the resonance frequency and the width at half maximum Δf as seen in Fig. 14. Also, the dynamic value of the Young's modulus E can be determined by the same method from the relation;

$$E = 38.32\,\rho L^4 f_o^2 / a^2 \dots \tag{3-15}$$

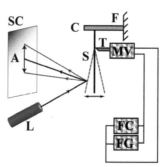

Fig. (15). Dynamic Resonance System. F; fixed support, C; clamp, MV; electro-mechanical vibrator, FG; electric current function generator, FC; frequency counter, L; low-power laser source, SC; screen, S; sample (solid rode or ribbon), T; transducer, and A is the reflected amplitude. Specimen can be contented in a temperature-controlled furnace or in an isolated cryogenic chamber [2,634].

Calculations and experiments show that Q^{-1} and dissipated energy are inversely proportional to the grain size and the relaxation time of any polycrystalline metal or alloy [636]. According to Valiev *et al* and Blanter *et al*; almost all bulk fine-grained and nanostructured metals (except those produced from the amorphous state) exhibit an internal friction peak which is not often found in well annealed (coarse-grained) or single-crystal metals, as in Fig. 16 [309,636]. The nature of this peak is still not entirely clear; some authors report a thermally activated, reversible anelastic grain boundary relaxation consistent with the Kê approach [632], while others attribute the effect to irreversible structural changes like recovery and recrystallisation, connected with short-range grain boundary diffusion in non-equilibrium grain boundary. Most of studies, for bulk fine-grained metals and alloys, in agree with the statement that the total temperature dependant internal friction can be expressed as;

$$Q_b^{-1} = Q_b^{-1}(T) + \Sigma Q_r^{-1}(T) \ ... \tag{3-16}$$

where the terms $Q_b^{-1}(T)$ and $\Sigma Q_r^{-1}(T)$ represent the *damping background* of internal friction (that is significant at very low temperatures ≤ 10 °K) and the *superposition* of different anelastic relaxation peaks, respectively [636]. First term is closely related to composition, microstructure, dislocation motion, annealing time, the amplitude and frequency of the imposed oscillatory strain. Second term is time-independent but frequency-dependent. In fact, there is no common agreement for the internal friction behaviour for submicron-grained bulk metals and alloys that proved to have a prominent vibrational attenuation power larger than the coarse-grained counterparts. Because of the lack of a combined study of nanostructured metals by different mechanical spectroscopy techniques, it is not easy to distinguish between different anelastic relaxation mechanisms during tests. The situation is more complex than was expected. For instance, in Cu a peak of the *Kê-type* has been observed by Woirgard *et al* and Rivière [637,638] even in single crystals, and therefore it must be associated with dislocations and other rheological properties of metals at high stress, rather than with grain boundary sliding. Also, the change in elastic modulus *vs.* Q^{-1} through isothermal annealing steps and between them while increasing the temperature monotonically [639]. These anomalies are associated not only with small grain sizes but also with specific defect structure connected with non-equilibrium grain boundaries. To our knowledge, there is no study relating the internal friction to different melt-spinning wheel speeds. It's still an open field of research.

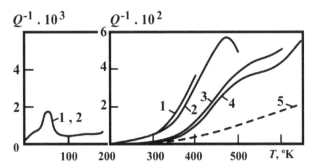

Fig. (16). Temperature dependencies of internal friction of nanostructured copper measured at consequent cycles of heating (the low temperature region is given with 10x magnification of an internal friction scale): 1,2,3 and 4 are internal friction during heating from 4 to 400, 500, 600, 650 °K, respectively, and 5 is the IF behavior for the single crystal one [309,636].

Finally, internal friction can generally be correlated with superplastic properties, because of the common mechanisms, and thus it can be used for determining the optimum temperature for superplastic deformation. Another extended controversy, treatments and material bibliography can be found in the literature [636].

3.3 CONSOLIDATION OF MELT-SPIN PRODUCTS

The unique properties of metastable materials (*e.g.*; amorphous, nanocrystalline phases, quasicrystalline or a mixed crystalline-amorphous state/phases ... etc) may be fully exploited if they are converted into powders and then densified into bulk parts that retain the initial metastable features. Numerous applications, particularly structural materials, require fully dense parts of certain size, geometry and final properties. In addition, dense specimens with metastable microstructures are important for magnetic, electrical or electronic functions such as permanent magnets, substrates in electronic circuitry, magnetic recording heads, capacitors, electrodes, and varistors. The key characteristic of metastable powder consolidation processes is to achieve densification with minimal microstructural coarsening and/or undesirable microstructural transformations. This requirement places significant restrictions on the consolidation process, particularly high temperature exposure which must be controlled *more carefully* than in the case of regular powders to constrain the diffusion mechanism through the boundaries of powder particles.

Fig. (17). Reduced-density Al-Li 2195 alloy is used widely for manufacturing the structural parts in the Space Shuttle, especially the gigantic external tank, making it super light-weight. (**a**) Mobile Launcher Platform and Crawler Transporter, Space Shuttle Atlantis creeps toward the launch complex at about 1 mph; compare its size by the near car at the under side of the picture. (**b**) The space shuttle while flight before the two boosters' rejection, while the non-recoverable external tank will be ejected, falling and breaking apart in ocean, in the later stage, as in (**c**). The RSP, P/M and SPF/DB techs are used extensively in every stage of the space shuttle manufacture [640].

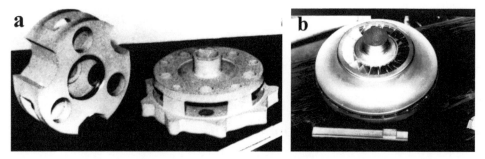

Fig. (18). A high-precession advanced products made from rapidly solidified Ti-base alloys, with the use of P/M consolidation followed by SPF/DB techs, for aeronautical industry. (**a**) A Ti-base alloy compacts for cases of a gear train and satellite gear. (**b**) A VT25u® Ti-alloy enclosed impeller of the centrifugal compressor, with inner complex components, for aircraft engine application. The 420 mm diameter impeller has internal guide channels formed by 32 thin-walled complex-shaped blades of two types. Complex geometrical shape in combination with a high property level inaccessible for foundry technology characterizes this impeller as a unique component [646].

Although lowest temperatures are desirable, full interparticle bonding and pore-free structures must be achieved. These restrictions to the sintering process are dependent on the type of microstructural metastability that needs to be retained. For instance, retention of chemical homogeneity is less restrictive on the sintering process than retention of metastable phases. So, the consolidation or compaction technique must be chosen carefully for any specific metastable material [313,641].

Consolidation of the small crystals have been performed at low as well as elevated temperatures under static (*e.g.*, uniaxial pressure, hydrostatic pressure, laser-sintering, laser reactive sintering routes) or dynamic conditions (*e.g.*, sinter-forging, shear, explosive consolidation routes). Laser reactive sintering is applicable preferentially to nanocomposits, especially for vacuum remelting and advanced coatings. The strategy of this method is to select mixtures (*e.g.*, a nanocrystalline ceramic and a metal) so that one component has a significantly lower melting point than the other(s). The objective is to achieve enhanced sintering from the low-melting component during pulse heating by a laser beam that is *scanned over* the pre-consolidated material [642]. In consolidating or sintering nano- or microcrystalline powders, it is noticed that the nanometer-sized (larger than 10 nm and smaller than 100 nm) powders densify at much lower temperatures and lower pressures than coarse grained powders with the same chemical composition, *i.e.*, the smaller the powder particles size, the lower the sintering temperature [641]. Due to the high density of interfaces, single-component nanostructured materials frequently exhibit crystal growth during sintering. One approach to *minimise crystal growth is to limit the time spent at the consolidation temperature.* This may be achieved by *sinter-forging* or *explosive* consolidation, as we will see later. Another approach is to use grain growth inhibitors such as pores, oxides, second phase particles …etc [643-645]. The latter method has been utilized, for instance, to sinter nanostructured WC (tungsten carbide) alloys by cobalt powders or by electrothermal or physical vapor deposition (PVD) of Co atoms over the movable WC powder. Grain growth was inhibited by transition metal carbides [643,645,647]. Also, powder consolidation is done by vacuum hot pressing and has also been done by the hot or cold isostatic pressing methods in vacuum or argon atmosphere. The packing density or

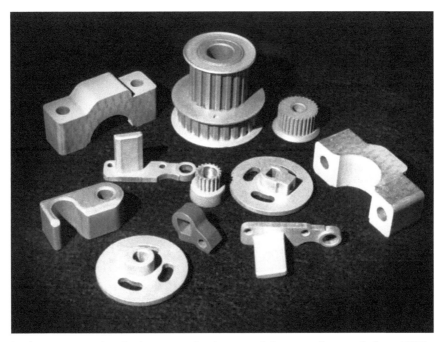

Fig. (19). Various aluminum parts produced using conventional press and sinter metallurgy techniques [648].

voidage is important in that it determines the bulk density of the material, and hence the volume taken up by a given mass. It affects the tendency for agglomeration of the particles, and it critically influences the material stiffness, creep and fatigue resistances and its resistance to the percolation of fluid through it. The consolidation process, densification or sintering is aimed at transforming the initial powders into bulk materials with a minimum amount of pores, or no pores. The driving force for densification is the tendency to reduce the free surface energy associated with individual powder particles. This driving force increases with the reduction of the initial particle size and becomes significant for powders in the nanometer range. Although there is a large driving force, sintering usually involves exposures at temperatures above room temperature for a certain time. These conditions may diminish, or sometimes, alter the initial metastable state of the powders. Since at least some departure from metastability will occur during sintering, it is important to define this metastability and the conditions upon which it is lost in order to utilize the adequate consolidation route for the concerned metastable powder [645].

3.3.1 Sintering Mechanism

Sintering is a processing technique used to produce density-controlled materials and components from metal and/or ceramic powders by applying thermal energy [649]. Unlike other processing technologies, various processing steps and variables need to be considered for the production of such parts (see Fig. **20**). For example, in the shaping step, one may use simple die compaction, isostatic pressing, slip casting, injection moulding … *etc.*, according to the shape and properties required for the end product. Depending on the shaping techniques used, not only the sintering conditions but also the sintered properties may vary considerably. In the sintering step too, there are various techniques and processing variables, so that variations in sintered microstructure and properties can result. Thermodynamically, fine-grained powders are unstable due to large surface area [126,313,641]. This morphological metastability is highly dependent on particle size, with nanopowders being the most unstable. The surface area increases dramatically with the reduction in powder diameter. The specific surface area (per unit mass), S, is given as:

$$S = \frac{6}{\rho_{th}\, d} \quad \cdots \tag{3-17}$$

where ρ_{th} is the materials' *theoretical density* and d is the particle diameter or size. For comparison, the surface area of 5 µm copper powder is 0.134 $m^2\ g^{-1}$ while for 5 nm powder is 134 $m^2\ g^{-1}$. The sintering process is driven by the tendency to reduce this surface area and, therefore, fine particles will have an enhanced tendency for sintering. The driving force for sintering is proportional to (γ/r), where γ is the surface energy, generally considered isotropic (but, in fact it is not because the actual powder particles are not ideal spheres), and r the particle size (also, known as particle "crystal" size, apparent or mean grain size). The dependence of the sintering kinetics on grain size may be illustrated using the equation for the densification or the shrinkage rate (dL/Ldt, L is the powder sample/compact linear size) developed by Johnson and co-workers [650] for all stages of sintering:

$$-\frac{dL}{L\,dt} = \frac{\gamma\,\Omega}{kT}\left[\frac{\delta\, D_b\, \Gamma_b}{d^4} + \frac{D_v\,\Gamma_v}{d^3}\right] \quad \cdots \tag{3-18}$$

where γ is the surface energy, Ω is the atomic volume, δ is the grain boundary width, D_b and D_v are the grain boundary and bulk diffusivities, Γ_b and Γ_v are functions of density, kT has the usual meaning, and d is the grain size. From this equation, it is seen that a decrease in grain size by three orders of magnitude (*e.g.*, from µm to nm) could enhance sintering rates by up to 12 orders of magnitude. Consequently, sintering of nanopowders and/or nanophases may be accomplished at significantly lower temperatures and shorter durations than for conventional powders. This has been noticed for numerous real nanoparticle sintering [313,641]. Due to the *accelerated* sintering at lower temperatures in the nanoregime, full densification may take place entirely during a pressureless heat-treatment (*i.e.*; *solid state sintering*) [651]. Sintering aims, in general, to produce sintered parts with reproducible and, if possible, designed microstructure through control of sintering variables. Microstructural control means the control of grain size, sintered density, and size and distribution of other phases including pores. In most cases, the final goal of microstructural control is to prepare a fully dense body with a fine grain structure. Main part of powder process technologies are formulated specifically to yield a product as close to *full* or *theoretical density* as possible. This contrasts significantly with the previous conventionally processed products where attainment of full density was not the primary goal. The full-density processes include *powder forging* (P/F), *metal injection molding* (MIM), *hot isostatic pressing* (HIP), *roll compaction*, (*unidirectional*) *hot pressing* (HP or *pressure-sintering*) and *powder extrusion* [649,652]. Basically, sintering processes can be divided into two types: *solid state sintering* and *liquid phase sintering*.

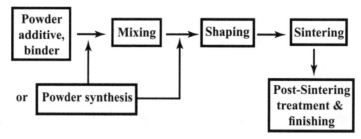

Fig. (20). General fabrication pattern of sintered parts [649].

Solid state sintering occurs when the powder compact is densified wholly in a solid state at the sintering temperature, while liquid phase sintering occurs when a liquid phase is present in the powder compact during sintering [649]. The solid state densification process consists of solid particle bonding or neck formation followed by a continuous closing of pores from a largely open porosity to essentially a pore-free body, as in Figs. **21** and **22** [653]. For simplicity, solid state densification is considered to be accomplished through three stages: *initial*, *intermediate* and *final*.

Multiple mechanisms are involved throughout these stages, viz., evaporation-condensation, surface diffusion [313,654-657], grain boundary diffusion [658-660] bulk diffusion, viscous flow and plastic deformation [641,661-665]. Each transport process exhibits a particular dependence on external pressure, the particle or grain size and defect density, where sinks and sources for material (or vice versa for vacancies) are the surfaces (*i.e.*; pore-solid interface), grain boundaries and dislocations stimulated by surface (tension) energy or, in other words, by *capillary forces*. The highest sensitivity on particle size is that of surface and grain boundary diffusion, *i.e.*; a scale-dependence [313,641,653]. Since nanocrystalline materials contain a very large fraction of atoms at the grain boundaries, the numerous interfaces provide a high density of short-circuit diffusion paths. Consequently, they are expected to exhibit an enhanced diffusivity in comparison to single crystals or conventional coarse-grained polycrystalline materials with the same chemical composition [666]. This enhanced diffusivity can have a significant effect on mechanical properties such as creep and superplasticity, ability to efficiently dope nanocrystalline materials with impurities at relatively low temperatures, and synthesis of alloy phases in immiscible metals and at temperatures much lower than those usually required for coarse-grained materials. The measured diffusivities in nanocrystalline copper are about 14-20 orders of magnitude higher than lattice diffusion and about 2-4 orders of magnitude larger than grain boundary diffusion. For example, the measured diffusivity at room temperature is 2.6×10^{-20} m^2 s^{-1} for 8-nm grain-sized copper samples compared to 4.8×10^{-24} for grain boundary diffusion and 4×10^{-40} for lattice diffusion [667]. Similarly enhanced diffusivities were also observed for solute diffusion in other metals. It may be mentioned in this context that some investigators [668] ascribe this increased diffusivity to the presence of porosity in the consolidated samples. For example, if the presence of porosity is properly taken into account, the diffusivity of nanocrystalline materials has been found to be comparable to that of grain boundary diffusivity. The increased diffusivity (and consequently the reactivity) leads to increased solid solubility limits (*e.g.*; the solid solubility of Hg in nanocrystalline Cu has been reported to be 17 at.% against an equilibrium value of < 1 at.% [669]), formation of metastable intermetallic phases (like the formation of Pd_3Bi at 120 °C, a temperature much lower than normally required for coarse-grained materials [670]) and sometimes new phases, and increased sinterability of nanocrystalline powders. Thus, increased diffusivity of nanocrystalline powders leads to increased sinterability; sintering of nanocrystalline powders can occur at temperatures much lower than those required for sintering coarse-grained polycrystalline powders (at about 0.3-$0.5T_m$) [671].

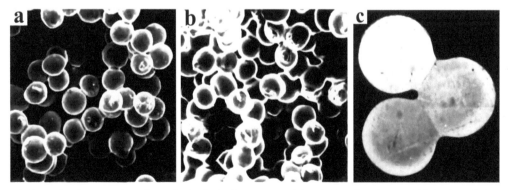

Fig. (21). Neck formation between loosely packed spherical copper particles during sintering at 1300 °K for 1 h (**a**) and 8 h (**b**), 150x. In (**c**), metallographic cross-section through an arrangement of three copper particles sintered at 1300 °K for 8h. Note the grain boundaries in the necks, 500x [653].

For conventional powders, it is known that surface diffusion does not lead to densification but to grain coarsening and surface diffusion is the most sensitive to particle size. Therefore, enhanced surface diffusion with reduced low-temperature densification should be observed in nanoparticle sintering. Also, it was found that sintering temperature (to full density) is particle size dependent and smaller is the particle size, lower is the sintering temperature at which full density is attained [651]. In addition to solid state and liquid phase sintering, other types of sintering, for example; transient liquid phase sintering and viscous flow sintering, can be utilized. Viscous flow sintering occurs when the volume fraction of liquid is sufficiently high, so that the full densification of the compact can be achieved by a viscous flow of grain-liquid mixture without having any grain shape change during densification. Transient liquid phase sintering is a combination of liquid phase sintering and solid state sintering. In this sintering technique, a liquid phase forms in the compact at an early stage of sintering, but the liquid disappears as sintering proceeds and densification is completed in the solid state. During sintering, however, the liquid phase disappears and only a solid phase remains because the equilibrium phase under the given sintering conditions is a solid phase. In general, compared with solid state sintering,

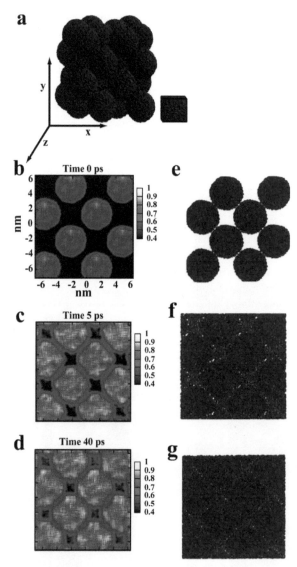

Fig. (22). A Molecular Dynamics Simulation of the generation of a Ge-nanoglass that is obtained by sintering nanometer-sized (5 nm) glassy spheres of Ge (**a**) by applying a pressure of 50 kbar at -73 °C (= 200 °K). Figs. (**e**)-(**g**) on the right side display the atomic structure of the nanoglass as the sintering process goes on. For the sake of simplicity, only those atoms are shown in these figures that lie within a thin slab (1.5 nm thickness) of Ge that is cut out of the sintered nanoglass block, at the right side of (**a**). The density distribution in the nanoglass slab of Ge as a function of the sintering time is displayed in (**b**)-(**d**). The scales on the right side of these figures indicate the correlation between the colors shown in the figures and the density of Ge. By comparing the colors in (**b**)-(**d**) it can be seen that the nanoglass consists of glassy regions with high density connected by glass-glass interfaces with reduced density. As the sintering process goes on, the density of the glass-glass interfaces increases, becoming wider and wider [672,673].

liquid phase sintering allows easy control of microstructure, produces full-density or poreless compact and makes reductions in processing cost, but degrades some important properties, for example, mechanical properties. In contrast, many specific products utilize properties of the *grain boundary phase* and, hence, need to be sintered in the presence of a liquid phase. Elimination of isolated pores is more time consuming and utilizes almost all of the sintering time. To ensure a poreless product, a pressure-sintering route must be used [649]. By filling in necks at the points of contact between the powder particles and, at a later stage, the pore space, sintering increases density and strength of the compacted powders. Depending on the size of the powder particles or the amount and dispersion of porosity in a compact, the total excess energy of the surface amounts to 0.1-100 J per mole of solid, where; the smaller number applies to coarse powders (~ 100 μm diameter) and to low-porosity material and the larger number to submicron powders or highly dispersed porosity.

3.3.2 Effect of Pores

Densification occurs when material is removed from the volume between the particle contacts. In numerous experiments, the decisive role of grain boundaries, as sinks for vacancies arriving from the neck surface or pores, has been demonstrated [653,674]. Pores shrink or collapse only when attached to or located very close to grain boundaries and particle centres approach each other only when the particles are separated by at least one grain boundary. Grain boundaries are usually assumed to be perfect vacancy sinks at the supersaturation levels caused by capillarity. The effectiveness of a grain boundary as a source for atoms or sink for vacancies is determined by the number and the mobility of the grain boundary dislocations, which may be reduced by solute atoms exerting a viscous drag, or by grain-boundary particles pinning the dislocations motion which act as a vacancy source sinks, and move non-conservatively by climbing [653].

After the neck forms by a *dislocation slip*, the adjacent particles rotate to achieve minimum grain boundary energy [662]. TEM studies of zirconia and copper nanoparticles indicate that such a rotation process occurs during early sintering [675]. If one started with nanopowders, by late sintering stages they will lose some of the more obvious initial nanofeatures and, therefore, their densification behavior converges towards that of regular grain-sized materials [313]. Furthermore, *sinter-HIP, i.e.*; a combination of pressureless sintering with or without a liquid phase (closing open porosity) and of subsequent hot isostatic pressing (so eliminating residual porosity), has found widespread technical application for high-density products and thus improving the mechanical properties [676,677].

One of the main difficulties which exist in the compaction forming process for powders includes a non-homogenous density distribution which has wide ranging effects on the final performance of the compacted part(s). The variation of density results in cracks and also in localized deformation in the compact, producing regions of high density surrounded by lower density material, leading to compact failure. The lack of homogeneity is primarily caused by large pores, friction (due to inter-particle movement), as well as relative slip between powder particles and the die wall. A similar phenomenon can occur when using dissimilar powders, whiskers or platelets. The die geometry and the sequence of movement result in a lack of homogeneity of density distribution in a compact. In summary; *the success of compaction forming depends on the ability of the process in imparting a uniform density distribution in the engineered part*. This is largely dependent on process variables, such as: the tool and the part geometry, friction between the powder and the tooling as well as friction between the particles themselves, besides the response of powder to external pressure. Similar to conventional powders, full density or rapid sintering of nanocrystalline powders is achieved when the green (unsintered) structure contains a *narrow pore size distribution*. Conversely, densification is retarded or inhibited when pore distribution is wide. In this case, big pores become larger and only small pores shrink. The removal of large pores is a lengthy process and requires higher temperatures. This way, the overall effect of large pores is to slow densification and induce undesirable coarsening and non-homogeneous density product [641]. Large pores usually originate from agglomerated powders.

The removal of large inter-agglomerate pores (Fig. **23**) based on vacancy diffusion requires significantly higher temperatures and longer sintering times. To reach high density-levels, it is necessary to avoid the *pore coarsening* process, which is the main obstacle for achieving full densification. Using non-agglomerated powders to generate a large array of small and uniform pores is the most common method to avoid pore coarsening. In conventional sintering, grain growth is suppressed due to grain boundary pinning by open pores. Growth of large pores at the expense of the small pores diminishes the pinning effect and grain growth is usually observed beyond the intermediate stage. Therefore, intense grain coarsening is usually observed in the late sintering stages because of larger pores that are difficult to eliminate in the late stages of sintering [678]. A small pore size throughout the sintering process is also critical in controlling the final grain size. For these purposes, a small and uniform pore population is desired in the green compact. Most often, such a pore distribution is associated with a high green density in non- or weakly agglomerated powders. The benefits of a small pore size and narrow size distribution in reaching high densities and reduced grain coarsening have been shown by many researchers [641,679]. A common method for powder *agglomeration prevention* is the use of *process control agents* (PCAs). PCA is added to the powder mixture during milling or pulverization, especially when the powder mix involves a substantial fraction of a ductile component. The PCAs are mostly organic compounds, which act as surface-active agents and evaporate or decompose at elevated temperatures prior to the sintering temperature, especially with the aid of a vacuum pump. The PCA adsorbs on the surface of the powder particles and minimizes cold welding between powder particles and thereby inhibits agglomeration. The surface-active agents adsorbed on particle surfaces interfere with cold welding and lower the surface tension of the solid material [680].

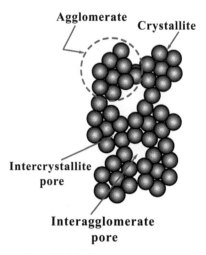

Fig. (23). Schematic diagram of an agglomerated powder [681].

It becomes clear from the above description that the analysis and simulation of powder compaction is important because the compact green density influences the further densification of the samples during sintering and hence the performance of the final part. These variations in density of the green compact will cause shape distortion after sintering along with regions of elevated stress and stress concentration. Thus, the control of compact density is very important for the manufacture of reliable products [682].

3.3.3 Powder Intrinsic Pressure

Also, it is known as the powder *capillary pressure*. Sometimes the row powder shows some degree of self-strength in room temperature even without any previous pressing. This effect is mainly attributed to the intrinsic pressure or, definitely, the intrinsic curvature-driven sintering stress σ ($\sigma = \gamma/R$, R is the particle curvature radius). According to Groza and Skandan [641,683], the effect of the applied stress is noticeable only if it exceeds the intrinsic curvature-driven sintering stress. When powders are sintered under an applied pressure, higher densification is achieved at the same temperature. The pressure has a mechanical role as well as an intrinsic role. Mechanically, the pressure has a direct effect on particle re-arrangement and the destruction of agglomerates, particularly in the case of nanometric powders. The intrinsic (sintering) effect of the pressure can be assessed from the driving force for sintering:

$$\frac{d\rho}{(1-\rho)dt} = B\left(g\frac{\gamma}{d} + P \right) \dots$$

(3-19)

where ρ is the fractional or relative density, B is a term that includes diffusion coefficient and temperature, g is a geometric constant, γ is the surface energy, d is the grain/particle size, t is time, and P is the applied external pressure. The first term on the right-hand side of Eq. (3-19) represents the intrinsic driving force for sintering while the second term represents the intrinsic contribution to the driving force by the applied pressure. The significance of the pressure on sintering thus depends on the relative magnitudes of the two terms. When the particle size (related to γ) is small, the relative contribution of the pressure is small, but becomes significant as the particle size increases. The point at which the two contributions are equal (*i.e.*; the threshold point) is represented by:

$$P = g\left(\gamma/d\right) \;\ldots$$
(3-20)

Eq. (3-20) provides the determination of the critical particle size above which the contribution of the pressure to the densification driving force becomes dominant and was verified by researchers [684,685]. So, an increased driving force for sintering, as a result of the application of a pressure, leads to a decrease in the sintering temperature and a limitation of grain growth [686]. Therefore, the *effective* applied pressure is dependent on the particle size while the intrinsic sintering pressure increases when the particle size is reduced and may reach high values as the particles become increasingly small (total sintering pressure is the sum of the capillary and external pressures). At this particle size, the applied pressure has to be higher than the curvature driven pressure. Consistent with this threshold effect, no external pressure effect was observed in the densification of powders unless a certain pressure level was achieved. The intrinsic curvature stress in nanopowders may be on the order of hundreds of MPa. Of course, the threshold pressure is specific for a certain material with a certain grain size and differs from matter to matter. It depends upon the materials' electronic structure, valence electrons of the diffusing atoms, entropy and chemical potentials [649,685,687,688]. Grain coarsening may occur by the time the pressure is applied, thus lowering the intrinsic sintering stress, where the migration of the grain boundary is then controlled by its intrinsic mobility (boundary control). Many times, pressure application is intentionally carried out only at high temperatures where the resistance to deformation is less and the pores are more likely to collapse. Large pores are more difficult to fracture, even at very high pressures, when the temperature is low. If pressure is applied at an elevated temperature, densification of the compact is much easier. Similarly, density improvements have been noticed when pressure was applied after heating vs. application of pressure at room temperature in sintering the same alloyed metal or ceramic powders. HIP induces only local shearing stresses and at the lowest levels among the pressure enhanced sintering methods. However, even these small deviations from a purely hydrostatic stress in HIP have been shown to contribute to enhanced sintering rates [641].

3.3.4 Theoretical Density

Theoretical density (or; pore-free and defect-free density) is calculated from crystal lattice parameters. Using the concept of a unit cell together with data on the atomic mass of constituent atoms, it is possible to derive a theoretical value for the density of a pure single crystal. The lattice parameter a, for the bcc cell of pure iron at room temperature, is 0.28664 nm. Hence, the volume of the unit cell is 0.02355 nm^3. The bcc cell contains two atoms, *i.e.*: (8 x 1/8 atom) + 1 atom. Using the Avogadro's constant N_A, we can calculate the mass of these two atoms as 2 x $(55.85/N_A)$ or 185.46 x 10^{-24} kg, where 55.85 is the relative atomic mass of iron. The theoretical density (mass/volume) is thus 7875 kg m^{-3}. The reason for the slight discrepancy between this value and the experimentally-determined value of 7870 kg m^{-3} will become evident when we consider the crystal imperfections (pores, interstitials, dislocations, polycrystallinity, grain boundaries,. *etc.*). Practically, in the field of powder metallurgy, density may be displayed as a physical value or a percentage of the theoretical density.

The theoretical density value of nanocrystalline materials is controversial. Some argue that a lower density value in nanocrystals may arise due to a large volume fraction of grain boundaries which, in turn, may have lower densities than adjacent crystals. It is known that the uniqueness of submicro- or nanomaterials originates from their enhanced ratio of atoms at the grain boundary to atoms contained in the grain interior (bulk or single crystal) and the interface properties. For example, a 5 nm (or lesser) grain-sized material may contain more than half of its atoms at grain boundaries [509,672,673], *i.e.*; $f_{gb} \approx f_{sc}$. If these grain boundaries have different properties than the grain itself, then these differences ought to be reflected in the overall properties of these materials. One such property shall be the theoretical density. A simple calculation of this final density, ρ, may be performed based on *the rule of mixtures* that has been shown to hold for some-microns and nano-grained materials [344,509,689-691]:

$$\rho = f_{gb}\,\rho_{gb}^{n} + \left(1 - f_{gb}\right)\rho_{sc}^{n} \;,\; -1 \le n \le 1 \; \dots \tag{3-21}$$

where ρ_{sc} and ρ_{gb} are the single crystal grain (bulk) and grain boundary densities, respectively, and f_{gb} is the grain boundary volume fraction, assuming a cubic grain of inner crystal size d and grain boundary (grain outer shell) of thickness δ, $\delta \cong 1$ nm (for typical metals) and contains 3 to 4 layers of atoms with atomic diameter of ~ 3 Å at least. The exponent n value is taken equal to 1 in most cases that is working with grain sizes not less than 100 nm, discarding the crystal morphology imperfections densities. The main problem in applying Eq. (3-21) to nanomaterials is the grain boundary disordered structure influencing both ρ_{gb} and f_{gb}. Besides, the (crystal) grain or particle size is not the same but real powder grains are different in shape, geometrical symmetries and actual size, as in Fig. **24**. Changing the assumed crystal morphology (rather than the cubic one) would certainly change the grain size scale and thus the parameters in Eq. (3-21). Some TEM studies showed that, in case of gold grains, the density of this grain boundary was shown to be 4% lower than the bulk material, since the spacing of atoms at grain boundaries is usually somewhat larger than the spacing of atoms within the single crystal (*i.e.*, the atoms are more loosely packed), so the polycrystaline samples are usually slightly-lower in density than single crystal ones. The currently accepted value of grain boundary thickness is 0.5-1 nm based on careful measurements in *High Resolution Transmission Electron Microscopy* (HRTEM) studies, whether the grain size is 1 μm or 100 μm, *i.e.*; there is no change in grain boundary thickness for any certain powder material [692]. However, grain boundary thickness is a sensitive function of the amount of impurities, boundary misfits and boundary junctions/corners densities [509,690]. At small grain sizes, other complexities arise. In view of the cubic morphology, as in Figs. **25a** and **26a**, we may explain the values of parameter n in Eq. (3-21). The domains that we have been discussing at this point, in terms of the lower density at the grain boundaries, represent planar regions where two crystal facets meet. The regions, where these planes themselves meet, are either nearly-linear regions of finite size that we will call *triple junctions* (tj); or cuboidal regions at cube corners that are *corner junctions* (cj). Note that triple junctions get their name because in most microstructures they are formed by the intersection of three planes. It is likely that the densities ρ_{tj} and ρ_{cj} of the material at triple junctions and corner junctions respectively are even lower than that of the grain boundaries, because of the increased lack of compatibility between the atoms in the crystal structure at these regions. Thus, the effective density of the polycrystalline material is given by;

Screen Aperatures

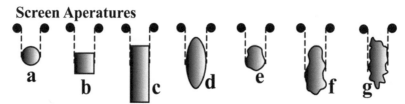

Fig. (24). Various shapes that would all be classified under the same sieve-aperture diameter. This aperture diameter is the well-known mean or apparent (crystal) grain/particle size of the produced powder by this sieve. Of course, the *real* crystal size is different from this mean or apparent size and they are related to each other by the *shape factor*, as in x-ray analysis and by statistical microscopic calculations. Grain shapes are symmetric as in (**a**)-(**d**); less asymmetric, moderately-asymmetric and highly-asymmetric as in (**e**), (**f**) and (**g**), respectively. All have the same apparent/mean grain size [344,691,693].

$$\bar{\rho} = \left(1 - f_{gb} - f_{tj} - f_{cj}\right)\rho_{sc} + f_{gb}\,\rho_{gb} + f_{tj}\,\rho_{tj} + f_{cj}\,\rho_{cj} \; \dots \tag{3-22}$$

and;

$$f_{sc} + f_{gb} + f_{tj} + f_{cj} = 1 \; \dots \tag{3-23}$$

Therefore, any mechanisms that are triggered at such junctions are likely to become very important *at small grain sizes*. The lower limit of the density would now be ρ_{cj}, where the atoms are typically the most disordered and making the value of *n* less than 1.

In fact, however, the density of many nanocrystalline materials is measurably less than the density of the corresponding conventional grain size material even for relatively large grain sizes of order 100 nm. The primary reason is because the processes that are used to make nanocrystalline materials typically result in the generation of significant quantities of defects in the material, particularly *porosity*. The effect of the pores on the density can be very easily-understood by considering the pores to represent a third phase of the material of density ρ_{defect}, where $\rho_{defect} = 0$. The porosity φ of the material is essentially the volume fraction of the pores, $\varphi = f_{defect} = V_{pores} / V_{total}$, and $\rho_{pores} = 0$. With this fourth constituent (the pores defect) included, the overall density of the polycrystalline material becomes now:

$$\bar{\rho} = \left(1 - f_{gb} - f_{tj} - f_{cj} - f_{defect}\right)\rho_{sc} + f_{gb}\rho_{gb} + f_{tj}\rho_{tj} + f_{cj}\rho_{cj} + f_{defect}\rho_{defect} \cdots \qquad (3\text{-}24)$$

Since $\rho_{defect} = 0$ when the defects are pores, it follows that;

$$\bar{\rho} = \left(1 - f_{gb} - f_{tj} - f_{cj} - \varphi\right)\rho_{sc} + f_{gb}\rho_{gb} + f_{tj}\rho_{tj} + f_{cj}\rho_{cj} \cdots \qquad (3\text{-}25)$$

The effect of the porosity can be quite dramatic, because the porosity φ itself can be substantial, depending on the processing route used to make the nanocrystalline material. Much of the literature considers material with a porosity

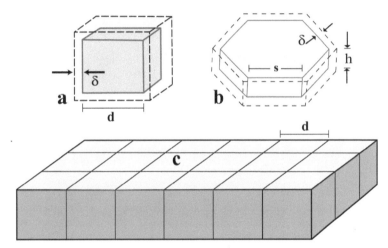

Fig. (25). (**a**) A cubical grain of size *d* with a grain boundary domain of thickness δ. (**b**) Another possible space-filling morphology, using hexagonal tiles of side *s*, height $h = \alpha s$, where α is the aspect ratio of the tile and the effective grain size in the plane of the hexagon is d = 2s. (**c**) The packing of cubic grains to form a material [509].

of 3-5% to be *fully dense* even though this level of porosity has a substantial impact on mechanical properties [509]. The situation will be relatively different when considering grain morphology rather than the simple cubic morphology, *e.g.*; the hexagonal morphology. The effects of grain morphology on the behavior of a material can be very extensive at small grain sizes. For example, consider the alternative space-filling hexagonal prism morphology shown in Figs. **25b** and **26b** for a hexagon of side *s* and height *h*, with aspect ratio $\alpha = h/s$. The effective grain size in the plane of the hexagon is $d = 2s$, but note that if the aspect ratio is either much less than 1 or much greater than 1, a single grain size is no longer a reasonable description of the polycrystal structure. However, if $\alpha = 0.1$, corresponding to a plate morphology (*e.g.*, from a rolling process), $d_{50} = 12$ nm; if $\alpha = 10$, corresponding to a rod morphology (*e.g.*, from an extrusion process), we obtain $d_{50} = 1.6$ nm. So, even the present morphology may have special cases that lead to different values for different types of volume fractions and densities, as in Fig. **26**, because of the very confusing variant values of the grain size of any specified powder. Besides, the density of the material near the grain boundaries and at the triple junctions is generally not known, because there is a very wide range of atomic arrangements possible at the boundaries, but they are lesser than ρ_{sc} as a general. Nevertheless, what single grain size number should be used for a long rod shaped crystal? Should it be the diameter of the rod or the length of the rod? One possibility, sometimes used in the literature, is to define the grain size as the size of the cube of identical volume to the grain in the actual polycrystal [509]. Thus, the concept of *effective* or *apparent* (crystal) grain size arises, which is derived from the x-ray analysis, powder sieving measurement methods and grain morphology geometry for any certain powder, as one can find in the literature [344,691,694,695] to get an applicable value for n (for the present specified powder). For instance, the effective grain size $\overline{d}$, to be used when considering the space filling hexagonal prism morphology shown in **Fig. 25b**, would be given by $\overline{d} = \left(3\alpha\sqrt{3}/2\right)^{1/3} s$. We should thus expect the rod morphology and the plate morphology to produce very different rates of change of behavior as the grain size is decreased. The sensitivity to morphology is evidently related to another specific issue, which is that the grain size is not a well-defined quantity for highly asymmetric structures, as in Fig. **24g**.

Accordingly, the overall density of the material is then predicted, using Eq. (3-25), to decrease *substantially* as the grain size of the polycrystals decreases below 50 nm or so, reaching 95% of ρ_{sc} at $d \sim 3$ nm for the cubic morphology or at $d \sim 5$ nm for the hexagonal morphology with $\alpha = 1$ (assuming fixed values for ρ_{gb}, ρ_{cj} and ρ_{tj}). Hence, the apparent density of fine-grained materials should be expected to be somewhat lower than that of conventional large-grain polycrystalline versions, and this may account for some of the situations where lower densities are observed but no pores are visible even in the TEM [696].

Any real powder has its own grain size distribution, namely; it has an upper and lower limit of its grain size values. Accordingly, any real powder has its own upper and lower limits of theoretical density. Due to the differences in chemical compositions among the various samples and the inevitable uncertainties in measurements (due mainly to thermal and surface effects, pore presence and specimen preparation and size), it was necessary to convert the values of consolidated powder to relative/fractional values (% theoretical density) in order to compare them. Given that exact phase volume fractions are not known, it is possible only to determine a range of percent theoretical density values for each powder compact. So, by averaging the relative densities obtained with the upper and lower limits of theoretical density, one can get the percent theoretical density (ρ%) as [697] :

$$\rho\left(\%Theoretical\right) = \frac{1}{2} \times \left[\left(\rho_{meas}\Big/\left(\sum_i \rho_i\, f_i\right)_{Upper\, Limit}\right) + \left(\rho_{meas}\Big/\left(\sum_i \rho_i\, f_i\right)_{Lower\, Limit}\right)\right] \times 100 \; \dots \; (3\text{-}26)$$

where ρ_i is density, ρ_{meas} is the measured value (using any of Archimedes principle methods) and f_i is the phase volume fraction.

Direct electron microscopy studies were also performed and revealed that grain boundaries in nanocrystals produced from single grained nanoparticles have a similar thickness as coarse grained materials [698]. A reduced atomic density in the grain boundaries down to 75% has also been inferred from Mössbauer spectroscopy in nano-Fe with 6 nm grain size [699]. Till now, a well understanding of grain boundary structure is under developing, but the present theoretical density calculation will be based on the existing data on grain boundary density and width for both conventional and nano-powders.

However, caution must be exercised when either theoretical or experimental density values are considered for nanocrystalline materials (*i.e.*; bulk nanomaterials or nanopowders). The accuracy of the rule-of-mixtures approach scales with the volume fraction of grain boundaries, f_{gb}. For example, the difference of the calculated theoretical densities between two grain boundaries of 10% and 20% of the bulk density of a powder grain, respectively, is very small, while this difference becomes significant when the grain size becomes 10 nm or lesser. This depends on grain boundary width which can have large variations by crystal faults. Also, uncertainties in the reduced value for grain boundary density complicate the problem to a great extent. Even for normal grain size materials, the decrease in grain boundary density is not commonly accepted. Besides, the precise experimental density or even grain size measurements are known to be difficult and thus the experimental verification is not easy at all [344,690,691].

3.3.5 Powder Contamination

Surface contamination is inevitable for any powder material and may become significant in nanopowders due to their whole large surface area. Metastable powders obtained by pulverized melt-spun ribbons, mechanical alloying or milling have a special case. Their structure is either amorphous or nanocrystalline but the powder particle size is in the micron range. For example; Trudeau *et al* showed highly non-stoichiometric surface oxides on nanocrystalline catalysts [700]. In this case, the surface area is small for nanometer powder particles but contamination is due to a large impurity intake during the processing steps. Oxides, nitrides and other compounds are often found in consolidated parts made of attrition-milled nanopowders [126,701,702]. Despite of any present contamination, these compounds maybe used to advantage, when present, as fine dispersions to prevent grain coarsening upon further densification.

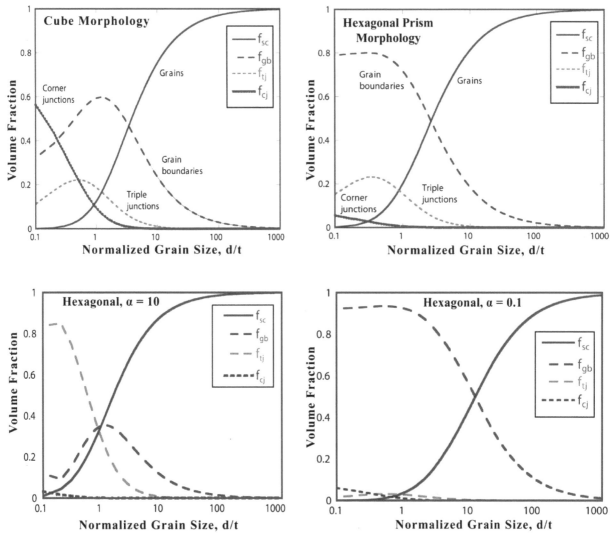

Fig. (26). Variation of grain volume fraction, grain boundary volume fraction, triple junction volume fraction, and corner junction volume fraction with normalized grain size d/t for: (**a**) the cubic, (**b**) the hexagonal grain morphology at $\alpha = 1$, (**c**) hexagonal morphology at $\alpha = 10$ (rod-like grains) and (**d**) hexagonal morphology at $\alpha = 0.1$ (plate-like grains) [509].

The influence of contamination on sintering was studied by *in situ* TEM under controlled oxygen levels as compared to ultrahigh vacuum conditions for both metal and ceramic nanopowders [656,657,703]. All results indicate that ultra-clean nanoparticles sinter very rapidly even at room temperature. In contrast, when sintering of nanoparticles takes place in the presence of controlled oxygen traces, very little or no neck growth was observed [704]. This oxygen contamination decreases the surface energy and slows down the sintering kinetics.

Although some caution is in order, Payne also showed that boundary interfaces display electronic states and, therefore, different properties in comparison to bulk crystals [705]. The interfacial energy may be modified by any distortion of a surface structure such as segregation or adsorption of impurities. Impurities dictate the thermodynamics of surface and surface behavior of micro- or nanoparticles. Macroscopically, stabilization of particle surfaces, *i.e.*; lower surface energy values when oxides are formed, results in a decrease of the driving force for sintering. This has been observed in nanosized particles which typically contain large amounts of adsorbates (*e.g.*; high oxygen in TiN [706]). In addition to the inherent morphological metastability related to fine grain size, nanocrystalline materials (bulk or powder) often display topological (*e.g.*; alternate crystal structures than at equilibrium) or compositional (*e.g.*; extended solid solubility or amorphous phase) metastabilities [641]. More specifically, TEM studies showed that sintering of Fe-Ni nanoparticles starts only when oxide layers are reduced [313,707]. When vacuum was used, no sintering took place up to 452 °C. To avoid further contamination, the as-produced powders may be either consolidated *in situ* or handled in controlled environment prior to sintering. Fig. **27** shows the closed system developed by Tohoku University for oxygen and moisture-free processing of light alloy powders from gas atomisation through to canning prior to degassing, sealing off and extrusion. This adds large costs to the actual component production. Melt-spun ribbons are produced at cooling rates of 10^6 °C s^{-1} or higher, which are then pulverized (by conventional routes, not the low or high energy milling routes) into flakes for subsequent consolidation. The powders or pulverized ribbons are processed by canning, degassing and hot pressing additionally to produce solid billet for subsequent forming by forging or extrusion. In a manufacturing environment, powders are consolidated in a remote place from the powder producing area. Therefore, contamination of small particles is likely to occur. Alternatively, consolidation processes that are less susceptible to the contamination levels are recommended (*e.g.*, high pressure consolidation and severe plastic deformation) [313,641].

The problem of process atmosphere is serious and has been found to be the major cause of contamination in many cases. In fact, it has been observed that if the container is not properly sealed, the atmosphere surrounding the container (usually air that contains moisture, nitrogen, oxygen and hydrogen) leaks into the container and contaminates the powder. Thus, when reactive metals like titanium and zirconium are processed or milled in improperly sealed containers, the powders are contaminated with nitrogen and oxygen. It has been reported that flushing with argon gas will not remove oxygen and nitrogen absorbed on the internal surfaces. Pickup of impurities during pulverizing would reduce the pressure within the container, allowing outside atmosphere to continuously leak into the container through an ineffective seal. In practice, it has been noted that if it is difficult to open the container lid due to the vacuum present inside, it is an indication that contamination of the powder is minimum. Attempts have been made to improve the container seal integrity to prevent the outside atmosphere leaking inside. Use of high-purity argon (99.998%) atmosphere and improvements of the seal quality resulted in the processing of high-quality titanium alloy powder with as little as 100 ppm oxygen and 15 ppm nitrogen [680,708,709]. This process, however, may not be economically viable and hence may not be feasible on a large industrial scale [701,702].

As an example for avoiding contamination, degassing of aluminum powder and its alloys prior to consolidation can be performed to improve the final properties of the sintered compact. The degassing should be carried out at elevated temperatures, in order to fully desorb the water vapor and decompose the hydride that forms on the surface, and thus ensuring an effective evacuation of the volatile species in the powder [710]. Degassing times and temperatures are dependent on the desired amount of degassing achieved, experience and the economics of the treatment. Complete degassing is very difficult to achieve, even at high temperatures (550 °C) and long times. The degassing is carried out in a partial or fine vacuum. Temperatures vary, but a range of 200-400 °C is effective. Many techniques have been developed to perform the degassing. A good discussion of several of these is contained elsewhere [295]. Degassing has been shown to convert the ductile aluminum hydroxide into a brittle form of alumina [711]. Once the hydroxide has been converted, it is stable for several days in air, possibly allowing batch degassing to be included in a production process. The brittle alumina is broken during pressing, allowing for a much larger number of metal-to-metal contact areas. This results in improved compressibility and improved strengths after sintering. Green compacts densities and green strengths are also improved dramatically by prior degassing [648], with strength improving by greater than 100% of the strength of the green compact.

3.3.6 Cold and Hot Compaction

The cold compaction forming stage can be accomplished using different techniques, such as: die compaction, cold isostatic pressing, rolling, injection molding and slip casting. The main stages of densification during cold compaction can be itemized as: particle rearrangement, plastic yield deformation and fragmentation. That is, the initial transition with pressurization is from a loose array of particles to a closer packing. Subsequently, the point contacts deform as the pressure increases leading to plastic deformation, and finally to fragmentation of the powder body. Essentially, these various techniques differ from the mode of application of the external force and the compaction is accomplished without the application of heat, depending on the used powder. The most conventional and widely used is uniaxial die compaction using mechanical or hydraulic press systems. This can be achieved by single or multi-level acting punches, depending on part complexity. However, many (hard) powders require

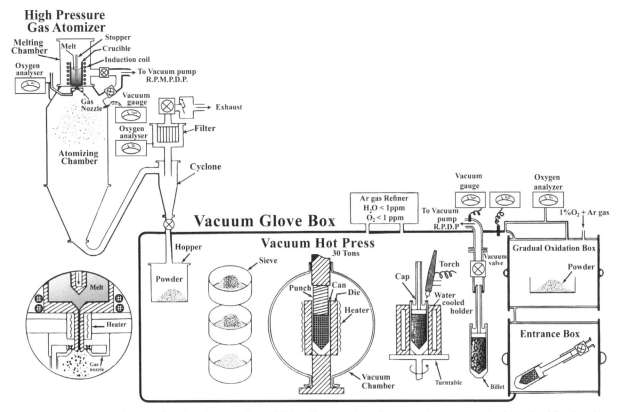

Fig. (27). A sequential System for closed processing of light alloy powders, incorporating gas atomisation, classification, hot-pressing and canning, all prior to degassing, sealing-off and extrusion. The powder atomization unit can be replaced by a Melt-Spinning unit followed by a ribbon-pulverizing device, all are working in *dry and highly-pure* inert gas [18,712].

preheating to begin consolidation. In the case of a single acting system, the top punch closes the die and compacts the powder to the required shape that is followed by a punch release. For multi-level components, multiple punches are used which act independently, as shown in Fig. **28**. In the cold die compaction forming process a controlled amount of mixed powder is automatically gravity-fed into a precision die and is compacted, usually at room temperature, at pressures as low as 138 N/mm^2 or as high as 827 N/mm^2 depending on the density requirements of the part. Higher pressures limit tool life, especially with thin or fragile tool components. Powders that develop green strength on compaction can, for example, be filled in all rubber molds of complex or simple shapes. After sealing with suitably formed rubber lids and sealing bands, the molds are inserted in the pressure chamber of a cold isostatic press and compacted by pressurizing with a fluid. In this way, billets for subsequent hot working or sintering plus hot working have been ever manufactured since cold isostatic presses have been available. Since there is no *wall friction*, powders need no lubricant addition; powders without green strength can be binder-treated or handled in sheet metal containers before consolidation [713,714], as in Fig. **29**. Alternatively, the hot compaction mode of forming P/M parts is suitable for consolidating hard powders and can be classified into; (*i*) hot isostatic pressing, (*ii*) hot extrusion, (*iii*) hot die compaction and (*iv*) hot powder spraying, among others [295,713]. In the hot compaction forming process, both the external force of pressure and the heat are applied simultaneously in the compacting process.

To eliminate adsorbates and enhance interparticle bonding, a *warm compaction*, and prior to hot pressing and in adequate time interval, at temperatures up to 475 °C has been largely applied to the powder. Compacting the loose powder produces a coherent piece called *green compact*. Strength of green compacts results mainly from mechanical interlocking of *irregularities* on the particle surfaces (*i.e.*; the powder particles surface tribology and grain surface friction forces), which is promoted by plastic deformation during pressing. The green compact has a shape of the finished part, when ejected from the die and also has sufficient strength for in-process *careful* handling and transport to a sintering furnace. The resulting green compact has very little cohesion and for any further applications the green compact has to be sintered. Owing to low applied pressures and the fact that powder slides relative to the die wall, the density distribution in the green compact is non-homogenous. The typical density attainment during die compaction is limited to 60 to 70 percent of the theoretical density of the fully dense metal.

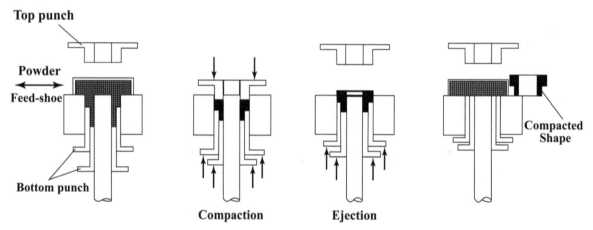

Fig. (28). Multiple punches for producing multi-level shaped component [682].

3.3.7 Hot Isostatic and Pseudo-isostatic Pressing

The most popular way to achieve 100% density in powder materials is with hot isostatic pressing. In the HIP processes, the compaction is typically performed in a gaseous (usually inert argon or helium) atmosphere contained within the pressure vessel. The HIP process consists of the powder to be compacted which is canned by a *mild steel* container of the desired shape which, in turn, is subsequently vacuumed. The canned powder mass is compacted by the simultaneous application of hydrostatic pressure (up to 300 MPa) and heat. HIP densification involves longer times and therefore grain growth is more likely to occur than in hot pressing, but by careful control of the HIP route parameters (mainly; temperature and pressure) a lesser grain growth can occur [715]. Common pressure levels which extend up to 103 N/mm^2 are combined with temperatures up to $0.7T_m$. Processing volumes currently attain sizes up to 1.2 m diameter by 2.7 m long. It is important to note that the end products from the HIP process do not need any further treatment and are ready for use. HIP is considered only a near-net-shape method. Owing to high operating pressure and temperature values and lack of die-tool setup during HIP, the end stage results in a more uniform density distribution and the achieved density of the end part is equal or nearly-equal to the theoretical density of the metal or alloy [682,714]. However, due to its very low production rate, costly equipment and unique tool requirements, the HIP process is normally relegated to expensive materials such as tool steels, superalloys, titanium, and so forth. The process also requires high-purity powders that should be generally spherical or semi-spherical in shape (as for production of complex-shaped components in cans *via HIPing*) due to their ability to mutually redistribute and fill the thinnest elements inside cans. High flowability of spherical granules opens good possibilities for consistent precise filling, including automated ones, during canning

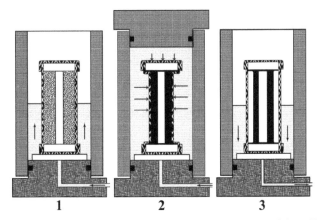

Fig. (29). Wet bag isostatic tube compaction. **(1)** Inserting mold into vessel. **(2)** Pressurizing. **(3)** Opening vessel after pressure release [713].

[646]. The powders are vibrated in place in a metal or ceramic container, which is then evacuated, sealed, and then placed in the HIP vessel, as in Figs. **27, 30,** and **31** [18,177,713]. In addition to sintering, HIP is also used in attempts to achieve full densification or to cure flaws remaining after sintering and this application is, in fact, more general than its role in pure sintering. In this case, since the pores are isolated in the sintered compacts, encapsulation is usually not necessary. If a high pressure is applied to a compact containing interconnected pore, on the other hand, HIP cannot completely densify the compact because the high pressure gas in the vessel is entrapped within the pores at the moment of pore isolation. Full densification by HIP can be achieved only after pore isolation with zero apparent porosity (interconnected porosity). On the other hand, when isolated pores are entrapped within grains due to pore boundary separation, full densification is also impossible by HIP.

Another approach to reduce the cost of HIPing is the true isostatic gas pressure forging [677,716-718] where encapsulated powders or *containerless preforms* without interconnected porosity are heated to hot-working temperature, inserted into a hot pressure vessel and exposed to a gas pressure pulse up to 1000 MPa by injecting a suitable fluid into the vessel. In contact with the hot environment, the liquid evaporates explosively and consolidates the material to full density. Other processes which usually do not employ a true isostatic state of pressure have at least reached the ready-to-use stage, *e.g.*; *pseudo-HIP*. This technique utilizes a solid powder, like the commercial casting sand SiO_2 of about 100 μm average grain size and known as a pressure transmitting medium, as the working fluid to pseudo-isostatically press a green compact. The word "pseudo-" is used because the pressure-transmitting hot medium is not completely rigid but has a non-zero static shear component and the preform is placed in the center of the pressing powder. At first, the preform is immersed in a *sol-gel suspension* of refractory powder ceramic to make a non-diffusing coating after drying.

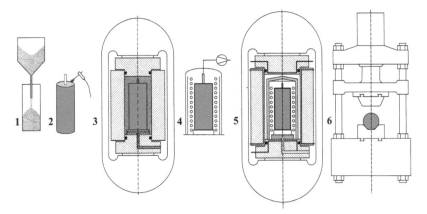

Fig. (30). Hot isostatic pressing. **(1)** Fill can. **(2)** Close can by welding. **(3)** Optional cold isostatic pressing for better heat transfer. **(4)** Heat and evacuate. **(5)** Hot isostatic consolidation and **(6)** hot work [713].

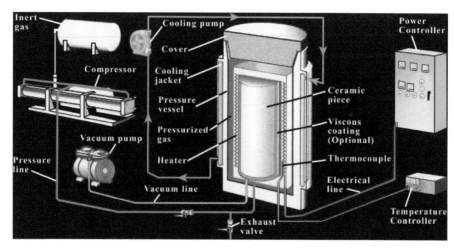

Fig. (31). Hot isostatic pressing (HIP) system for advanced refractory alloys and ceramic arts manufacturing [177].

The coated preform then is inserted in the center of the compressing powder and cold-pressing the system. To remove gases the system should be preheated to $0.3T_{m}$ for 1 h, while evacuating the chamber to ~ 5 Pa (*i.e.*; fine vacuum) using a vacuum pump. Then, pressure is applied by a uniaxial pressing while raising the temperature to the sintering degree for the required time.

Also, there are other variants of this route, namely: the STAMP, Rapid Omnidirectional Compaction (ROC), Ceracon and Ceramic Mold Processes [85,713]. Fig. **32** shows an example of the pseudo-HIP that can be used effectively in manufacturing many-parts preforms with complex designs in consolidation temperatures as elevated as 2200 °C and for multiple preforms at the same time too [718,719]. Recently, a new HP technique, where a pulsed electric field is applied, has been developed, the so-called *Spark Plasma Sintering* (SPS) [720-722] or *Pulsed Electric Current Sintering* (PECS) [723]. The experimental set-up is similar to that of HP with a graphite die and graphite punches. The heating of the sample, however, is achieved by resistance heating caused by a pulsed current (typically a few thousand amperes) under a pulsed DC voltage (typically a few volts ~ 10 V). A high heating rate of more than a few hundred °C min^{-1} can be achieved, as in Fig. **33**. In SPS the densification is very fast, taking a few minutes, while grain growth is very limited at process temperatures which are measured to be lower than HP temperatures [649]. Many benefits of the SPS technology include higher consolidation densities, smaller grain sizes, cleaner grain boundaries, high heating rates, the application of pressure, and the effect of the current activated sintering by the boundary mass transport and increasing the point defect concentration. The general explanation for these observations is the role of presumed plasma forming between particles in the sintering compacts. Finally, for more information about the HIPing, see references [295,713,725-727].

3.3.8 Sinter-Forging Technique

Sinter-forging is the simultaneous densification and deformation of a porous green compact at elevated temperature under the action of a uniaxial stress. Also known as Powder Hot Forging; a sintered blank of appropriate geometry is

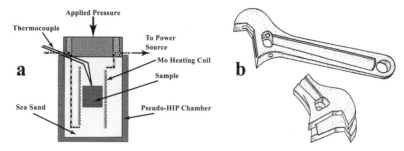

Fig. (32). (a) Pseudo-HIP apparatus consisting of a stainless steel pressure vessel 110 mm inside diameter, 180 mm outside diameter and 145 mm deep, filled with commercial casting sand (SiO$_2$, average particle size is 100 μm). **(b)** A many-part-preformed wrench [718,719].

hot-forged in a closed die to the required shape and dimensions. It uses the superplasticity of the material itself by closing pores with the aid of plastic flow. To eliminate large pores and prevent grain coarsening, large strains must be applied before small pores are eliminated by diffusion, *i.e.*, during the intermediate sintering stage. For this purpose, a two-stage sinter-forging (high, then low, strain rate) has been developed and resulted in the highest densities and smaller grain sizes [641]. Generally, the stress levels required for densification by sinter-forging are lower than in hot pressing or HIP. This method received substantial attention, both theoretically and experimentally, as applied to ceramic densification [641,714,728-731]. The most attractive benefit in using the sinter-forging technique is the capability of densifying green compacts with large inter-agglomerate pores. The method involves a uniaxilly applied pressure/stress with or without moulding die in temperatures below $0.5T_m$ in order to suppress the grain growth, and, as a result, a finer grain size values can be obtained. Sinter-forging is carried out as a two-step process, as in Fig. **34**. The first step of this process includes the cold compaction of the metal powder under appropriate pressure at room temperature for some or many hours in a compressing cylinder or moulding die. Following is a hot-uniaxial press, with lower pressure than the first stage (lesser than 1GPa), after transferring the preform into a slightly *wider* pressing cylinder or die but with the same inner volume, at suitable temperature, where a degassing is applied with a slow and gradual increase in temperature till its final value (commonly below $0.5T_m$), and then the pressing chamber is back-filled with inert gas. Walls of the die must be lubricated with a suitable high temperature lubricant (like C-blacks or graphite fine powder in a mineral oil; the so-called graphite paste) in order to decrease the friction between the compact and die walls. So, a lesser pressure is lost, leading to make the *radial plastic deformation* of the preform uniform in all directions. This creates stress states favorable for *pore closure* and bonding across collapsed pore interfaces and allowing easier compact ejection after consolidation, compared with single step hot forging/pressing. Thus, no *flaking*, *bulging* or cracking appear around the edge in the final compacted workpiece [714,731,732]. The time of the uniaxial pressing/stress differs from alloy system to another, depending upon the conditions of temperature and pressure used that do not lead to a grain growth but attaining the compact fine microstructure, and the duration is commonly 1 minute to 1 hour. Typically, time of this process is lesser than the

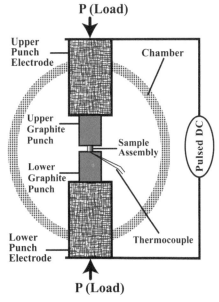

Fig. (33). Sumitomo model 1050 SPS apparatus. The unit is capable of producing a DC current up to a maximum of 5000A at a maximum voltage of 10 V [724].

HIP and leads to full density compaction. During the sinter-forging process, the compact deformed significantly and eventually conformed to the die walls under the effect of the stimulated intrinsic superplasticity of the compacted powder. Apparently, this technique is more effective than not only pressureless sintering, but also straight hot pressing and HIPing. The applied high pressure and the sinter-forging geometry enhance shearing and plastic strain controlled pore closure and at the same time suppress grain growth. The stress also plays a significant role in stress-assisted densification (diffusion) mechanisms. Possible stress effects on deformation and densification can be seen from the established formulas for creep, due to grain boundary sliding and plastic flow under the uniaxial stress [729,733]. In addition, the extensive plastic deformation under high stress results in disruption of possible continuous oxides on powder particle surfaces that may degrade bonding and properties. Precautions should be considered to avoid metallic specimen fracture at high strain rates and low temperatures. Some limitations in using pressure consolidation are related to metallic and ceramic specimen fracture at high strain rates (at low temperatures) that restrict the internal heat generated due to plastic working to dissipate quickly and thereby, reduce the resistance of sintered materials against deformation [731]. Here, volumetric constancy is no longer valid for establishing the compatibility conditions, as density of the compacts increases simultaneously with the deformation due to closing of inter-particle pores. Hence, yielding of sintered powder compacts is sensitive to hydrostatic stress component. The investigations during plastic deformation of sintered powder preforms revealed that compressive forces gradually close down the interparticle pores leading to decrease in the volume and increase in the relative density. Although a reduction in temperature or time during consolidation would be preferable to further constrain grain growth, it would require higher applied pressures to achieve full density and hence introduce undesirable complexities in the process. Our current choice of the consolidation parameters may have *compromised* some grain growth for ease in operation [732].

3.3.9 Severe Plastic Deformation Consolidation (SPDC)

The key strategy in enhancing the nanocrystal number density and thereby to improve both property performance and microstructure stability is to promote the nucleation density of nanocrystals while minimizing the change of the amorphous matrix phase. One new opportunity for enhancing *the number density of nanocrystals* is presented by severe plastic deformation of rapidly quenched marginally glass-forming alloys (*i.e.*, producing nanocrystalline material by the *metglass devitrification*) [309,734,735]. In addition to nanostructure formation, the deformation treatment serves also as a consolidation step, which is important for producing bulk shapes more cheaply. Many Studies show representative examples of partially nanocrystallized metallic samples after thermally-induced or deformation-induced nanocrystallization. The comparison clearly indicates the enhanced nanocrystal number density that can be obtained by combining different nonequilibrium processing pathways sequentially. These initial results together with results from combining different plastic deformation treatments indicate an entire range of advanced processing routes for obtaining bulk nanostructured materials or bulk nanocomposites that still waits to be explored [569,736]. During the process, raw powder particles with a size of several microns experience severe plastic deformation (*i.e.*; undergo a repetitive cold welding by the collisions and frictions of the powder particles) and fracturing mechanism. These processes lead to (*i*) milling of softer amorphous particles by harder intermetallic and semi-amorphous particles, (*ii*) deformation and crystallization of the surface layers of hard particles due to friction between these particles, (*iii*) welding amorphous particles to each other and to intermetallic particles and the formation of soft amorphous layers on the surfaces of the intermetallic particles, (*iv*) sticking of intermetallic

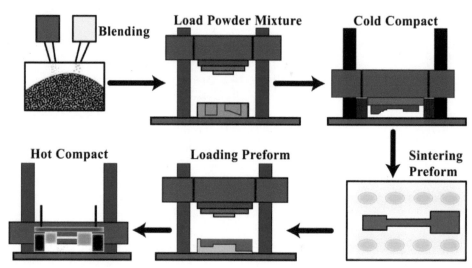

Fig. (34). Schematic of sinter-forging process [714].

particles to each other through the soft amorphous layers and (*v*) tearing of these amorphous layers leading to separation of the hard particles during their further rotations. Almost full consolidation occurs in regions predominantly consisting of amorphous and semi-amorphous powder particles at a true strain of ~ 2. However, further excessive deformation leads to strain localization and shear crack formation in these regions but consolidation can be enhanced by cold compaction, prior to the process, and using flakes rather than spherical powder particles to prevent the particles rotations [565,737].

Apart from consolidation of nanopowders, *severe plastic deformation* (SPD) shows promise to deform bulk structures having coarser grains into submicron products with a mean grain size of about 100 nm. SPD can produce large amounts of bulk samples without residual porosity for mechanical testing. However, the final structures are metastable and therefore susceptible to grain growth at high temperatures. There are two major procedures for severe plastic deformation processing, namely; *severe plastic torsion straining* (SPTS, also known as the *high pressure torsion* HPT) and *equal-channel angular pressing* (ECAP) or *extrusion* (ECAE) [309,565,738]. In the SPTS process, a sample in the form of a disk is subjected to very large shear torsion straining under the applied high pressure of several GPa at room temperature. One of the sample holders rotates and surface friction deforms the material by shear (Fig. **35a**). In ECAP, the sample is pressed through a die in which two channels, equal in cross section, intersect at an angle $\Phi \geq 90°$, with an additional angle Ψ defining the outer arc of the curvature where the two channels intersect (Fig. **35b**).

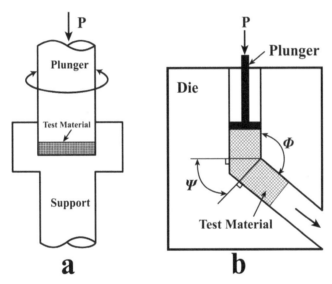

Fig. (35). Principles of SPD methods. (**a**) Torsion under high pressure and (**b**) ECAP [309].

For the melt-spun ribbons, they are packed into copper cans without preliminary milling (because the ECAP process acts as the mechanical ball milling, so ribbons can only be chopped or sliced to small slices or flakes of few mm size) and cold compacted to near 75% theoretical density to conform to the die of the SPD route cross-section. The cans will then be sealed with or without prior degassing of the compacted material. Prior to extrusion the die and can were heated to the selected extrusion temperature, T_{extr}, and equilibrated for few minutes. Extrusion is done at a rate of few mm/s with back-pressure applied to the exit channel of the die. T_{extr} should be carefully-selected for the optimized conditions of producing a full amorphous or nanostructured material and should be well below the transformation temperature of the existing metastable phases. This transformation temperature can be obtained by the differential thermal analysis (DTA) of the melt-spun ribbons prior to the consolidation processing. The can with chopped and cold-compacted ribbons change its cross-sectional shape during the extrusion pass, and the cross-section remained unchanged during subsequent passes, as in Fig. **36**. At least one extrusion pass is applied in the range of 0.2 GPa to several GPa as the alloy system and material demand. The sample is pressed through the die using a plunger or a pressing punch and subjected to shear as it passes through the shearing plane at the intersection of the two channels. As deformed, dimensions are identical to the initial ones, thus it is possible to repeat this process for many cycles to accumulate a large plastic strain. In this regard, ECAP is more attractive than SPTS because it can be used to produce not only laboratory samples but bulk submicron-grained billets directly for industrial applications. Both of these procedures are known to induce a high density of dislocations that arrange subsequently into metastable subgrains of *high-angle boundaries*. In pure metals, the application of SPTS usually results in the formation of an ultrafine-grained structure with an average size of about 100 nm and lesser, while the application of ECAP yields a grain size of 200-300 nm. A particular crystallographic texture is formed in ECAP samples. However, it is not necessary to obtain high density precompact since ECAP is much more effective in increasing density. The as-processed samples have high dislocation density, high internal stress and high angle non-equilibrium grain boundaries [565,738,739]. As an example, the submicron-sized grains of Al and its alloys processed by ECAP are generally unstable at temperatures above 250 °C except when the precipitates are present. The precipitate particles can inhibit the grain growth of the selected Al-alloys. In this case, grains remain within the submicrometer range up to temperatures in the vicinity 400 °C. The stability of ultrafine grains due to precipitate formation at elevated temperatures can lead to the occurrence of superplasticity, as in Al-2024, Al-7034 and Al-Mg$_{3-4}$-Sc$_{0.2}$ alloys processed by ECAP [309,740,741].

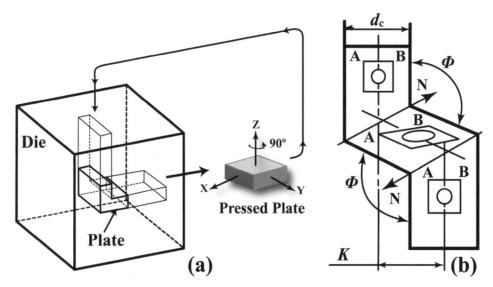

Fig. (36). The ECAP SPD process. in (**a**), pressing a plate in a horizontal configuration that can be repeated many times and the die can be surrounded by heating elements. In (**b**), a multi-pass facility for ECAP, where N is in the shear direction, d_c is the channel diameter, K is the displacement between the two channels, Φ is the angle of intersection between the two parts of the channel and the internal shaded areas depict the shearing as the sample traverses the shearing zone. Note the change in dimensions of inner grains (represented by circles) forced by SPD that leads to quenching-in of large grains into fine ones with the collapse of all pores and agglomerates [309,565].

Another variant of the ECAP route that uses the severe plastic deformation techniques for consolidation of rapidly quenched metals and alloys, as fully amorphous structures, is the new process of the *twist extrusion* (TE) that may enable fabrication of large cross-section materials. It was used for the first time for consolidation of amorphous melt-spun ribbons and it involves passing a billet through a die with a twisted channel (Fig. **37**). Cyclic torsional deformation can be applied in subsequent passes where the torsional strain is applied in either the same direction or the opposite direction relative to the initial torsion. In such a way, the deformation in TE is similar to that in high-pressure torsion processing of thin disks, where the billet cross-section remains unchanged throughout the extrusion. Although the details are different, ECAP processing similarly allows an accumulation of strain by subsequent extrusion passes [742-744].

In particular, the deformation gradient in TE processing is quite steep in the cross-section area and this method is characterized by intense flow of material being deformed within cross-sections of the billet and can be fully consolidated by only one pass with a 360° full twisting. Also, the TE route may be conducted before applying the ECAP procedure, leading to a reduction in the grain size to a level that is more amenable to easy refinement to produce a true ultrafine-grained microstructure.

As a general, SPD methods resulted in overcoming of a number of difficulties connected with residual porosity in compacted samples, impurities from ball milling, processing of large scale billets and practical application of the given materials. Thus, the principle of processing of bulk amorphous and nanostructured materials using SPD methods is an alternative to the existing methods of powder compacting. Other SPD methods can be found in literature [565].

3.3.10 Shockwave Consolidation

Also it is known as the dynamic or explosive consolidation technique that proceeds with the passage of a *large-amplitude compressive stress* or a *shock-wave* generated by plate impact or explosion without any external heating. The peak pressure values may be on the order of tens of GPa up to few hundreds of GPa in some routes and set-ups, thus providing densification by plastic yielding for both metals and ceramics with no occurrence of long-range diffusion. It is a short time sintering method with application of very high pressure, for densifying and compacting ceramic, metal and intermetallic materials. In this regard, shock wave compaction is attractive because it can be used to fully densify powdered materials without inducing thermal activated microstructural and compositional changes that are normal in conventional thermomechanical processing. Such densification is possible because of the interparticle bonding due to localized melting at the interfaces between the particles.

The high rates of heating and subsequent cooling (10^8-10^{11} °C s^{-1} for metals) which occur during the explosive compaction process (*i.e.*; densification, particle surface melting, rapid surface re-solidification and cooling due to the high difference in temperatures between the undeformed interior and the deformed surface area of the powder particles) mean that grain growth is greatly minimised because of the very small diffusion distances (~ 10-100 nm) [85,641,714,746-750]. Localized heating, possibly up to melting temperatures due to *particle interfriction*, occurs and enables good interparticle bonding. In nanosized powders, the heat may transfer throughout the entire particle,

Fig. (37). The twist extrusion (TE) processing technique, where a sample is passed through a twisted channel. The sample cross-sectional area and shape do not change during this process, but its spatial dimensions with grain size refinement [742,745].

thus providing an advantage over coarser materials where the heating is only superficial. Best results are achieved when high temperatures are reached before the shockwave passes. If particles are heated, they may deform rather than fracture when the stress is applied. This very short, high-temperature exposure provides *the best means to retain* fine grain size or out-of equilibrium conditions such as amorphous structures, hard quasicrystalline powders or supersaturated solid solutions. The major drawback is the difficult coordination of these short stress and heat application events which result in frequent specimen fracture (cracking), especially in ceramic materials due to their high values of hardness, melting temperatures and viscosity [641,747,751-753]. Metallic or alloy powders are usually processed through this route due to the fact that plastic deformation in metal is quite easier compared to ceramics. A typical laboratory shock-wave setup and a schematic of the equipment are shown in Figs. **38** and **39**. If a metal powder at atmospheric pressure, ambient temperature, and initial packing density is subjected to a shock wave of energy E, then consolidation is all about how much energy is being used along the interfaces of the particles in order for the lot to be densified properly. As in Fig. **40** [754]; energy is absorbed by the packing material or spent towards, creating a loud noise! Net shock energy spent in the interfaces is inversely proportional to the surface area of the particles [756,757]. It is obvious again that less energy is needed to consolidate nanoparticles compared to coarse-grained particle, which is a merit indeed. Different models have been proposed to describe the solid-state changes and reactions involved during shock wave processing and several experiments have been conducted to correlate those models with the experiments [757-759]. The critical processes, occurring during the microsecond time duration of shock compression loading, involve the heterogeneous deposition of shock energy, resulting in interparticle bonding, and configurational changes in the particles due to the annihilation of voids *via* plastic flow and dispersion of fragments. Hence, with shock compression, it is possible to fabricate nanocomposite bulk magnets without grain growth since the heating is limited to the particle surface regions and it occurs in microseconds. Microstructural modifications and a high density of defects produced under such conditions generate a highly activated dense-packed state. This can result in improvements in properties of bulk shock consolidated materials, or even alter the thermodynamics and kinetics of post-shock crystallization and precipitation treatments to permit the retention of metastable phases or nanoscale size of crystallites or precipitates [714,755]. Also, the explosive flyer compaction shockwave can be used for exciting or triggering the Self-propagating High-temperature Synthesis (SHS) reactions used for producing advanced and ultrahigh-temperature bulk ceramics like WC, ZrO_2 and TiC-TiB_2 and ZrB_2-SiC cermets and composites, as an example [753,760-762], for space rockets and extraterrestrial orbiters and all hypersonic engines.

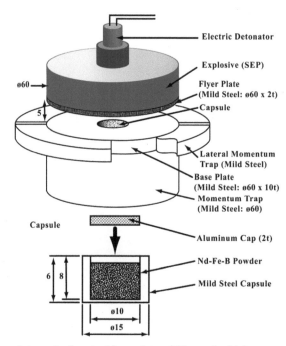

Fig. (38). Compaction by the flyer plate method, a steel base plate of 10 mm in thickness and 60 mm in diameter, and a center hole of 15 mm in diameter and 10 mm in depth is bored on the base plate. Momentum traps are glued to the base for removing excessive shock wave reflection. φ for diameter and t for thickness. SEP™ explosive is a Japanese trademark [755].

For the process equipment, the container must be strong enough at the high consolidation or reaction temperature (2000 °C and more) to keep sample pieces from flying away. Mild steel has been found to satisfy this requirement; a highly conductive and highly ductile material. The container must also be able to allow the escaping, high temperature gases to exit after achieving consolidation. For the SHS reaction, another requirement is that the thermal conductivity and heat capacity of the container be low enough so that the heat generated by the reacting sample is not drained away to the container walls and, thus, quenching the reaction. This requirement is met by making the steel container out of a thin annular ring or sleeve made of mild steel also (surrounding the powder preform) and backing this ring with a thermally insulating material such as alumina plaster. All of these components are inside the thick-walled mild steel container and below the pressing flyer plate. The fact that the plaster is used for the bulk of the containment vessel satisfies another requirement, namely; the compressibility of the container by the explosively generated pressure shockwave becomes roughly equivalent to that of the powder sample. Finally, a high temperature and relatively inert material must be placed between the sample and the steel container to prevent the iron from diffusing into the hot or reacted sample. Grafoil and Zirconia sheets have been found to satisfy this condition [753]. Many types of explosive materials are used in this technique with limited amounts for the use in this route depending on the optimized conditions present for producing high quality products.

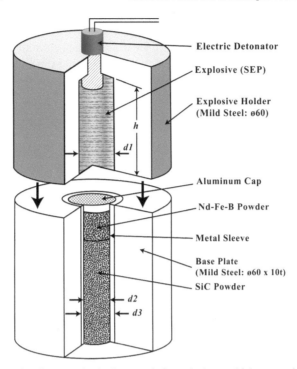

Fig. (39). Explosive compaction by the direct method. The metal sleeve is 1 mm thickness and made of mild steel. The unit can be used for the SHS production methods and for compacting many different powder preforms at the same time by one explosive charge [755].

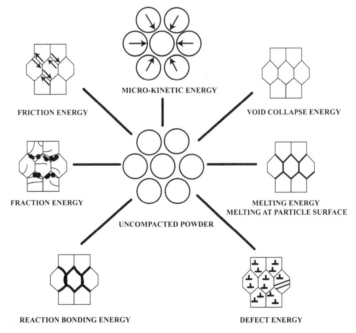

Fig. (40). Various modes of energy dissipation in shock compression of powders [754,757].

For the common cylindrical setups, the shock pressure on the powder can be varied by varying the amount and the type of the explosive. By increasing the detonation velocity from 3500 m/s (which is the lowest range of Ammonium Nitrate and Fuel Oil or ANFO, Nitrammite, Ammonium Perchlorate and Baratol explosives) to 7000 m/s and higher (for plastic explosives like PETN, PBX, Cyclotol, compositions A, A-3, B, B-3, C-4 …. etc), the peak shock pressure is four times higher since the pressure varies with the square of the detonation velocity. Of course, these are extremely high dangerous materials and anyone is not allowed to use it freely. For the social and people safety, many countries impose tight restrictions for using these high explosive and military materials for the industrial usage, so one must get permissions from his local governmental authorities for this serious industrial technology. The entire fixture is then buried under a suitable pile of sand, which served to protect the case and contents from excessive *thermal shock* and sharp vibrations. This thermal shock is mainly due to the fast heating rates, accompanying vibrations, the type of materials and the non-homogeneous density distribution of the green compact. After the fixture and its contents returned to ambient temperature, the compacted billet/preform is extracted. Once the processing is completed, extreme care needs to be exercised to look for unexploded explosive material. The pressure pulse converges toward the central axis of the cylinder, and, if excessive, a hole is generated along the cylinder axis, known as *Mach stem* and stimulates macro- and micro-cracks. The Mach stem is due to the so-called *Mach reflections* effect of the explosive charges and it is used world-widely in blast warheads when a bomb is pre-detonated at some distance above the ground target. So, the reflected and incident waves overlap, producing a third wave of higher pressure and impulse than the original incident one with vertical wavefront, and finally increasing considerably the radius of effectiveness of the bomb damage. Mach stem can be eliminated by putting a solid metal rod (mandrel) along the axis or by adjusting the shock and detonation conditions [761,763,764].

For the cylindrical set-up (Fig. **41**), which results in the convergence of the shock-waves at the central axis, it is responsible for five types of cracks [764,765-769]:

1. *Helicoidal cracking*: this type of cracking is due to the compressive shear-stress; the diameter of compacted material has to decrease as it moves towards the centre. If the ductility of the material is low, helicoidal cracks are generated.

2. *Mach-stem formation*: this is due to the compressive stress; as a result of overcompaction, in the middle of the tube. The generated pressure is too high, *i.e.*; excessively large shock energies are used and, accordingly, the material turbulates. Further, it melts and can be blown out.

3. *Circumferential cracking*: this is due to the tensile stress, where the reflected and transmitted shock-waves meet each other, even inside the material itself, and superpose causing circumferential cracks.

4. *Radial cracking*: this is due to the tensile stress also. Just before the pressure release the material near the metal tube is under higher compressive stress than the material in the centre. As a result, radial cracks are formed initiating at the centre of the tube.

5. *Transverse cracking*: this is due to the tensile stress resulting from: (*i*) impedance mismatch between the compacted material and the bottom plug, (*ii*) stretching of the cylinder, and (*iii*) thermal stresses during cooling.

The use of lower detonation velocity explosives (so making an intermediate shockwave compaction) and preheating the powder (*i.e.*; hot shock compaction) avoid or minimise cracking of compacts. High temperature can induce additional ductility to the powders and reduce their strength and hardness by recrystallization. Another advantage of predetonation heating is that the shock energy required to melt the powder surfaces is decreased. Also, by making the preforms of uniform density and *uniform pore-size distribution* the sintering process is done without cracking as the sintering rate increases with increasing green uniformity and density, thus reducing the sintering times and/or temperatures required [771].

Another approach for the crack prevention is the *converging underwater shock wave* compaction method with temperatures even more 1000 °C [750,772-776]. Explosive compaction, with the shock wave transmitted into the powders through a water medium, provides the ability to employ a more dispersed shockwave of prolonged duration, which in some cases is beneficial for attaining better consolidation by providing an increased time for interparticle bonding [777]. In an experiment [772], a stainless steel cover plate was placed on the top of the powder bed, on which a water container of 5 mm column height was placed. Water acts as a shock-transmitting medium and

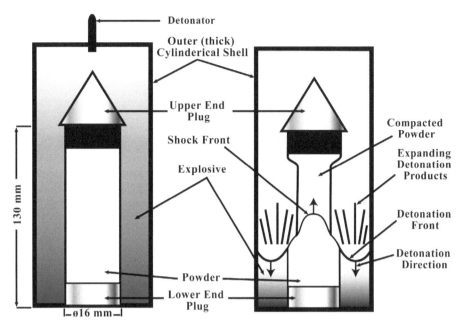

Fig. (41). Axisymmetric setup and process of the cylindrical explosive compaction that is used commonly for producing High-T_c Superconductors [763,770].

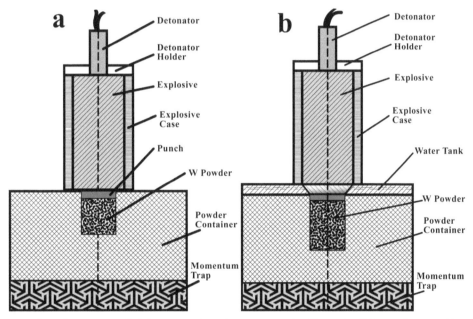

Fig. (42). Two types of test die-setups applied for compaction of tungsten powder. The main difference between (**a**) and (**b**) is a water tank that added to (**a**) as an interface between explosive and tungsten powder. All components were made of mild steel [773].

pressure modulator. An explosive container was then placed on the top of the assembly to create a *planar shock wave*, which was transmitted through the water container prior to compressing the powder. With this geometry, described in more detail elsewhere [778], the pressure acting on the powder can be controlled by varying the height of the water column, *i.e.*; a control of shock loading conditions. As in Fig. **42**, the difference between (**a**) and (**b**) is the water tank, which is machined into a truncated cone (a hollow taper component) and is able to converge the shock wave and reflect it on the wall of the water tank. A momentum trap is *glued* into the lower part of the setup, made of high internal friction and ductile material such as hardened lead alloy or a high ductility mild stainless steel, to attenuate the explosion vibrations. Thus, the set-up is able to generate a higher and longer uniform, axisymmetric pressure pulse and acts as a shockwave modulating lens. Therefore, the application of underwater shock wave increases the thickness of the melting layer in powder surface, which enhances the bonding between particles [750,773].

Of late, gas guns are widely used in performing shockwave consolidation. Gas gun looks similar to a canon, where the shockwave generation was achieved using aluminium projectiles (length 50 mm and diameter 25 mm for some setups), with an embedded flat stainless impactor, launched and accelerated using a single stage smooth-bore light gas gun, as shown in Fig. **43**. The gun uses either a powder propellant breech or a compressed gas. Projectile then hits the flat and parallel stainless flyer plate which acts as the velocity transducer into the stainless powder die, making the compacting planar shockwave, as in Fig. **44** [779-781], although some modifications have been made to accommodate higher driver gas pressures (up to 15 MPa) and a longer barrel (3.0 m) which provide projectile/impactor velocities up to 720 m/s by the release of the compressed gas that causes a shockwave impact of about 20 GPa. In some gun setups, the projectile can be accelerated to ~ 8 km s^{-1}, then generating an impact shockwave of 600 GPa; 6 million times the pressure of air at earths' surface [714,748,771,782,783]!

3.4 ELECTRICAL PROPERTIES OF RSP PRODUCTS:

Over the last two decades, considerable effort has been directed toward obtaining an understanding of the nature of electronic transport in disordered systems. As amorphous, liquid, or glassy alloys lack long-range periodicity, *the concept of the Brillouin zone is no longer applicable* [784]. The understanding of electronic transport and phonon transfer in such disordered systems therefore demands a totally new approach compared to that used for crystalline materials. The main distinguishing characteristic of the resistivity in MS metallic products is that it is greater than that of the corresponding alloy systems in the crystalline state, *becoming several times larger in the case of metallic glass products*, and the temperature change in resistivity is usually small and *often negative*.

There exist a number of anomalies which have initiated a great deal of research, including the occurrence of a resistivity minimum at low temperatures for alloys with a positive *temperature coefficient of resistivity* (TCR), and also a *positive Hall coefficient* in some amorphous alloys. Although much has been achieved in understanding the

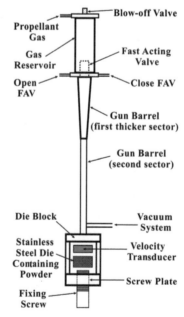

Fig. (43). A setup of single stage light gas gun, with a 3 m length barrel. The projectile is made of aluminium with a maximum speed of 720 m s^{-1} that results in a consolidation shock wave pressure of 8-10 GPa [771].

possible mechanisms that may play a role in electronic conduction in disordered systems, as yet clear explanations to the above anomalies are lacking. One common feature in the resistivity behavior of these materials is the $\sqrt{T}$ -

dependence with a *negative TCR* at lowest temperatures [785,786], as in Fig. **45**. Alloys with positive TCR correspond to cases in which electron-phonon interactions dominate [787]. Cochrane and Strom-Olsen [788] have demonstrated that this is a general behavior related to the high resistivity of the system and not to the amorphous state since the effect is observed in systems such as amorphous Nb-Si [789], crystalline Si-P [790], and granular Al [791] as well as in metallic glasses such as $Y_{77.5}Al_{22.5}$ and $Fe_{80}B_{20}$.

Also, the positive Hall coefficient in the metallic glasses with temperature dependence [792-797] indicates clearly that this is an electronic structure-related phenomenon rather than a structural one. However, much remains to be done to confirm the interesting new explanations put forward for the cause of the positive sign and its temperature dependence. Moreover, *superconductivity* is evident in RSP metals and alloys and has taken great amount of researches up till now. A comparative specific heat study for cubic, icosahedral and amorphous $Mg_3Zn_3Al_2$ showed T_c's of 0.32, 0.41 and 0.75 °K, respectively [798]. As a representative for metastable crystalline superconductors containing no transition metals, $Al_{100-x}Si_x$ alloys will be briefly discussed. These alloys (up to x = 30) can be obtained by rapid quenching or by pressure quenching. In the latter process, the samples are heated to 725 °C under a pressure of 4 GPa followed by a rapid quench, yielding a rather homogeneous Al-Si solid solution with Si occupying regular sites of the Al FCC lattice [799]. In these $Al_{100-x}Si_x$ alloys, T_c goes up to ~ 6 °K (compared to T_c = 1.2 °K for pure Al). Metastable BCC $Zr_{100-x}Si_x$ alloys can be prepared by rapid quenching for $8 \leq x \leq 12$ [800]. For $12 \leq x \leq 24$, the alloys are amorphous and show superconductive behaviour. Immediately after the discovery of high-T_c superconductivity by Bednorz and Müller [801], many routes to obtain high-T_c RSP superconductors were followed. For instance, $(YBa_2Cu_3O_{7-\delta})$ has a T_c of ~ 92 °K (≈ -184.16 °C) [802]. The standard procedure consists of mixing and pressing powders of CuO, Y_2O_3, and $BaCO_3$ and giving specified heat treatments in air. In melt quenching [803], these powders are melted together at 1400 °C. The melt is then poured quickly on an iron plate and pressed to a thickness of ~ 1-2 mm. A final heat treatment at 900 °C results in samples with T_c = 90 °K. The twin roller technique was used to obtain ~ 50 μm-thick samples [804] that consist of metastable crystalline and amorphous regions [804]. After annealing in air, superconductivity with T_c = 93 °K (midpoint of resistive transition) was observed. In contrast to the former method, densities close to the microscopic density of $(YBa_2Cu_3O_{7-\delta})$ were

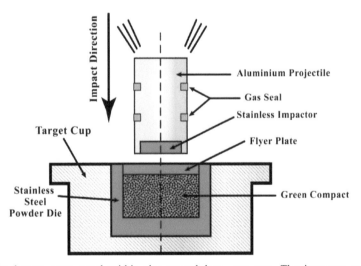

Fig. (44). The impact shockwave setup used within the smooth-bore gas gun. The impactor and flyer plate should be axisymmetric and/or have the same diameter for homogeneous plane shockwave generation.

found [803].Experiments on the physical properties of high-T_c superconductors require well-characterized materials, preferably single crystals, in order to elucidate the mechanism responsible for superconductivity. Another feature is the resistance to radiation of *superconducting metallic glasses*, which could lead to applications in large superconducting magnets required for fusion reactor technology [805]. Partial crystallization of a metallic glass has been reported [806-808] to improve markedly superconducting properties while retaining good mechanical ductility. Rapid solidification and subsequent annealing have nevertheless proved to be a reliable preparation process. Of

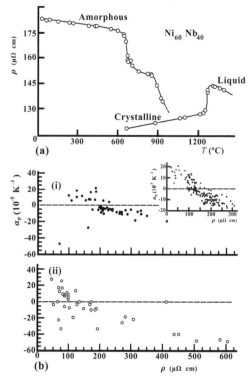

Fig. (45). (**a**) Temperature dependence of the resistivity of $Ni_{60}Nb_{40}$ in amorphous, liquid and crystalline states. (**b**) The Mooij correlation between the temperature coefficient of resistivity α_ρ, and the value of the resistivity for: (***i***) most metallic glasses except those containing Ca, and (***ii***) metallic glasses containing Ca as one component. Shown in the inset is the correlation discovered by Mooij for amorphous and disordered crystalline metals [792,793].

course, only the future can show if this route is of importance for technical applications of high-T_c superconductors [809]. Finally, a *fact must be in mind is that*; non-equilibrium solid phases produced by rapid solidification are often metastable noncrystalline fine-grained or mixed crystalline-noncrystalline state, and an attempt to characterize and understand the electronic structure of these solids appears to lead to a foggy path. Owing to recent advances in experimental and theoretical methods, however, significant progress has been made in this field, and we are beginning to accomplish more quantitative description of the structure of non-crystalline solids to interpret the visible RSP materials properties [810-813]. The key for a successful experimental structure determination is to measure the scattering intensity accurately up to high values of scattering vector, using high-energy x-rays or neutrons. This usually requires the use of special facilities such as synchrotrons, rather than a short trip to the x-ray laboratory down the hall. Therefore, an accurate characterization of glassy solids is a major research project by itself. But it is worth the effort to know such a structure, since, unlike in the gaseous state, atoms in the glassy states are not really randomly distributed in space but are *highly correlated* and exhibit strong local order, which determines the properties of the solid. The recent discovery of quasicrystalline solids has significantly changed our attitude toward the ordered state. It is now clear that the crystalline state is not always the state with the lowest energy, and once the restriction of periodicity (*which, by the way, is a severe restriction*) is removed, a large variety of atomic configurations such as icosahedral clusters become attainable. Although many fundamental issues remain unsolved and await future research, a clear understanding of some of the important structural features of non-crystalline solids is not beyond reach.

3.5 MAGNETIC PROPERTIES OF RSP METALS AND ALLOYS

It is known that 3d or bulk metal-based amorphous alloys obtained by rapid quenching of the melt are excellent soft magnetic materials including inductive and *magnetoresistive* ones. The report by Duwez and Lin [814] in 1967 of typical soft ferromagnetic behaviour for ($FeP_{15}C_{10}$, at.%) metallic glass heralded a major new field of research in magnetism.

Conventional metallic soft magnetic materials range from low-cost common iron and silicon iron used in large volumes mainly for cores of power transformers, generators and motors, to relatively costly nickel and cobalt irons (*permalloy*; $Fe_x Ni_{100-x}$, $10 \leq x \leq 50$, $\mu_r = 70,000$, and *supermalloy*; $Ni_{78}Fe_{17}Mo_5$, $\mu_r = 1,000,000$ and all by wt.%) used in smaller volumes in electronic devices, especially hard disks and magnetic thin films.

Of prime importance for applications in cores are a high permeability, low magnetic losses, and a low coercivity. Such costly fabrication procedures contrast with the relatively simple and inexpensive techniques of melt-spinning and planar-flow casting which produce ferromagnetic metallic glasses in the form of continuous ribbons and wide tapes at a high speed in one step directly from the melt. In addition, their high electrical resistivity, very high permeability, very high magnetization and low coercivity reduce core losses under alternative current (AC) excitation to levels typically one-third those of the best oriented silicon iron. Also, soft magnetic materials are widely employed for the fabrication of magnetic recording heads. These materials must have a high saturation magnetization in order to produce a large gap field. A high permeability is required in order to ensure high efficiency and a small magnetostriction to produce low medium-contact noise. The coercivity has to be low in order to ensure a low thermal noise and a high electrical resistivity in order to reduce eddy currents, especially in read-and-write magnetic heads. To ensure good reliability and a long operating life, the materials must exhibit a good thermal stability and a high resistance to wear and corrosion [815]. Energy savings resulting from low core losses are the main incentive for substituting metallic glass for silicon iron in power devices, while cost savings are the main attraction for applications in electronic devices. Also, the uniformity of field-emission from metallic glasses suggests uses as electron sources in electron-optical equipment [816], while their reproducible glass transition temperature might be useful in an electrical fuse material [19,817].

The usual approach to the atomic structure of ordered magnetic solids assumes lattice periodicity. Such a framework idea is not applicable for metallic glasses, which are defined as solids in which the orientation of local symmetry axes fluctuate with a typical correlation length $l \approx 10$ Å. Therefore, it has been, and still it is, an important challenge within the scientific knowledge of magnetic materials to unravel the fundamental physics that underlies their excellent technical properties [815,818-821].

In a crystalline environment, frequently, one crystallographic direction is found in which the atomic magnetic moments have a lower energy than in other directions. Such a direction is called the *easy magnetization direction* or, equivalently, the preferred magnetization directions. The magnetization is not completely free to rotate but is linked to distinct crystallography directions. Different compounds may have a different easy magnetization direction. In most cases, but not always, the easy magnetization direction coincides with one of the main crystallographic directions. There are some fundamental questions related to the existence of well-defined magnetic order in a material having structural disorder. In fact, if we consider ferromagnetic interactions of the magnetic moments we immediately think of a ferromagnetic structure. *Magnetic anisotropy effects have been neglected in this quick impression.* It is important to note that, in a single crystal, magnetic anisotropy is perfectly compatible with ferromagnetic order. Magnetic moments tend to arrange their orientations parallel to each other *via* exchange interactions; this is the way they do when lying along a magnetic easy axis which is in the same direction at every point in the single crystal or grain. However, if the easy-axis orientation fluctuates from site to site, a conflict between ferromagnetic coupling and anisotropy arises. As long as we imagine lattice periodicity, a ferromagnetic structure is a consequence of ferromagnetic exchange interactions and the strength of the anisotropy is irrelevant. But we are assuming a major simplification; the direction of the easy axis is uniform throughout the sample. This is a simple example of the questions related to the influence of an amorphous structure on magnetic order.

Presently, we know that magnetic order stems from two contributions, exchange and local anisotropy. Magnetic anisotropy also originates by the interaction of the local electrical field with spin orientation, through the spin-orbit coupling. Therefore, magnetic anisotropy is also a local concept. Nevertheless, the structural configuration of magnetic solids exerts an important influence on the macroscopic manifestation of the local anisotropy. The most significant contribution to magnetic anisotropy in a $Ni_{80}Fe_{20}$ permalloy films (also known as mu-metal or μ-metal alloy by the notation of some references) arises from Fe-Fe pair ordering. Such ordering is induced upon deposition in a magnetic field or upon annealing at a temperature that is high enough to allow the required atomic jumps to take place in the bulk. The latter option is not usually chosen because the concomitant grain growth will take place with an increase in coercive force. Other contributions to magnetic anisotropy may arise from anisotropic stress in the film plane, and from anisotropic distributions of dislocations, stacking faults, twins and grain boundary voids [822].

As outlined above, in a single crystal there is a macroscopic anisotropy with a macroscopic easy axis oriented in the same direction as all the local axes. However, when the local axes fluctuate in orientation as a consequence of the structural fluctuation (amorphous structure, micro-stresses and shears causing crystal dislocations, for example), calculation of the resultant macroscopic anisotropy becomes a difficult task. Now, it's known that local fluctuations (*i.e.*; the random distribution) of the easy axes orientation can result in ferromagnetic order with vanishing or minimized macroscopic anisotropy as well as a *frustration* of ferromagnetism in some cases [819]. In fact, the correlation length of the magnetic moments, for a constant correlation length of orientation fluctuations of local easy axes, depends on the strength of the local *magneto-crystalline anisotropy constant*, K_i, relative to that of the exchange interactions. Hence, frustration of order is expected for high local anisotropy strength and, in this case, the anisotropy overcomes the exchange, and the spins are oriented along the easy axes and therefore their orientations fluctuate with the same correlation length as that of the structure [815,818,819,823]. When anisotropy is relatively weak, the exchange interactions dominate and ferromagnetism is achieved with correlation length, L, which is much larger than the structural correlation length l [823]. In this case the macroscopic or large scale anisotropy undergoes a drastic lowering relative to the local anisotropy strength by a factor of $L^{-3/2}$. The correlation length l of such fluctuations, which is typically the correlation length of the structure, ranges from 10 Å (amorphous) to 10 nm (nanocrystals) to 1 mm (polycrystals).

For the nanocrystalline ferromagnets, the magnetic softness is a dramatic function of the grain size, l. Consider the well known alloy $(Fe_{73.5}Cu_1Nb_3Si_{13.5}B_9)$ [824], which can be rapidly quenched as an amorphous alloy ribbon. A subsequent annealing heat treatment (*i.e.*; devitrification) above its crystallization temperature produces an ultrafine randomly-oriented grain structure of α-Fe(Si) with grain sizes of $l \sim 10\text{-}20$ nm. This grain size is remarkably less than that of the grains usually obtained by recrystallizing of metallic glasses ($l \approx 0.1\text{-}1$ μm) [818]. The appearance of such small grains or nanocrystals has been associated with the presence of Cu and Nb in Fe based amorphous alloys. Cu helps nucleation of α-Fe(Si) grains whereas Nb limits grain growth. As annealing temperature increases, so does grain size, thereby causing the nanocrystalline structure to evolve towards a polycrystalline one. The ability to control the grain size (over a range of six orders of magnitude) by annealing at different temperatures has been successfully used to give evidence of the close relation between magnetic properties and microstructure (see also Fig. **46a**). In the amorphous state the initial permeability, μ_i, is 8000. After isochronal annealing of 1 hr at temperature, T_a, of 525 °C, μ_i increases up to 10^5. This value does not change with T_a ranging from 525 to 560 °C. Above this range of T_a, μ_i varies continuously and drastically decreases to 10^2 for $T_a = 610$ °C. It is worth noticing that a relatively narrow range of T_a (85 °C) gives rise to such remarkable variation in the magnetic softness [818]. Also, fast cooling to a sufficiently low temperature freezes in the directional order obtained. It leads to a uniaxial magnetic anisotropy where the easy axis of the magnetization direction becomes in the field direction. Hysteresis loops measured in this same direction are square (Fig. **46b**).

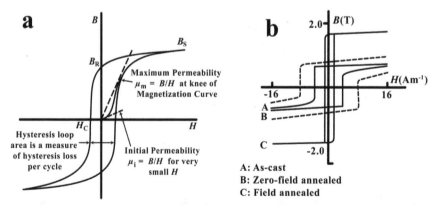

Fig. (46). (a) Hysteresis loop of a magnetic material in which several magnetic quantities relevant to soft magnetic materials are defined [825,793]. (b) Hysteresis loops observed for amorphous $(Fe_{72}Co_8Si_5B_{15})$ alloy treated in deferent ways [826].

The full explanation of the previous observed magnetic behavior can be summarized as follows. In the amorphous state there is no structural macroscopic anisotropy, however, internal stresses give rise to the appearance of magnetic anisotropy through *magnetoelastic coupling*, which is proportional to the saturation magnetostriction constant. After annealing at 520 °C, the onset of crystallization has been overcome and some nanocrystals coexist with the amorphous matrix. The grain size of the nanocrystals is smaller than the exchange correlation length, L, and therefore there is not any structural macroscopic anisotropy contribution associated with them. However, the magnetostriction constant of the α-Fe(Si) crystallites is (-6×10^{-6}), whereas in the amorphous structure its value is ($+20 \times 10^{-6}$) [818]. Hence, it is expected that for a ~ 0.23 volume fraction of the α-Fe(Si) the average magnetostriction should be very small, thus explaining the annealing temperature window for magnetic softness. Since magnetoelastic coupling can contribute to the anisotropy, soft magnets usually require very low magnetostriction as well.

3.5.1 Coercivity and Magnetostriction

For permalloys, the low coercivity alloy has a near zero value of the magnetostriction at the composition 81%Ni-19%Fe and, hence, this composition is the preferred one for thin film inductive recording heads, which almost always are under stress in the as-deposited condition. The magnitude of the detected signal increases with the saturation magnetization and permeability. The detection sensitivity of the inductive head increases with decrease in the coercivity and the noise. The magnitude of the detected signal and the head noise are also affected by the magnetic anisotropy and domain structure. Of these parameters, the saturation magnetization alone is insensitive to microstructure in the soft magnetic films [822].

For higher annealing temperatures, the grain size increases, ultimately reaching the exchange correlation length. Consequently, the effect of crystalline anisotropy becomes increasingly noticeable and the softness drastically drops. In fact, magnetostriction depends not only on material structure, but also on the fabrication routes, even on the cooling rate and structure relaxation (Fig. **47**). So, magnetostriction is the only factor governing the technical magnetism of the amorphous ferromagnets [815,818]. Unfortunately, there is no theoretical basis for predicting the effect of structure upon most of the magnetic properties, although many of them, such as coercivity, are sensitive to grain size, dislocation density, ribbon/film thickness microstructure and even geometry of the final product [827-832]. For example, to be able to predict the effect of structure on coercivity it is necessary to know the mechanism of the domain response to the applied magnetic field. Given this knowledge then, an attempt can be made to apply the applicable model for the effect of structure upon coercivity [833]. However, no general rule can be stated that covers all the permalloy parts and thin films in use. Nevertheless, in the regime of nanocrystalline grain sizes it appears that the coercivity decreases with decreasing grain size [834,835]. This correlation is likely to be a manifestation of *domain wall boundary motion* limitation of the magnetization reversal. Domain walls tend to be pinned by dislocations, grain boundaries, sharp gradients of stress, second phases, voids, surface roughness, and other discontinuities and sharp gradients in the structure. Noise in soft magnetic thin films is produced by sudden changes in magnetic domain configurations, as occasioned, for example by thermal de-pinning of domain wall motion by eddy currents.

All soft magnetic thin films must have low values of magnetocrystalline anisotropy energy and magnetostriction for reading and writing applications to prevent *Permeability Roll-off* caused by high frequency eddy currents. The decrease in permeability that occurs beyond some frequency of magnetization reversal is a consequence of eddy-current damping. The most effective method of eliminating the decrease in permeability with frequency is to make the magnetic layer thickness less than the eddy current skin depth and to use a multilayer assembly of such magnetic layers separated by insulating non-magnetic ones, with a sufficient number of magnetic layers to obtain the desired flux with increasing the thin film resistivity by decreasing the microstructure as possible. Another similar technique for minimizing eddy current losses, which works better for high-frequency applications, is to make the core out of iron or permalloy nano- or submicron powder instead of thin iron sheets. Like the lamination sheets, these granules of iron are individually coated in an electrically insulating material, which makes the core nonconductive except for within the width of each granule. Powdered iron cores are often found in transformers handling radio-frequency currents.

The differences in coercivity between different soft magnetic films may derive from their differences in the resistance to motion of domain boundaries. Amorphous films are likely to have the least resistance to domain wall motion because they are the most homogeneous of the films [822,823]. Fig. **48a** shows the correlation found between coercivity and dislocation density which is a manifestation of the barrier to domain wall motion offered by dislocations [836] which is a further evidence that domain wall motion governs the coercivity in such nanocrystalline ($Ni_{81}Fe_{19}$) permalloy films. However, in precipitation hardened permalloy containing Nb, it has been inferred that magnetization reversal occurs by coherent rotation. If the inference is valid then one would expect that in this material coercivity would vary inversely with grain size [837]. The fact is that; this permalloy film has a fine nanocrystalline grain structure. Moreover, the dependence of coercivity on grain size reverses below a certain critical grain size when the domain rotation process becomes thermally activated [838]. Perhaps this situation applies to the latter material. Such a reversal is shown for Ni in Fig. **48b**. It appears that any discontinuity in the structure or sharp gradient in it acts to hinder domain wall motion and thereby increase the coercivity when it is controlled by domain wall motion [815,822,823]. Besides, it has been reported that coercivity decreases as the thickness of permalloy films increases [839]. A theory of this effect has been developed by Néel for Bloch wall domain motion [840] which predicts that coercivity should vary as the inverse 4/3 power of the thickness. The experimental value of this exponent is hardly ever -4/3. Possible contributions to the deviation of the exponent from the predicted value are the variations of the average grain size and surface roughness with film thickness. The coercivity of $Ni_{80}Fe_{20}$ seems to be insensitive to texture in that an epitaxial film having (100) texture was found to have the same coercivity as a polycrystalline film of the same thickness having (111) texture [822,841].

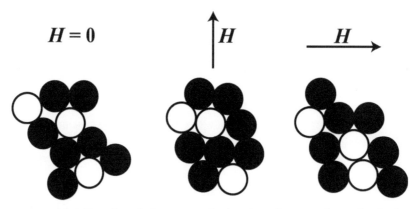

Fig. (47). Atomic rearrangement is taking place during structural relaxation of an amorphous alloy in the presence of magnetic fields in two different directions [815].

In many alloys, thermal or mechanical treatments that lead to increases in coercivity, *e.g.*, by precipitation annealing, also lead to increases in yield stress. However, because of the different dependence of domain-wall pinning and dislocation pinning on defect size, magnetic and mechanical hardness are not always concurrent. For example, solid-solution alloying elements in iron or nickel can produce mechanical hardening without producing magnetic hardening. An extreme example is that of amorphous alloys, which are mechanically very hard but magnetically very soft [832]. The outstanding magnetic softness and other favorable characteristics, such as frequency response of nearly zero magnetostriction alloys, make the amorphous material ideal for magnetic cores of transformers, electronic devices, stress and magnetic sensors. The only drawback of this kind of material is the relatively low saturation magnetization. Highly magnetostrictive alloys can be used as excellent sensing elements for detecting stresses. Also, amorphous alloys have been successfully incorporated as magnetic field detectors, in encoding devices and in the anti-theft systems industry, especially those alloys containing terbium and Tb-base alloys [842-844]. Reading heads in computer hard disks is made from nanostructured or amorphous permalloys that satisfy the basic requirements of this type of application, *i.e.*; large magnetoresistance, small magnetization, uniaxial anisotropy, small hard axis coercivity, low magnetostriction, low resistivity, wear and corrosion resistance. The nanocrystalline magnetic materials discovered more recently also open a wide scope of possibilities in various technical applications [818].

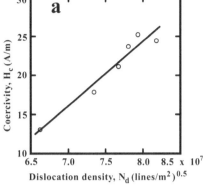

Fig. (48). (**a**) An apparent correlation between coercivity and dislocation density deduced from X-ray rms microstrain measurements. (**b**) Coercivity of Ni as a function of grain size [822,834,845].

3.5.2 Magnetic Shielding

Another, but not final, application of soft magnetic nanostructured and amorphous ribbons is the *magnetic shielding*. The first electronic application was of Fe-Ni-P-B metallic glass ribbons braided or woven into continuous fabric up to 2 m wide for use as magnetic shielding material (Fig. **49**) with the advantage over conventional permalloys of being formable into the required shape without impairment of shielding performance [816,817,846-850]. Magnetic shielding is a process that limits the coupling of a magnetic field between two locations. Magnetic shielding can be done with a number of materials, including sheet metal, metal fabrics, metal mesh, ionized gas or plasma. The purpose of magnetic shielding is most often to prevent magnetic fields from interfering with electrical devices. It is necessary because, unlike electricity, magnetic fields cannot be blocked or insulated. This is explained in one of *Maxwell's Equations*, Div(**B**) = $\nabla \cdot \mathbf{B}$ = 0 (all are vector quantities), which means that; there are no *magnetic monopoles*. Therefore, magnetic field lines must terminate on the opposite pole. There is no way to block these field lines; nature will find a path to return the magnetic field lines back to an opposite pole. This means that even if a nonmagnetic object, like normal glass, is placed between the poles of a horseshoe magnet, the magnetic field will not change. Instead of attempting to stop these magnetic field lines, magnetic shielding re-routes them around the shielded object (Fig. 50). This is done by surrounding the device to be shielded with a soft magnetic material. If the material used has a greater permeability than the object inside, the magnetic field will tend to flow along this material, avoiding the objects inside. Thus, the magnetic field lines are allowed to terminate on opposite poles, but are merely redirected [851-854]. This can be explained by considering the limit of a thin, high permeability shell for which b is the outer radius and inner one is a and $b = a + d$, made up of material of permeability μ and placed in a formerly uniform magnetic field B_o. For the conditions $d/a \ll 1$ and $\mu/\mu_o \gg 1$ (for a permalloy shell), then the induced magnetic field inside the spherical shell, B, can be given by [852]:

$$B \simeq \frac{3}{2} \frac{\mu_o}{\mu} \frac{a}{d} B_o \; \ldots \tag{3-27}$$

Thus, if $\mu = 10^5 \mu_o$ for a permalloy fabric shell, then we can reduce the magnetic field strength inside the shell by almost a factor of 1000 using a shell whose thickness is only $1/100th$ of its radius. Clearly, a little mumetal goes a

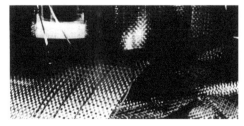

Fig. (49). Flexible magnetic shielding woven from metallic glass melt-spun ribbons [19,846,849].

long way! Note, however, that as the external field strength, B_o, is increased, the μ-metal shell eventually saturates, and (μ / μ_o) gradually falls to unity. Thus, extremely strong magnetic fields (typically; $B_o \sim 1$ Tesla) are hardly shielded at all by mumetal, or similar magnetic materials [852]. While the materials used in magnetic shielding must have a high permeability, it is important that they themselves do not develop permanent magnetization. Magnetic shielding is often employed in hospitals, where devices such as *magnetic resonance imaging* (MRI) equipment generate powerful magnetic flux. Magnetic shielding rooms are constructed to prevent this equipment from interfering with surrounding instruments or meters. Similar magnetic shielding rooms are used in electron beam exposure rooms where semiconductors are made, or in research facilities using magnetic flux. Smaller applications of magnetic shielding are common in home theater systems. Speaker magnets can distort a cathode ray tube (CRT) television picture when placed close to the set, so speakers intended for that purpose use magnetic shielding. Magnetic shielding is also used to counter similar distortion on computer monitors. Also in hard disks where the magnetic sensor head is shielded by a magnetic shielding shell. The stray field which is used to read the information of a bit is shielded between two soft magnetic layers to avoid interferences with stray fields from neighboring transitions [855]. These shields are based on Permalloy, Supermalloy or Sendust ($Fe_{84}Si_{10}Al_5$) alloys [856]. A number of companies will custom-build magnetic shields from a diagram for home or commercial applications. Also, magnetic shielding using superconducting magnets is being researched as a means of shielding spacecrafts from cosmic radiation.

3.6 PROPERTIES AND APPLICATIONS FOR RSP AMORPHOUS METALLIC STRUCTURES

The structure of an amorphous solid is never unique, unlike that of a crystal, and can depend on preparation method and/or thermal history. It is confirmed that the bulk metallic amorphous alloys (BMAs) or glasses (BMGs) exhibit various characteristics such as high mechanical strength, high elastic energy, high impact fracture energy, high wear resistance, high corrosion resistance, strongly high temperature-dependant electrical resistivity (roughly one or two orders of magnitude higher than that of the typical crystalline counterparts and equal to that of the liquid state), higher T_c-superconductivity, good soft magnetic properties, high frequency permeability, higher thermal expansivity with T^2-dependant thermal conductivity (with anomalous behaviour at $T \leq 2$ °K), fine and precise viscous deformability, good castability and high consolidation tendency into bulk forms. By utilizing these unique characteristics inherent to the BMAs, the new alloys are expected to open up new application fields in the near future [857-862].

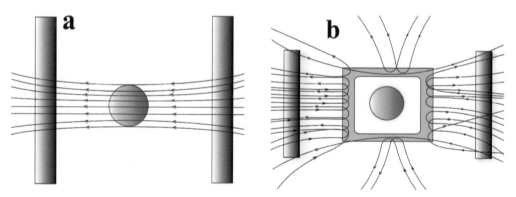

Fig. (50). The magnetic shielding of an object from stray magnetic lines that can cause electromagnetic coupling and electronic noises; (**a**) unshielded and (**b**) shielded, where the field is forced to travel within the walls of the box and not to penetrate the inside of it and thus stray magnetic lines are deflected away from the inside of the shield.

Amorphous alloys are now known for many compositions, mainly; (*i*) the (cheap) Late transition metal-metalloid (*e.g.*; $Fe_{100-x}B_x$ (x = 12-25) and $Co_{70.5}Fe_{4.5}Si_{15}B_{10}$ [863,864]), (*ii*) early transition metal-metalloid (*e.g.*; $Ti_{100-x}Si_x$ (x = 15-20) and $W_{60}Ir_{20}B_{20}$ [127,865]), (*iii*) early transition metal-late transition metal (*e.g.*; $TiFe_{20}Si_{10}$, $Zr_{57}Ta_5Cu_{20}Ni_8Al_{10}$ and $Nb_{100-x}Ni_x$ (x = 40-70) [866-871]), (*iv*) aluminum-based (light) alloys (*e.g.*; $Al_{100-x}La_x$ (x = 10, 50-80) and $Al_{85}Ni_5Y_8Co_2$ [483,872,873]), (*v*) lanthanide-based (*e.g.*; $La_{100-x}Al_x$ (x = 18-34), $La_{100-x}Ge_x$ (x = 17-22) and $Nd_{48}Al_{20}Fe_{27}Co_5$ [874-876]), (*vi*) alkaline-earth based (*e.g.*; $Ca_{100-x}Al_x$, $Mg_{100-x}Zn_x$ (x = 25-32) and $Mg_{65}Y_{10}Cu_{15}Ag_{10}$ [877-878]) and (*vii*) actinide-based (*e.g.*; $U_xCr_yV_z$ (x = 60-80; y or z = 0-40) [879-880]), all compositions in at.% [127,133,351,857,881]. Within a given alloy system, the range of compositions which can be made amorphous depends on the production method and conditions, as mentioned earlier.

3.6.1 Amorphous Strength Applications

Amorphous alloys possess very high strength levels, exceeding 1,000 MPa, approaching those of dislocation-free single crystals or a hard-drawn piano wire. For example, Fe80B20 has tensile yield strength of 3.6 GPa and good ductility in bending or compression, but not in tension [882]. Excellent fatigue resistance of the amorphous alloys has also been recently reported [883]. However, a major disadvantage of amorphous alloys is their poor fracture behavior. This is a consequence of the lack of the dislocation mechanism equivalent to that of work-hardening in crystalline alloys to disperse slip and so extend the time required for nucleation of fatigue cracks. Due to their high strength, high wear and corrosion resistances and isotropic two-dimensional forms, amorphous alloy ribbons (by melt spinning) have been used for reinforcement of mortar [884], polymers, long-life razor blades or surgical scalpel [885,886], and metallic matrix materials and many planar composites [887-889]. Amorphous Ni-Cr-P-B alloy layers on mild steel, produced by laser/ion beam fusion of a prior deposit of the required composition, had a corrosion resistance comparable to that of melt-spun alloy [495,510,890-892]. Table 1 summarizes the fundamental characteristics and application fields of the bulk amorphous alloys.

In addition to their elastic elongations of 2%, Fe-, Ni-, Cu-, Pd-, and Pt-based metal-metalloid BMGs also have a good ductility as evident from the achievement of plastic elongation. Conversely, Co-based metal-metalloid and metal-metal BMGs and Fe-based metal-metal BMGs exhibit no plastic elongation. The 2% elastic elongation property has been recognized for other BMGs such as Zr-, Mg-, and Ln (lanthanide metal)-based alloy systems. The elastic limit of a crystalline alloy is typically less than 0.65% due to the presence of dislocations. BMGs do not contain dislocations and ideally should exhibit an elastic limit of 2%. However, some BMGs do not achieve this 2% elastic limit due to cast defects. Therefore, the 2% elastic limit reflects the random atomic configuration of an ideal glass and is an essential factor for the achievement of high fracture strength [194].

In fact, recent studies have shown that a metallic glass with very low unconstrained plasticity can show large plasticity under geometry constrained conditions. Under hydrostatic pressure, metallic glasses can exhibit large plasticity as a result of geometry constraint of shear bands [893-895]. To improve the mechanical properties of BMGs, bulk-metallic glass-matrix composites with ductile metals, refractory ceramic particles, and fibers have been developed [896-897]. According to Nieh [893], there are two composite approaches used to improve the plasticity and toughness of BMGs; extrinsic and intrinsic [898]. The extrinsic approach is to artificially introduce second phases into a glass matrix system, whereas the intrinsic approach, in contrast, is to design a chemical composition such that second phases are formed in situ during processing. The composite concept is borrowed from the intensive development of glass/ceramic matrix composite in the 1980s. In fact, it is similar to the strength and toughness improvement in silicon carbide fiber-reinforced glass matrix (lithium alumino-silica) composites [899]. In the case of extrinsic composites, Conner *et al* [900] demonstrated that fracture strain as well as the toughness of a $Zr_{41.25}Ti_{13.75}Cu_{12.5}Ni_{10}Be_{22.5}$, the *Vitreloy*, metglass can be significantly improved by reinforcing the glass matrix with either continuous tungsten fibers or 1080-steel wires. The increase in compressive toughness was suggested to come from the fibers restricting shear band propagation, promoting the generation of multiple shear bands (with very fine grains) and additional delamination of interfaces between fiber and matrix. The extrinsic approach has also been extended to Mg-based BMGs [901].

For joining or welding of BMGs, it has demonstrated the success of several welding processes, namely, friction welding, spark welding and electron-beam welding [902-905]. Results confirmed that with *careful control of the process parameters* the joints remain amorphous and the strength is comparable to the bulk of the metallic glass [906].

3.6.2 BMAs Wear Resistance

Also, amorphous alloys and related materials are of interest for a diverse range of applications. The tribological properties must often be taken into account, and in some applications, they are of primary interest. For example, amorphous alloys have been proposed as coatings in dry or unlubricated bearings for use in space [907]. The viscous

Table 1. Fundamental characteristics and application fields of BMGs [510,857].

Fundamental characteristic	Application field
High strength	Machinery structural materials Reinforced rubber tires Spring components Metal-matrix composites
High hardness with high manufacturability	Optical precision material Cutting tools Abrasive wheels Penetrators (with the self-sharpening property)
High impact fracture energy	Tool materials Armor penetrators
High elastic energy (high resiliency)	Cutting materials Tennis racquets Baseball bats Composite car breaks and structural sections Exceptional mechanical energy absorption BMG-Foams
High corrosion resistance	Corrosion resistant materials Anti-weathering surface coats
High wear resistance	Ornamental materials Anti-friction surface coats Brilliant antiscratch hand-watch frames Writing appliance materials Medical devices materials and human body implants transmission gears and actuators
High viscous flowability	Composite materials Reinforced plastics Acoustic attenuation systems MEMS
Good soft magnetism	Writing appliance materials Strong permanent ferromagnets Magnetic shielding for electronics
High magnetostriction	Sporting goods materials Soft magnetic materials and sensors High magnetostrictive materials

flow of amorphous alloys in the supercooled liquid state region and the ability of making elongations exceeding 300%, at suitable *low strain rates* and temperatures, have been exploited to make not only small machine parts but also ultrafine mechanical devices (MEMS), *e.g.*; gear wheels and micro-motors/actuators [191,194,351,908-912]; for these also the friction and wear properties are of direct concern. These alloys at supercooled liquid state behave like a Newtonian fluid and can fill cavities or flow through extrusion dies with sizes in the order of 100 μm [231,906]. For example, a tensile elongation of 20,000% has been reported in a $La_{55}Al_{25}Ni_{20}$ alloy in the supercooled liquid region [893]! It is generally observed in the study of homogeneous deformation that BMGs behave like a Newtonian fluid at low strain rates, but becomes non-Newtonian at high strain rates [913-916]. In some cases, the BMAs wear resistance can be of primary concern, in others it becomes an important secondary property. The distinctive mechanical properties of amorphous alloys make their wear resistance of fundamental interest also [351]. As discussed above, the lack of work hardening and consequent shear instability in amorphous alloys may make them more attractive as coatings than as bulk materials. In addition, the good tribological properties but relative expense of amorphous alloys may favor applications as coatings. In this section, amorphous and related partially or fully crystallised coatings produced specifically for wear resistance are considered. Rapid solidification at a surface can be achieved by transient melting using a pulsed or a moving heat source. Using a scanned laser or electron beam, a surface can be glazed, *i.e.*; the rapidly solidified surface can be made amorphous [85,714].

While achieving the required quench rate is not difficult, there can be problems with surface roughness and with melting in one area causing partial crystallisation (or; *devitrification*), of a neighbouring already-amorphised area [917]. So precursor alloys for amorphous coatings should be chosen carefully in order to improve lasting good hard coatings after induced recrystallisation. Essentially, all practical metallic materials are metastable in some degree. Yet the departure from equilibrium of amorphous alloys appears particularly extreme as they preserve the high temperature structure of the liquid alloy. It is expected that they will undergo structural changes on annealing, or during wear itself. Crystallisation can lead to property changes similar to those induced by structural relaxation, though structurally the two processes are entirely distinct. Often, though not quite universally [918], crystallisation leads to hardening and embrittlement. The property changes can be beneficial for applications especially when discussing the fully devitrified alloys and nanophase composites. The crystalline materials obtained by transforming amorphous alloys have some characteristic features. They may contain ultrahard phases, such as borides and carbides, because of the amorphous alloy compositions. Typically, the microstructure is ultrafine, compared to typical solidification microstructures.

Table 2. Some fundamental properties of the soft-magnetic late transition metal (LTM)-based BMGs [194].

Fe-based	Soft magnetism (glass, nanocrystal)
	Hard magnetism (nanocrystal)
	High corrosion resistance
	High endurance against cycled impact deformation
Co-based	Soft magnetism (glass, nanocrystal)
	Hard magnetism (nanocrystal)
	High corrosion resistance
	High endurance against cycled impact deformation
Ni-based	High strength, high ductility
	High corrosion resistance
	High hydrogen permeation
Cu-based	High strength, high ductility (glass, nanocrystal)
	High fracture toughness, high fatigue strength
	High corrosion resistance
Pd-based	High strength
	High fatigue strength, high fracture toughness
	High corrosion resistance
Pt-based	Very low glass transition temperature, T_g
	Very low liquidus temperature, T_L
	High glass forming ability (High GFA)
	High corrosion resistance

Besides, the best static and dynamic soft magnetic properties are presently achieved as well in amorphous cobalt-base as in nanocrystalline iron-base alloys (see also Table **2** that summarizes some features of some BMAs for magnetic and wear resistance applications). Both alloy systems reveal isotropic near-zero magnetostriction with the so-called giant magnetoimpedance effect [919]. Add to that, amorphous magnetic properties can be changed systematically by substituting different elements without changing any kind of structure [920]. Apart from its higher saturation induction, however, the nanocrystalline material shows a much better thermal stability of its magnetic properties than its amorphous counterpart and, especially, when it is based on the inexpensive raw materials as iron and silicon. Accordingly, nanocrystalline alloys provide an invaluable supplement to the existing soft magnetic materials manifested in a steadily increasing number of applications [921]. Yet, the variability of their soft magnetic properties, as well as their form of delivery so far and thermal sensitivity, is still restricted compared to amorphous or other soft magnetic materials. Thus, amorphous alloys may reveal good soft magnetic properties already in the as-quenched state or after moderate annealing. They can be delivered as a semi-finished, ductile product useful for flexible magnetic screening, for example, or for sensor applications, most noticeably in electronic article surveillance [922,923]. Accordingly, the major drawback of the nanocrystalline materials is the severe embrittlement upon crystallization, which requires final shape annealing and restricts their application mainly to toroidally wound cores. Yet, the situation is similar for highly permeable amorphous alloys due to the necessary stress relief treatment, which also causes embrittlement [130].

3.6.3 Amorphous Magnetic Applications

For magnetic applications, two interesting families of BMGs exist: (*i*) bulk (3-d) metalloid (metalloid = B, Si, C, *etc.*), and (*ii*) early transition element-late transition element (ETM-LTM). Here, the Fe-Zr and Co-Zr systems are famous; both systems show an unusual magnetization behavior. The amorphous alloys are produced with the general formula (metal)$_{80}$(metalloid)$_{20}$ (metal = Fe, Co, Ni, metalloid = B, Si, Al, C, *etc.*). The metalloid is necessary for lowering the melting point of the alloy, thus stabilizing the amorphous state. The most promising systems are the Fe-Co-based amorphous alloys [920].

Amorphous materials are generally obtained as thin ribbons (thickness 20-50 μm) by the CBMS technique [924,925]. The absence of grains and grain boundaries leads to excellent soft magnetic properties [926]. High permeability and good frequency response of ferromagnetic amorphous alloys make them obvious candidates for replacing conventional soft magnetic materials in magnetic cores in various transducers.

BMAs are uniquely suited to some sensing applications [927-932]. These sensors are used in recording tapes heads, read/write HDD heads, magnetic compasses, geological prospecting, and even submarine detection. Fluxgate magnetometers, which depend upon the nonlinearity of the *B-H* loop of their cores as they enter saturation, have been made from Co-based amorphous alloys [850,933-938].

As an outstanding application, glass disks or Ni-P-coated Al-Mg (4 at.%) alloy substrates are used for HDD, especially for *perpendicular data storage* (the glass disks are harder and stiffer). A thick layer (10 μm) of amorphous Ni-P is electrolessly plated onto the Al-Mg disk. Ni-P can be polished mirror-like (with surface roughness ~ 5 nm) and a surface texture can be added. Ni grains are superparamagnetic in the Ni-P layer so they do not contribute any magnetic signal. The magnetic storage layer will be isolated from the substrate by an underlayer (*the guiding soft underlayer*), which induces crystallographic texture, and a seed layer which controls the granular size distribution. Its top surface is protected by depositing a mechanically hard capping layer (diamond-like carbon or dense SiC) and finally a lubricating layer is added [856].

3.6.4 BMAs for Hydrogen Storage

Finally, metal hydrides find applications in *hydrogen storage systems*, *hydrogen pumps*, and *heat pumps*. Traditional crystalline materials used for hydrogen storage are intermetallic compounds (*e.g.*; LaNi$_5$), which are brittle and easily disintegrate into very small particles during hydrogen sorption cycles. Metal hydrides have been attracting very serious attention because they are materials for safe storage of hydrogen and can store hydrogen with a higher volume density than liquid hydrogen alone. For a material to be seriously considered for hydrogen storage, it should have [702]: (1) high hydrogen storage capacity, (2) fast kinetics of storage and removal (reversible hydrogenation and dehydrogenation), (3) low hysteresis, (4) possibility of hydriding/dehydriding at relatively low temperatures and pressures, preferably at room temperature and atmospheric pressure, (5) long lifetime, (6) high chemical stability and thermally easily decomposable, (7) no or minimal need to activate the material, (8) easy availability, (9) relative immunity to impurities, and (10) low cost (including that of fabrication).

However, RS unhydrided alloys are strong and ductile and thus are not likely to disintegrate. In an early investigation [939], it was shown that, under similar conditions of temperature and pressure, amorphous alloys have greater hydrogen absorption capacity than their crystalline counterparts. Subsequent work [940] has shown that no plateau exists in the pressure-composition isotherms of amorphous Ni-Zr alloys, in contrast with the crystalline counterpart, indicating that no hydride coexists with a saturated solid solution of hydrogen. It has also been shown [941,942] that amorphous TiCu absorbs 35% more hydrogen than crystalline TiCu and that even though the maximum hydrogen absorption capacity is determined by the electronic structure of the compound, the crystal structure, *i.e.*; the type and size of interstitial sites in the lattice, may not always permit maximum absorption to take place readily. A critical comparison of the hydrogen sorption behavior in the crystalline and amorphous alloys has already been presented [942-944]. Other outstanding features are found in some light magnesium alloys. Among the different alloy systems investigated so far, magnesium alloys have the pronounced advantage because of their large hydrogen capacity (7.6 wt.%), low cost, and easy availability [127,702,945-951]. Elements such as Ce, Nb, and Ti; which form hydrides, or Fe, Co, or Cr; which do not form hydrides themselves, can all serve as *hydrogen pumping agents* when added to the Mg-base alloy. These catalysts force the reaction of hydrogen adsorption by the nano- or amorphous powder to complete in less than 1 min. rather than three or 10 min. or more without the catalyst, even at reduced temperatures [952-954]. Briefly, amorphous alloys offer a larger capacity, wide composition range, and higher ductility than crystalline materials for hydrogen storage. Potential applications include fuel for space rockets, vehicles (zero emission vehicles), heat pumps, and air conditioners [510,702,955,956].

3.7 PROPERTIES AND APPLICATIONS FOR QUASICRYSTALS

Quasicrystal (QC) formation has been reviewed very briefly in chapter (1) and discussed widely in many useful literatures [957-965] for many studied QC alloy systems. As a rule in RSP, a stable crystalline phase formed when the melt was cooled at low cooling rates and an amorphous phase formed at the highest cooling rates. A QC phase formed at intermediate cooling rates [225,702,959,966]. In some cases, the QC formation is the equilibrium phase selection; *e.g.*; Al-Cu-Fe, Al-Pd-Mn, Zn-Mg-R (R = Y, and lanthanides/lanthanoids) and $Sc_{12}Zn_{88}$ systems [967-971]. These QCs form as equilibrium phases at certain composition and temperature regions, and then the growth of single-grain samples has become possible [972,970]. The physical properties, namely; electronic, magnetic, and mechanical properties, of a QC have been investigated, and the following have been discovered: the presence of pseudo-gap at Fermi level leading to a large electrical resistivity with negative temperature coefficient, large thermoelectric power, low optical reflectivity and high optical absorptance (especially in the infrared region), positive temperature dependence of *Pauli paramagnetism*, short-range magnetic ordering satisfying the icosahedral symmetry with different magnetic behaviours [973], excess specific heat at high temperature beyond the *Dulong-Petit law*, low thermal conductivity, *phason hopping/flipping*, very low friction coefficient, very low water-surface adhesion ability (*i.e.*; very good *non-sticking* behavior), outstanding catalysis, superior hydrogen sorption, superior corrosion, wear and scratch resistances [217,351,961,970,974]. Most notable disadvantages are the extreme brittleness, especially at room temperature, and low fracture impact resistance with poor resistance to abrasive wear for *pure* QCs.

Mechanical properties of QCs also relate to their first potential practical applications; they are in some respects similar to those of metallic glasses, where, in particular, they do not show work hardening (see the review in Ref. [975]). Firstly, QCs are exceptionally brittle. They have few dislocations, and those present have low mobility. Since metals bend by creating and moving dislocations, the near absence of dislocation motion causes brittleness up to three-quarters of their melting temperature and become very ductile above. Second, the stress-strain curves exhibit a pronounced yield drop followed by a stage of very low (sometimes negative) strain hardening. Third, although it is well known, since the work of Wollgarten *et al* [976], that deformation proceeds by dislocation motion, specific mechanisms are expected on account of the lack of translational periodicity [977].

On the positive side, the difficulty of moving dislocations makes QCs extremely hard. They strongly resist deformation. This, in fact, is due to the complex deformation mechanism in QCs as a result of their structure symmetry with un-recurring patterns/cells (or; a short distance recurring patterns) in the long range order. The phonons vibrations (causing dislocations) are not alone here, but another accompanying vibrations exist, namely; *the phasons vibrations* [977,978]. Phasons are (point) lattice vibrations that take place in different directions, relative to phonons, and can make relatively large lattice displacements (known as *phason flips*) [979]. These point defects, which are typical of

QCs, can lead to broadened X-ray diffraction peaks [980,981], and are thermally-activated causing the observed high-temperatures plastic behaviour of the QC materials [975,982]. They are characterized by low activation volumes and high activation energies which do not yield any clear indication on the deformation rate-controlling process. More recent investigations, however, showed that dislocation motion takes place by climb motion [983,984]. Conversely, glide appears almost impossible, probably on account of *the very high corrugation of dense planes*, which is an intrinsic property of QCs. Under such conditions, the low-temperature brittleness can easily be interpreted by the absence of glide, which is the usual mode of deformation of conventional alloys at low temperature [977]. Actually, it has been clarified that QCs are plastically deformable at high temperatures and that the deformation is brought by a dislocation process; QCs shear deformation by dislocation glide is possible at high temperatures. An important difference from crystals is that dislocation glide does not restore perfect order, but a *phason wall* (which is a special type of stacking fault) is left behind each gliding dislocation. Such *coupling* between conventional defects (dislocations) and phasons might also give rise to relaxation effects [636,978,985]. Generally, the phason flip is a thermally activated process and thus relaxation of a phason strain, whose elementary process is the phason flip, must proceed relatively slowly like atomic diffusion in solids [636,978,986-988]. This is in sharp contrast to the conventional strain, which can be relaxed instantaneously *via* displacive phonon modes [989,990-994].

These mechanical properties make QCs excellent candidates for *high-strength surface coatings*. For example, this is achieved by coating the surface by a quasicrystalline precursor powder in a *sol-gel emulsion* form, and, after drying, the coated surface is then subjected to a sweeping laser beam, electronic beam or any efficient heat source of enough power to melt the powder coating *in situ* and forming the hard quasicrystalline coating layer. Other methods include thermal spray technology [85,714,995]. Indeed, the first successful application of QCs was as a surface treatment for aluminum frying pans where the non-sticking ability or the low surface energy is proportional to the material hardness. Add to that, QCs are poor heat conductors with very good thermal stability up to or near their melting temperatures [217,351,974,996]. The thermal conductivity of QCs is lower than that of copper by a factor of 200, several times smaller than that of stainless steel, and is comparable to that of dielectric glasses like fused silica [996]. Also, QCs are excellent thermal and diffusion barriers due to their low thermal conductivities, high-temperature induced ductility and their intrinsic thermal expansion that is comparable to that of metals, making them very adequate for low friction and wear resistant turbine blades coatings [996-999]. Especially interesting is the $Al_{71}Co_{13}Fe_8Cr_8$ compound, which shows a hexagonal unit cell nearly identical to that of Al_5Co_2 and melts peritectically at 1100°C, very close to the liquidus temperature at 1170°C. In comparison to the Al-Cu-Fe icosahedral QCs, the range of working temperatures is increased by about 200°C and matches the situation found in most combustion engines (although, as a matter of fact, designers of future aircraft engines are looking for even higher temperatures, in the range 1300-1400°C) [1000]. Coating of the composition $Al_{71}Co_{13}Fe_8Cr_8$, deposited by PVD magnetron sputtering on a Ni-based superalloy substrate, shows an inter-diffusion layer that is already a few micrometres in thickness [974,996]. The same alloy has been used for the antifriction thermal barrier for internal combustion engines too. The QC coated pistons and sealing rings heat slowly, decreasing the engine piston temperature and fuel consumption, and thus saving energy [1001]. The same also for machine parts made from Al-base bulk alloys containing QC-phases. Presence of QC-phases enhances the thermal stability and high temperatures creep resistance. To enhance the room temperature abrasive resistance of QC-coatings, the QC-powder should be mixed with a ductile phase, so increasing the applicability of coatings, especially against weathering and icing wears encountered in aeroplanes [351]. The inclusion of only 1 vol.% of ductile Fe-Al in the QC-powder increases its sliding wear resistance by a factor of 2 to 3. QC-icosahedral powder were plasma-sprayed by De Palo *et al* [1002] and Lugscheider *et al* [996, 1003, 1004]. They obtained coatings having low friction coefficients in the range $\mu = 0.1$-0.6. Also, bulk QC samples, made by sintering, have mechanical and wear-resistance properties that depend largely on the processing conditions [1005], and should contain ductile phases for applicability in different engineering fields. As a general, the differences in the strength and ductility data are primarily attributed to the dependence of the mechanical properties on the volume fraction, the size and the shape of the particles and, hence, on the alloy composition. QC-reinforced Al-alloys were synthesized by RSP techniques and powder metallurgy methods, which can keep up with or excel the performance of conventional solid solution or precipitation hardened Al-alloys. The application of different processing techniques for different alloys resulted in the formation of composites with considerably different microstructures. The classifying differences in the microstructures lead to similarly different mechanical properties. A representative collection of characteristic mechanical data for the QC-composite alloys is shown in Table **3** [1006]. Note the superior mechanical properties of MS-samples over other mentioned techniques within this table.

Another category appears here, strongly related to the CBMS-tech, and is the nanoquasicrystalline bulk (3-D) samples. Nano-quasicrystals (NQCs) can form by CBMSing of QC-forming alloy systems or by suitable annealing/devitrification of the amorphous ribbons or bulk forms of these alloys [192,193,201-205,486,553]. Even the mechanical alloying of QC-alloy compositions can form NQCs, but within moderate mechanical alloying intensities. It appears that the formation of NQCs have resolved the demerits of conventional QCs. High hardness, a high Young's modulus, a high elevated temperature strength, low coefficient thermal conductivity, and high resistance against oxidation have all been reported in icosahedral QCs originating from the three-dimensional (3-D) quasiperiodic lattice. The structure controls (such as: volume fraction, grain size, and distribution of icosahedral particles) must be required for the improvement of these properties, and can create new unique properties. It is expected that they can be achieved by the formation of a nanometer-scale icosahedral phase (nano-icosahedral QC). The formation of a nano-icosahedral phase is effective by quasicrystallization from the amorphous or glassy structure because the homogeneous structure consisting of icosahedral and amorphous or glassy phases can be obtained, and the volume fraction and grain size of the icosahedral particles are easily controlled by changing the annealing conditions [1007-1009]. It is noticed that the dispersion of the nanoicosahedral phase causes a dispersion

Table 3. Summary of characteristic mechanical data obtained from constant rate compression tests of quasicrystalline Al-Mn-Ce and Al-Cu-Fe quasicrystalline bulk composites prepared *via* different processing routes, all in at.%. (*E*; Young's modulus, σ_y ; yield strength, σ_f; fracture strength, ε_f ; fracture compression, ε_{pl} ; plastic compression) [1006].

Method/Sample	E GPa	σ_y MPa	σ_f MPa	ε_f %	ε_{pl} %
Arc-melted prealloy.					
$Al_{92}Mn_6Ce_2$	68	226	319	4.2	5.0
Cast rod					
$Al_{88}Mn_{12}$	69	353	492	5.5	4.8
$Al_{86}Mn_{12}Ce_2$	80	831	848	1.5	0.3
$Al_{90}Mn_8Ce_2$	79	565	825	3.5	1.6
$Al_{92}Mn_6Ce_2$	77	392	565	19.4	18.2
Melt-spun ribbons; pulverized and extruded.					
$Al_{91}Mn_7Ce_2$	94	766	975	6.7	5.4
$Al_{92}Mn_6Ce_2$	102	713	851	24.8	23.2
$Al_{93}Mn_5Ce_2$	98	758	844	17.0	16.1
Mechanically alloyed and extruded powders $Al_{100-x}(Al_{0.63}Cu_{0.25}Fe_{0.12})_x$.					
$Al_{20}(Al_{0.63}Cu_{0.25}Fe_{0.12})_{80}$	72	567	660	3.1	1.8
$Al_{40}(Al_{0.63}Cu_{0.25}Fe_{0.12})_{60}$	68	394	524	10.3	9.2
$Al_{50}(Al_{0.63}Cu_{0.25}Fe_{0.12})_{50}$	70	284	419	15.6	14.7
$Al_{60}(Al_{0.63}Cu_{0.25}Fe_{0.12})_{40}$	68	167	363	26.2	25.8

hardening as well as a *dispersion ductilization* [192,193]. The effects of the NQC phase with the existing amorphous phase do not lead to the observed improved dispersion strengthening and ductility only, but also to a significant superplastic deformation for 3D specimens. Ishihara and Inoue *et al* [192, 1010, 1011] have studied the superplastic deformation in the Zr-based bulk glassy alloys with nanoicosahedral particles dispersions. They reported that the

$Zr_{65}Al_{7.5}Ni_{10}Cu_{12.5}Ag_5$ and $Zr_{65}Al_{7.5}Ni_{10}Cu_{12.5}Pd_5$ bulk glassy alloys exhibit an elongation of 400-450 % at a strain rate of $\sim 10^{-3}$ s^{-1} near T_g (= 400 ºC), which is considerably larger than that ($\sim$ 270-340 %) in the $Zr_{65}A_{7.5}Ni_{10}Cu_{17.5}$ bulk glassy alloy at strain rates of 3×10^{-3} - 5×10^{-2} s^{-1} in the supercooled liquid region [1011]. These results show that: *the precipitation of the nanoicosahedral phase is effective for the enhancement of superplasticity.* Thus, the application of the mechanisms for QC dispersion strengthening and ductilization is expected to enable the future fabrication of new materials with improved engineering properties. Al-based NQC alloys have shown many better improvements too. Examples include Al-Mn-Ln (high strength alloys), Al-Mn-Cu-TM (high ductility alloys), and Al-Fe-Cr-Ti (high elevated temperature strength) alloy systems [193, 1012]. These examples exhibit much better combinations of almost all properties; including tensile strength, elevated temperature strength, Charpy impact fracture energy, cold workability, elongation with high Young's modulus, especially when comparing them to the commercial wrought 7075-T6 and AU4-G (or; AA2017A) Al-base alloys. The same for some Mg-base alloys that improved high thermal stability and strength; *e.g.*; $Mg_{32}(Al,Zn)_{49}$, $Mg_{95}Zn_{4.3}Y_{0.7}$ and Mg_4CuAl_6 alloys [483]. With these improvements, they are very acceptable competitors for the expensive Ti-base alloys. These excellent characteristics are quite attractive for future applications as a new type of Al-based alloys with high fracture strength, high ductility and high elevated temperature strength.

The electric properties of QCs have proved to be rather unusual. Unlike their constituent elements, which tend to be good electrical conductors, QCs conduct electricity poorly. For alloys of aluminum-copper-ruthenium, their conductivities differ by as much as a factor of 100. As the perfection of the quasicrystalline order grows, the conductivity drops. Such behaviour is consistent with the appearance of a gap in the electronic density of states at the *Fermi surface*, which is the energy level separating the filled electronic states from empty ones. Since it is only Fermi-surface electrons that carry current, vanishingly small density of such electronic states leads to low electrical conductivities in semiconductors and insulators. Such a gap in the density of states may also play a role in explaining the formation of quasicrystalline structures. This is known as the *Hume-Rothery* rule for alloy formation (for *diffusionless transformations* in alloys; electronic phases) [1013, 1014]. The same is found for amorphous forming alloys and amorphous phases [1015]. Since the Fermi-surface electrons are the highest-energy electrons, diminishing the number of such electrons may lower the overall energy. Also, QCs are showing one or another form of magnetism that can be divided into two main classes: those containing transition metals (TM) and those containing rare earth (RE) elements. While periodic crystals of similar symmetry and chemical composition usually exhibit comparable magnetic properties, their aperiodic counterparts belonging to the same structural family can show completely different magnetic behavior. Diamagnetism (which is the common QC magnetic behavior in a very broad temperature ranges, extending from about 50 ºK up to their melting point), paramagnetism, ferromagnetism, and spin-glass behavior have been reported in the literature for TM-containing QCs, whereas spin-glass and paramagnetic state as well as short and long-range antiferromagnetic correlations have been experimentally found in RE-based QCs [970,973,974, 1016]. The reason for the unconventional behavior of quasiperiodic structures is the combination of geometrical frustration, structural disorder, deformation mechanism, and complicated distance-dependent magnetic interactions. The most striking difference of QCs from their crystalline counterparts is their symmetry that is forbidden in conventional periodic solids. Therefore, ever since the report of long-range icosahedral atomic structure in rapidly solidified $Al_{86}Mn_{14}$ alloys, one of the fundamental issues has become the effect of this symmetry on the magnetic properties. Following the crystal field theory, magnetism in solids basically results from the lifting of degeneracy of d and f atomic states according to the crystal field symmetry present. The RE-QCs are strong paramagnets at room temperature and show glass-like freezing at low temperatures. However, they are fundamentally different from the canonical spin-glass and show complicated magnetic ordering at low temperatures. Besides, QCs show complicated behavior of low-field magnetoconductivity at low temperatures, reflecting the importance of the quantum coherence. Because of these complexities, the magnetism in QCs is still a subject of ongoing scientific debate [973, 1017-1019].

Optically, the strong suppression of the *Drude peak* in optical conductivity, *i.e.*, low absorption and emission in the IR region, and the high absorption coefficient in the visible spectral range (where solar radiation has a maximum intensity) are combined with the high thermal stability and anticorrosion properties of icosahedral QCs. So, absorbance is high, although with no selectivity, and the material may be used as a light absorber from infrared to visible range, provided that its emissivity is maintained as low as possible. This is feasible with QCs because they keep transparency down to thicknesses as high as a few tens of nm, in comparison to, for example, gold that becomes opaque to visible light for thicknesses above 10-12 nm [974]. Owing to the combination of these properties, QC films are promising for practical application in *selective solar radiation absorbers* with a multilayer coating containing a QC film as one of the constituting layers [996, 1020-1023].

Another remarkable feature, related to the unique electronic structure of QCs, is the high catalysis and hydrogen absorption and storage [483, 1024-1026]. Ti-base, Zr-base and Zn-Mg-Y QCs alloy systems have attracted more attention [1027-1031]. These materials may be in the form of fine or ultra-fine powders or have a nanometer size (NQCs) by making them by devitrification of their amorphous form [1027, 1032]. Ti-Zr-Ni QCs (especially $Ti_{45}Zr_{38}Ni_{17}$, $Ti_{45}Zr_{35}Ni_{20-x}Y_x$ and $Ti_{45}Zr_{35}Ni_{20-x}Pd_x$) store prodigious quantities of hydrogen, up to two hydrogen atoms for each metal atom, presumably due to the large number of tetrahedral sites in the quasilattice, and the favorable alloy chemistry with much higher reversible capacity. The low-cost materials and the relatively rapid absorption and desorption at temperatures between 300°C and 350°C, make these systems extremely promising for hydrogen storage applications [1033-1038]. The loading capacity of the QCs is competitive (more than 10 wt%), absorbing hydrogen to a higher weight percentage than the intermetallic $LaNi_5$, commonly used in hydride batteries, or TiFe, one of the most promising materials for stationary hydrogen storage applications. Potentially wider applications may be based on their unique electronic, thermal, optical and hydrogen storage properties. In Pd-coated QC $Zr_{69.5}Cu_{12}Ni_{11}Al_{7.5}$ (made by CBMSing), hydrogen absorption was accelerated by a factor of about 10 than the normal finely-powdered form and these alloys are promising negative electrode materials of Ni-MH (Ni-Metal Hydride) batteries due to good cycling stability, corrosion resistance and potential high discharge capacity [1036, 1039, 1040]. By contrast, the high brittleness of QCs allows us to prepare large amounts of ultra-fine powders through grinding, without any loss of the chemical homogeneity of the ingot (after proper heat treatment) [974].

The field of QCs is still wide and open for more research, explorations and applications, especially when one knows that most of QCs papers and reviews are focusing on the icosahedral-QC symmetry, rather than other QC-symmetries.

3.8 SUMMARY AND CONCLUSION

In the present chapter we have browsed many promising features and valuable applications of materials produced by the CBMS tech, mainly; bulk amorphous and fine-grained metals and alloys (including nanostructures, QCs and amorphous phases), through the following innovative P/M consolidation and SPF/DB methods. Approximately, they can be used in most fields of our lives. This technology has led to the developing of new metallic materials with significant improvements in mechanical and chemical properties that are exactly meeting our great and continuous need for new materials with exceptional compromised ensemble of useful properties. Some industries depend principally on this technology like the aerospace and automotive industries. New alloy systems for structural materials can not be produced without the RSP technologies, resulting in more lighter and tougher structural components. Even old metallic materials, they can be processed by the CBMS route to gain more improved mechanical properties, corrosion and wear resistances, especially when used for producing new composite products. CBMS is the only cheaper route for producing amorphous or mixed amorphous-microstructured bulk metals and alloys, for electronic and mechanical components, featured by high elastic, strength and wear resistance properties. Also, CBMS method is the best route for producing quasicrystalline materials for structures, very hard coatings (against wear and weathering) and hydrogen storage systems. The future strategy should now include novel compositions, combinations and the ability to produce materials at lower overall cost than can be achieved with conventional processing. There is no doubt that the RSP technology will continue to make a significant impact and will be a major part of the materials science and technology of the future. This chapter is an industrial guide for the benefits of the RSP by CBMS technology.

REFERENCES

[1]   D. A. Porter and K. E. Easterling, *Phase transformations in metals and alloys*, Chapman and Hall, 1993.

[2]   Tarek N. El-Ashram, "Rapidly Solidified Bearing Alloys and Solders", PhD Thesis, Physics Dep., Faculty of Science , 2002 , Al-Manşūrah (Al Mansoura) University, Egypt.

[3]   Mustafa Kamal, *Metal Physics and Metallic Alloys Technology* (in Arabic), Etrac for Printing, Publishing and Distribution, Cairo, Egypt, 2001.

[4]   R. A. Swalin, *Thermodynamics of Solids*, John Wiley and Sons, 1962.

[5]   B. Chalmer, *Principles of Solidification*, John Wiley and Sons, 1964.

[6]   A. H. Cottrell, *Theoretical Structural Metallurgy*, English Language Book Society (ELBS) and Edward Arnold (publishers) Ltd., 1964.

[7]   G. H. Geiger and D. R. Poirier, *Transport Phenomena in Metallurgy*, Addison-Wesley, Reading, MA, 1973.

[8]   M. Flemings, *Solidification Processing*, McGraw-Hill Book Co., 1974.

[9]   P. Cotterill and P. Mould, *Recrystallization* and *Grain Growth in Metals*, Surrey University Press, 1976.

[10]   W. Kurz and D. Fisher, *Fundamentals of Solidification*, 3rd Ed. (reprinted), Trans Tech Publications, Switzerland, 1992.

[11]   R. W. Cahn and P. Haasen, Eds., *Physical Metallurgy*, Vol. 1, North-Holland and Elsevier Science B.V., 1996.

[12]   S. Davis, *Theory of Solidification* ,Cambridge University Press, 2001.

[13]   J. Campbell, *Castings*, Butterworth-Heinemann and Elsevier Science Ltd., 2003.

[14]   G. Vander-Voort, Ed., ASM Handbook, Vol. 9; *Metallography and Microstructures*, 2004.

[15]   R. W. Cahn, in: *Rapidly Solidified Alloys: Processes , Structures , Properties* and *Applications*, H. Liebermann, Ed., Marcel Dekker Inc., 1993, and references therein.

[16]   P. Duwez, "Effect of Rate of Cooling on the Alpha-Beta Transformation in Titanium and Titanium-Molybdenum Alloys", *Trans. AIME* 191, pp. 765-771, 1951.

[17]   N. Ohashi, in: Suppl. Vol. 1 of the *Encyclopedia of Materials Science and Engineering*, R. W. Cahn, Eds., Pergamon, Oxford , pp. 88, 1988.

[18]   H. Jones, "A perspective on the development of rapid solidification and nonequilibrium processing and its future", *Mater. Sci. Eng.* A (304-306), pp. 11-19, 2001, and references therein.

[19]   H. Jones, "Review: The status of rapid solidification of alloys in research and application", *J. Mater. Sci.* 19, pp. 1043-1076, 1984, and references therein.

[20]   D. Moore and D. Gabe, "Electrodeposition of cadmium-lead alloys", *Trans. Inst. Met. Finish.*, 1983.

[21]   R. P. Allen, S. D. Dahlgren and M. D. Merz, in: *Rapidly Quenched Metals*, B. C. Giessen and N. J. Grant, Eds., MIT Press, Cambridge, MA, pp. 37, 1976.

[22]   P. Duwez, R. Willens and W. Klement, "Metastable Electron Compound in Ag-Ge Alloys", *J. Appl. Phys.* 31, 1137, 1960.

[23]   P. Duwez and R. H. Willens, "Rapid quenching of liquid alloys", *Trans. Met. Soc. AIME* 227, 362-365, 1963.

[24]   P. Krehl, *History of Shock Waves, Explosions and Impact: A Chronological and Biographical Reference*, Springer Berlin-Heidelberg, 2009.

[25]   G. Price, *Northern Architect* 10, pp. 222–223, 1963, cited in Ref. 18.

[26]   R. J. Bayuzick, W. H. Hofmeister and M. B. Robinson, *Undercooled Alloy Phases*, TMS, Warrendale, PA, , pp. 207–231, 1987.

[27]   G. Agricola, *De Re Metallica*, Froben, 1556.

[28]   W. Marriott, British Patent 3322 (1873), cited in Ref. 18.

[29]   P. F. Cowing, "Shot-Making Machine", US Patent 809,671, Jan. 9 1906.

[30]   M. U. Schoop, British Patent 5712 (1910), cited in Ref. 18.

[31]   A. R. E. Singer, "Principles of spray rolling of metals", *Met. Mater.* 4, 246-257, 1970.

[32]   H. Bessemer, British Patent 11317 (1846), cited in Ref. 18.

[33]   E. Small, "Apparatus for Making Wire-Solder", US Patent 262,625, Aug. 15, 1882.

[34]   N. R. Lyman, "Machine for Making Printers' Leads", US Patent 315045, Apr. 7, 1885.

[35]   E. H. Strange, C. A. Pim , "Process of Manufacturing Thin Sheets, Foil", Strips, or Ribbons of Zinc, Lead, or Other Metal or Alloy., US Patent 905,758, Dec. 1,1908.

[36]   W. Staples, "Metal-Strand Machine", US Patent 989,075, Apr. 28, 1911.

[37]   E. Dix, Jr., *Proc. Am. Soc. Test Mater.* 25,120-122, 1925, cited in Ref. 18 and in: M. H. Burden and H. Jones, "A metallographic study of the effect of more rapid freezing on the cast structure of aluminum-iron alloys", *Metallography* 3, pp. 307-326, 1970.

M. Kamal and Usama S. Mohammad

[38]   W. Hofmann, "X-ray Methoden zur Untersuchungvon Aluminiumlegierungen" (X-ray Methods for the Investigation of Aluminium Alloys), *Aluminium* 20, 865-872, 1938.

[39]   R. S. Busk, T. Leontis, "The Extrusion of Powdered Magnesium Alloys", *Trans. AIME* 188, 297-306, 1950.

[40]   A. Brenner, D.E. Couch , E.K. Williams , "Electrodeposition of Alloys of Phosphorus with Nickel or Cobalt", *J. Res. Nat. Bur. Stands.* 44, pp.109-122, 1950.

[41]   W. Buckel, R. Hilsch , "Supraleitung und elektrischer Widerstand neuartiger Zinn-Wismut-Legierungen" (Superconductivity and electrical resistance of novel tin-bismuth alloys), *Z. Phys.* 146, 27-38, 1956.

[42]   H. Jones, "Cooling rates in freezing finite slabs", *Mater. Sci. Eng.* 5, 297-299, (1969/1970).

[43]   H. Jones, "Cooling rates during rapid solidification from a chill surface", *Mater. Lett.* 26, pp. 133-136, 1996.

[44]   R. C. Ruhl, "Cooling rates in splat cooling", *Mater. Sci. Eng.* 1, , pp. 313-320. 1967.

[45]   H. Jones and C. Suryanarayana, "Rapid quenching from the melt: An annotated bibliography 1958-72", *J. Mater. Sci.* 8, pp. 705-753, 1973.

[46]   G. Falkenhagen and W. Hofmann, , *Z. Metallkunde* 43, pp. 69-81, 1952, cited in: P. Duwez, "Metallic glasses - a new class of materials: their scientific and industrial importance", *Sadhana* 2, pp. 117-132, 1979.

[47]   C. Hayzelden, J. Rayment and B. Cantor, "Rapid solidification microstructures in austenitic Fe-Ni alloys", *Acta Met.* 31, pp. 379-386, 1983.

[48]   B. Cantor, in: *Science and technology of the undercooled melt*, P. Sahm, H. Jones and C. Adams, Eds., Nijhoff, Dordrecht, the Netherlands, pp. 3-28, 1986.

[49]   A. Gillen, B. Cantor, "Photocalorimetric Cooling Rate Measurements on a Ni-5 wt% Al Alloy Rapidly Solidified by Melt Spinning", *Acta Metal.* 33, pp. 1813-1815, 1985.

[50]   B. Bewlay and B. Cantor, "Cooling rates in melt spun 316L stainless steel", *Int. J. Rapid Solidification* 2, pp. 107-123, 1986.

[51]   P. Cremer and J. Bigot, "An infrared thermography study of the temperature variation of an amorphous ribbon during production by planar flow casting", *Mater. Sci. Eng.* 98, pp. 95-97, 1988.

[52]   M. Ishra, "Annealing Kinetics of Some Amorphous Metallic Alloys", MSc. Thesis, Physics Dep., Faculty of Science, Al-Manşūrah University, Egypt, 1982.

[53]   M. Calvo-Dahlborg, "Structure and embrittlement of metallic glasses", *Mater. Sci. Eng.* A (226-228), pp. 833-845, 1997.

[54]   R. B. Pond and R. Maddin, "Method of Producing Rapidly Solidified Filamentary Castings", *Trans. Met. Soc. AIME* 245, pp. 2475-6, 1969.

[55]   M. Kamal, J. Piéri and R. Jouty, "Preparation et étude de l'évolution structurale des alliages metalliques amorphes Pb-Sb" (Preparation and Structural Evaluation of Amorphous Metallic Alloys Pb-Sb), Mémoires et Etudes Scientifiques Revue de Métallurgie, Mars 1983, pp. 143-148.

[56]   T. Anantharaman and C. Suryanarayana, *Rapidly Solidified Metals*, Trans Tech Pub., Switzerland, 1987.

[57]   P. Pietrokowsky, "Novel Mechanical Device for Producing Rapidly Cooled Metals and Alloys of Uniform Thickness", *Rev. Sci. Instr.* 34, pp. 445-446, 1963.

[58]   J. Carpenter and P. Steen, "Planar-flow spin casting of molten metals: process behavior", *J. Mater. Sci.* 27, pp.215-225, 1992.

[59]   P. Steen and C. Karcher, "Fluid Mechanics of Spin Casting of Metals", *Annu. Rev. Fluid Mech.* 29, pp. 373-397, 1997.

[60]   E. Breinan, B. Kear and C. Banas, "Processing Materials with Lasers", *Physics Today*, pp. 44-50, Nov. 1976.

[61]   R. Pond, R. Maringer and C. Mobley, in: *New Trends in Materials Processing*, ASM Metals Park, OH, pp. 128, 1976.

[62]   P. Ochin, A. Dezellus, Ph. Plaindoux, J. Pons, Ph. Vermaut, R. Portier and E. Cesari, "Shape memory thin round wires produced by the in rotating water melt-spinning technique", *Acta Mater.* 54, pp. 1877-1885, 2006.

[63]   H. Chen and C. Miller, "A rapid quenching technique for the preparation of thin uniform flims of amorphous solids", *Rev. Sci. Instrum.* 41, pp. 1237-38, 1970.

[64]   M. Moss, D. Smith and R. Lefever, "Metastable Phases and superconductors produced by Plasma Jet Spraying", *App. Phys. Lett.* 5, pp. 120-1, 1964.

[65]   S. Whang, "Rapidly Solidified High Strength, Ductile Dispersion-Hardened Tungsten-Rich Alloys", US Patent 4,908,182, Mar. 13, 1990.

[66]   I. Donald and B. Metcalfe, "The Preparation, Properties and Applications of Some Glass-Coated Metal Filaments Prepared by the Taylor-Wire Process", *J. Mat. Sci.* 31, pp. 1139-1149, 1996.

[67]   A. Cox, J. Moore and E. Van Reuth, in: *Superalloys: Metallurgy and Manufacture*, B. H. Kear, D. Muzyka, J. Tien and S. Wlodek, Eds., Claitor's Publishing Division, Baton Rouge, LA, pp. 123, 1976.

[68]   V. Anand, A. Kaufman and N. Grant, in: *Rapid Solidification Processing: Principles and Technologies* II, R. Mehrabian, B. Kear, and M. Cohen, Eds., Claitor's Publishing Division, Baton Rouge, LA, pp. 273, 1980.

[69]   Y. Yingxue, G. Shengdong and C. Chengsong, "Rapid prototyping based on uniform droplet spraying", *J. Mater. Proc. Tech.* 146, pp. 389-395, 2004.

[70]   K. Nagashio, H. Murata, K. Kuribayashi, "In situ observation of solidification behavior of Si melt dropped on Si wafer by IR thermography", *J. Crystal Growth* 275, e1685-e1690, 2005.

[71]   M. Watanabe, K. Higuchi, A. Mizuno, K. Nagashio, K. Kuribayashi and Y. Katayama, "Structural change in silicon from undercooled liquid state to crystalline state during crystallization", *J. Crystal Growth* 294, pp. 16-21, 2006.

[72]   A. Singer, in: *Advanced Fabrication Processes*, Proceedings of the AGARD (Advisory Group for Aeronautical Research and Development) Conference 256, Florence, pp. 19-24, 1978.

[73]   B. Rickinson, F. Kirk, and D. Davies, "CSD: A Novel Process for Particle Metallurgy Products", *Powder Metallurgy* 24, pp. 1-6, 1981.

[74]   M. Kim and H. Jones, in: *Rapidly Quenched Metals*, Vol. I , T. Masumoto and K. Suzuki, Eds., JIM (The Japan Institute of Metals, Sendai), pp. 85-88, 1982.

[75]   R. Alarashi, A. Shaban and M. Kamal, "Evaluation of electronic transport and premature failure in the melt-spun Pb–Sn–Sb–Ag rapidly solidified alloys", *Mater. Lett.* 31, pp. 61-65, 1997.

[76]   M. Narasimhan, "Continuous Casting Method for Metallic Strips", U.S. Patent 4142571, Mar. 6, 1979.

[77]   R. E. Maringer, "Method of and Apparatus for Casting Metal Strip Employing Free Gap Melt Drag", U.S. Patent 4614222, Sep. 30, 1986.

[78]   R. Maringer and C. Mobley, "Casting of metallic filament and fiber", *J. Vac. Sci. Technol.* 11, pp. 1067-1071, 1974.

[79]   B. Kear, J. Mayer, J. Poate and P. Strutt, in: *Surface Treatments Using Laser, Electron and Ion Beam Processing Methods*, Metallurgical Treatises, J. Tien and J. Elliott, Eds., Met. Soc. AIME, pp. 321-343, 1981.

[80]   W. Walton and W. Prewett, "The Production of Sprays and Mists of Uniform Drop Size by Means of Spinning Disc Type Sprayers", *Proc. Phys. Soc.* 62, pp. 341-50, 1949.

[81]   R. Fraser, J. Jang and J. Mollen-Dorf, *J. Inst. Fuel* 36, pp. 316, 1963, cited in Ref. 19.

[82]   H. Lubanska, "Correlation of spray ring data for gas atomization of liquid metals", *J. Metals* 22, pp. 45-49, 1970.

[83]   M. H. Kim, "Formation and decomposition of solid solutions extended by rapid solidification", Ph.D Thesis, Sheffield, 1982.

[84]   C. Yolton and J. Beckman, "Powder metallurgy processing and properties of the ordered orthorhombic alloy Ti-22at.% Al-23at.% Nb", *Mater. Sci. Eng.* A (192-193), pp. 597-603, 1995.

[85]   E. Lavernia and T. Srivatsan, "The rapid solidification processing of materials: science, principles, technology, advances, and applications", *J. Mater. Sci.* 45, pp. 287-325. 2010.

[86]   H. Henein, "Single fluid atomization through the application of impulses to a melt", *Mater. Sci. Eng.* A 326, pp. 92-100, 2002.

[87]   Y. Yingxue, G. Shengdong and C. Chengsong, "Rapid prototyping based on uniform droplet spraying", *J. Mater. Proc. Tech.* 146, pp. 389–395, 2004.

[88]   F. Gillessen, "Keimbildungs kinetik in unterkiihlten, glasbildenden Metallschmelzen" (Nucleation Kinetics of Undercooled, Glass-Forming Metallic Melts), Ph.D. Thesis, Ruhr-Universität Bochum, Germany, 1989.

[89]   D. Herlach, P. Galenko and D. Holland-Moritz, *Metastable Solids from Undercooled Melts*, Pergamon Materials Series, Elsevier B.V., 2007.

[90]   D. Turnbull, "Under what conditions can a glass be formed?", *Contemporary Physics* 10, pp.473-488, 1969.

[91]   J. H. Perepezko and I. Anderson, in: *Synthesis and Properties of Metastable Phases*, E. Machlin and T. Rowland, Eds., TMS-AIME, Warrendale, PA, p-31, 1980.

[92]   J. Perepezko, in: *Rapid Solidification Processing: Principles and Technologies* II, R. Mehrabian, B. Kear and M. Cohen, Eds., Claitor's, Baton Rouge, LA, p-56, 1980.

[93]   C. Adam and L. Hogan, "Crystallography of the Al-Al$_3$ Fe eutectic", *Acta Metall.* 23, pp. 345-354, 1975.

[94]   I. Hughes and H. Jones, "Coupled eutectic growth in Al-Fe alloys", *J. Mater. Sci.* 11, pp. 1781-1793, 1976.

[95]   W. J. Boettinger and J. H. Perepezko, in: *Rapidly Solidified Alloys: Processes, Structures, Properties and Applications*, H. Liebermann (Editor), Marcel Dekker Inc., pp. 17, 1993, and references therein.

[96]   H. Jones, "Rapid Solidification of Metals and Alloys", Monograph No. 8, Institution of Metallurgists, London, 1982.

[97]   K. Nagashio and K. Kuribayashi, "Rapid Solidification of Y$_3$Al$_5$O$_{12}$ Garnet from Hypercooled Melt", *Acta Mater.* 49, 1947-1955, 2001.

[98]   M. Cohen and R. Mehrabian, in: *Rapid Solidification Processing, Principles and Technologies III, R. Mehrabian*, Eds., National Bureau of Standards (Now; the U.S. National Institute of Standards and Technology, NIST), Gaithersburg, MD, , p. 1, 1983.

[99]   C. Levi and R. Mehrabian, "Heat flow during rapid solidification of undercooled metal droplets", *Metall. Trans.* A 13, pp. 221-234, 1982.

[100] R. Mehrabian, "Rapid Solidification", *Int. Met. Rev.* 27, pp. 185-208, 1982.

[101] T. Clyne, "Numerical treatment of rapid solidification", *Metall. Mater. Trans.* B, pp. 369-381, 1984.

[102] H. Biloni and W. J. Boettinger, in: *Physical Metallurgy*, Robert W. Cahn and P. Haasen, Eds., Vol. 1, North-Holland and Elsevier Science B.V., pp. 669, 1996, and references therein.

[103] D Turnbull, "Formation of Crystal Nuclei in Liquid Metals", *J. Appl. Phys.* 21, 1022-1028, 1950.

[104] J. H. Perepezko and J. S. Paik, in: *Rapidly Solidified Amorphous and Crystalline Alloys*, B. Kear, B. Giessen and M. Cohen, Eds., North-Holland, Amsterdam, pp. 49, 1982.

[105] S. LeBeau, J. Perepezko, B. Mueller and G. Hildeman, in: *Rapidly Solidified Powder Aluminum Alloys*, ASTM STP 890, M. E. Fine and E. A. Starke, Jr., Eds., ASTM, Philadelphia, PA, pp. 118-136, 1986.

[106] B. Mueller and J. Perepezko, in: *Aluminum Alloys: Their Physical and Mechanical Properties*, E. A. Starke, Jr. and T. H. Sanders, Jr., Eds., EMAS West Midlands UK, , pp. 201, 1986.

[107] B. Mueller and J. Perepezko, "The undercooling of aluminum", *Metall. Trans.* A 18, pp. 1143-1150, 1987.

[108] W. Yoon, J. Paik, D. La Court, and J. H. Perepezko, "The effect of pressure on phase selection during nucleation in undercooled bismuth", *J. Appl. Phys.* 60, pp. 3489-94, 1986.

[109] J. Perepezko, B. Mueller and K. Ohsaka, in: *Undercooled Alloy Phases*, E. W. Collings and C. C. Koch, Eds., TMS, Warrendale, PA, pp. 289-320, 1987.

[110] J. Perepezko and W. Boettinger, in: *Surface Alloying by Ion, Electron and Laser Beams*, ASM, Metals Park, OH, pp. 51, 1987.

[111] H. Jones, "Observations on a structural transition in aluminium alloys hardened by rapid solidification", *Mater. Sci. Eng.* 5, pp. 1-18, 1969.

[112] S. C. Huang and K. P. LaForce, in: Rapidly Solidified Metastable Materials, R. H. Kear and B. C. Giessen, Eds., Elsevier, New York, pp.125-151, 1984.

[113] C. Levi and R. Mehrabian, "Microstructures of rapidly solidified aluminum alloy submicron powders", *Metall. Trans.* A 13, pp. 13-23, 1982. See also Ref. [99].

[114] D. Turnbull and B. Bagley, in: *Treatise on Solid State Chemistry*, Vol. 5, N. Hannay, Ed., Plenum, New York, pp. 513, 1975.

[115] H. Jones and W. Kurz, "Growth temperatures and the limits of coupled growth in unidirectional solidification of Fe-C eutectic alloys", *Metall. Trans.* A 11, pp. 1265-1273, 1980.

[116] J. Perepezko and W. Boettinger, in: *Alloy Phase Diagrams*, L. H. Bennett, T. B. Massalski and B. C. Giessen, Eds., Elsevier, New York, p-223, 1983.

[117] C. White, D. Zehner, S. Campisano, and A. Cullis, in: *Surface Modification and Alloying by Lasers, Ion and Electron Beams*, J. Poate, G. Foti and D. Jacobson, Eds., Plenum, New York., pp. 81, 1983.

[118] S. Picraux and D. Follstaedt, in: *Surface Modification and Alloying by Lasers, Ion and Electron Beams*, J. Poate, G. Foti and D. Jacobson, Eds., Plenum, New York, pp. 287, 1983.

[119] J. Baker and J. W. Cahn, *The thermodynamics of solidification*, ASM Seminar Series on Solidification, ASM, Metals Park, OH, pp. 23-58, 1971.

[120] W. Boettinger, in: *Rapidly Solidified Crystalline and Amorphous Alloys*, B. H. Kear and B. C. Giessen, Eds., Elsevier North Holland, NY, pp. 15, 1982.

[121] J. W. Cahn, in: *Rapid Solidification Processing: Principles and Technologies* II, R. Mehrabian, B. H. Kear, and M. Cohen, Eds., Claitor's, Baton Rouge, LA, pp. 24, 1980.

[122] J. Aptekar and D. Kamenetskaya, "Diffusionless transformations in steels", *Fiz. Metal. Metalloved.* 14, pp. 123-131, 1962.

[123] H. Biloni and B. Chalmers, "Predendritic Solidification", *Trans. Met. SOC. AIME* 233, pp. 373-379, 1965.

[124] J. W. Cahn, S. R. Coriell and W. J. Boettinger, in: *Laser and Electron Beam Processing of Materials*, C. W. White and P. S. Peercy, Eds., Academic Press, NY, pp. 89, 1980.

[125] C. White, D. Zehner, S. Campisano and A. Cullis, in: *Surface Modification and Alloying by Lasers, Ion, and Electron Beams*, J. Poate, G. Foti and D. Jacobson, Eds., Plenum Press, NY, pp. 81, 1983.

[126] C. Suryanarayana, "Mechanical alloying and milling", *Prog. Mater. Sci.* 46, pp. 1-184, 2001.

[127] R. W. Cahn and L. Greer, *Metastable States of Alloys, in: Physical Metallurgy*, R. Cahn and P. Haasen, Eds., Vol. 2, North-Holland and Elsevier Science BV, pp. 1724, 1996, and references therein.

[128] C. Suryanarayana, Rapid Solidification Processing, in: *Encyclopedia of Materials: Science and Technology*, K. Buschow, R. Cahn, M. Flemings, B. Ilschner, E. Kramer, S. Mahajan, and P. Veyssière, Eds., Elsevier Science Ltd., pp.1-10, 2008.

[129] D. Turnbull, "Metastable structures in metallurgy", *Metall. Trans.* A 12, pp. 695-708, 1981.

[130] G. Herzer, Amorphous and Nanocrystalline Materials, in: *Encyclopedia of Materials: Science and Technology*, K. Buschow, R. Cahn, M. Flemings, B. Ilschner, E. Kramer, S. Mahajan, and P. Veyssière, Eds., Elsevier Science Ltd., pp. 149-156, 2008.

[131] R. Cahn, in: *Glasses and Amorphous Materials*, J. Zarzycki, Ed., Vol. 9 of Materials Science and Technology, VCH, Weinheim, Germany, pp. 493, 1991.

[132] T. Egami, in: *Bulk Metallic Glasses: An Overview*, M. Miller and P. Liaw, Eds., Springer Science+Business Media, LLC, pp. 27, 2008, and references therein.

[133] S. Takayama, "Review: Amorphous structures and their formation and stability", *J. Mater. Sci.* 11, pp. 164-185, 1976.

[134] R. Schwarz, in: *Rapidly Solidified Alloys : Processes, Structures , Properties and Applications*, H. Liebermann, Ed., Marcel Dekker Inc., pp. 157, 1993.

[135] S. Kavesh, in: *Metallic glasses*, J. Gilman and H. Leamy, Eds., ASM, Metals Park, OH, pp. 36, 1978.

[136] J. Hollomon, and D. Turnbull, "Nucleation", *Prog. Met. Phys.* 4, Interscience, New York, pp. 333-388, 1953.

[137] J. Perepezko, "Nucleation Kinetics and Grain Refinement", in: *Casting*, J. Davis, Ed., ASM Handbook, Vol. 15, ASM International, 2008, pp. 276-287.

[138] J. W. Christian, *The Theory of Transformation in Metals and Alloys*, Pergamon, Oxford, 1975.

[139] F. Spaepen, "A structural model for the solid-liquid interface in monatomic systems", *Acta Metall.* 23, 729-743, 1975.

[140] W. Tiller, in: *Physical Metallurgy*, 2nd edn., R. W. Cahn, Ed., North-Holland, Amsterdam, pp. 403, 1970.

[141] J. Lee and H. Aaronson, in: *Lectures on the Theory of Phase Transformations*, H. Aaronson, Ed., The Metallurgical Society, pp. 83, 1975.

[142] J. Perepezko and M. Uttormark, "Nucleation-Controlled Solidification Kinetics", *Metall. Mater. Trans.* A 27, pp. 533-547, 1996.

[143] W. Boettinger and J. Perepezko, in: *Rapidly Solidified Crystalline Alloys*, S. Das, B. Kear, and C. Adam, Eds., The Metallurgical Society, pp. 21, 1986.

[144] J. Perepezko, "Nucleation in undercooled liquids", *Mater. Sci. Eng.* 65, pp. 125-135, 1984.

[145] R. Lacmann, A. Herden and C. Mayer, "Review: Kinetics of Nucleation and Crystal Growth", *Chem. Eng. Technol.* 22, pp. 279-289, 1999.

[146] A. Mersmann, "General prediction of statistically mean growth rates", *J. Cryst. Growth* 147, pp. 181-193, 1995.

[147] A. Mersmann, "Crystallization and precipitation", *Chem. Eng. Proc.* 38, pp. 345-353, 1999.

[148] R. Wagner, R. Kampmann and P. Voorhees, in: *Phase Transformations in Materials*, G. Kostorz, Ed., Wiley-VCH Verlag GmbH, pp. 309, 2001.

[149] R. Doherty, in: *Recrystallization in Metallic Materials*, F. Haessner, Ed., Riederer-Verlag, Stuttgart, pp. 23, 1978.

[150] E. Butler and P. Swann, "In situ observations of the nucleation and initial growth of grain boundary precipitates in an Al-Zn-Mg alloy", *Acta Metall.* 24, , pp. 343-352, 1976.

[151] H. Schubert, A. Mersmann, "Determination of heterogeneous nucleation rates", *Trans. I. ChemE.* A 74, pp. 821-827, 1996.

[152] X. Zhang and S. Tsukamoto, "Theoretical Calculation of Nucleation Temperature and the Undercooling Behaviors of Fe-Cr Alloys Studied with the Electromagnetic Levitation Method", *Metall. Mater. Trans.* A 30, pp. 1827-1833, 1999.

[153] B. Chakraverty and G. Pound, "Heterogeneous nucleation at macroscopic steps", *Acta Metall.* 12, pp. 851-860, 1964.

[154] C. Sholl and N. Fletcher, "Decoration criteria for surface steps", *Acta Metall.* 18, , pp. 1083-1086, 1970.

[155] M. Atzmon, J. Verhoeven, E. Gibson and W. Johnson, "Formation and growth of amorphous phases by solid-state reaction in elemental composites prepared by cold working", *Appl. Phys. Lett.* 45, pp. 1052-1054. 1984.

[156] J. Kim, Y. Choi, S. Suresh and A. Argon, "Nanocrystallzation during nanoindentation of a bulk amorphous metal alloy at room temperature", *Science* 295, pp. 654-657, 2002.

[157] R. Enrique and P. Bellon, "Compositional Patterning in Systems Driven by Competing Dynamics of Different Length Scale", *Phys. Rev. Lett.* 84, pp. 2885–2888, 2000.

[158] A. Lund and C. Schuh, "Molecular simulation of amorphization by mechanical alloying", *Acta Mater.* 52, , pp. 2123-2132, 2004.

[159] A. Lund and C. Schuh, "Topological and chemical arrangement of binary alloys during severe deformation", *J. Appl. Phys.* 95, pp. 4815-4822, 2004.

[160] S. Mayr, "Activation Energy of Shear Transformation Zones: A Key for Understanding Rheology of Glasses and Liquids", *Phys. Rev. Lett.* 97, pp. 195501-4, 2006.

[161] R. Doherty, in: *Physical Metallurgy*, Robert W. Cahn and Peter Haasen, Eds., Vol. 2, North-Holland and Elsevier Science B.V., pp. 1363, 1996, and references therein.

[162] R. Doherty, D. Hughes, F. Humphreys, J. Jonas, D. Jensen, M. Kassner, W. King, T. McNelley, H. McQueen and A. Rollett, "Current issues in recrystallization: a review", *Mater. Sci. Eng.* A 238, pp. 219-274, 1997, and references therein.

[163] J. Perepezko, K. Kimme and R. Hebert, "Deformation Alloying and Transformation Reactions", *J. Alloys Compounds* 483, pp. 14-19, 2009.

[164] G. Purdy and Y. Bréchet, in: *Phase Transformations in Materials*, G. Kostorz, Ed., Wiley-VCH Verlag GmbH, 2001.

[165] M. Larson and J. Garside, "Solute clustering and interfacial tension", *J. Cryst. Growth* 76, pp. 88-92, 1986.

[166] F. Spaepen, "Homogeneous nucleation and the temperature dependence of the crystal-melt interfacial tension", *Solid State Phys.* 47, Academic Press, Inc., Boston, pp. 1-32, 1994.

[167] J. Perepezko, in: Casting, ASM Handbook 15, ASM International, 9th Ed., 1988, pp. 101-108.

[168] A. Zewail, "Micrographia of the Twenty-First Century: from Camera Obscura to 4D Microscopy", *Phil. Trans. R. Soc.* A 368, pp. 1191-1204, 2010.

[169] A. Yurtsever and A. Zewail, "4D Nanoscale Diffraction Observed by Convergent-Beam Ultrafast Electron Microscopy", *Science* 326, No. 5953, pp. 708-712, 2009.

[170] A. H. Zewail, "4D Ultrafast Electron Diffraction, Crystallography, and Microscopy", *Annu. Rev. Phys. Chem.* 57, pp. 65-103, 2006, and references therein.

[171] B. Barwick, D. Flannigan and A. Zewail, "Photon-induced near-field electron microscopy", *Nature* 462, pp. 902-906, 2009.

[172] H. Park, O. Kwon, J. Baskin, B. Barwick and A. Zewail, "Direct Observation of Martensitic Phase-Transformation Dynamics in Iron by 4D Single-Pulse Electron Microscopy", *Nano. Lett.* 9, pp. 3954-3962, 2009.

[173] K. Nagashio, M. Adachi, K. Higuchi, A. Mizuno, M. Watanabe, K. Kuribayashi and Y. Katayama, "Real-Time X-Ray Observation of Solidification from Undercooled Si Melt", *J. Appl. Phys.* 100, pp. 1-6, 2006.

[174] W. King, M. Armstrong, O. Bostanjoglo and B. Reed, in: *Science of Microscopy*, P. Hawkes and J. Spence, Eds., Springer Science+Business Media, LLC, New York, pp. 406-444, 2007, and references therein.

[175] W. Yoon, J. Paik, D. LaCourt and J. Perepezko, "The effect of pressure on phase selection during nucleation in undercooled bismuth", *J. Appl. Phys.* 60, pp. 3489-3494, 1986.

[176] H. Fecht and J. Perepezko, "Metastable phase equilibria in the lead-tin alloy system: Part I. Experimental", *Metall. Trans.* A 20, pp. 785-794, 1989.

[177] Britannica Encyclopedia, Ultimate Reference Suite, a program (on DVD), Britannica Encyclopedia Inc., 2008.

[178] D. Shechtman, I. Blech, D. Gratias and J. Cahn, "Metallic phases with long-range orientational order and no translational symmetry", *Phys. Rev. Lett.* 53, pp. 1951-3, 1984.

[179] D. Herlach, "Metastable Phases Solidified from Undercooled Melts", Proceedings of the 22nd Risø International Symposium on Materials Science, Roskilde, Denmark, pp. 429-434, 2001.

[180] F. Frank, "Supercooling of Liquids", *Proc. Roy. Soc.* (London) A 215, pp. 43-46, 1952.

[181] H. Fecht, M. Zhang, Y. Chang and J. Perepezko, "Metastable phase equilibria in the lead-tin alloy system: Part II. Thermodynamic modeling", *Metall. Trans.* A 20, pp. 795-803, 1989.

[182] D. Shechtman and I. Blech, "The microstructure of rapidly solidified Al_6Mn", *Metall. Trans.* A 16, pp. 1005-1012, 1985.

[183] H. v. Löhneysen, in: *Rapidly Solidified Alloys: Processes, Structures, Properties and Applications*, H. Liebermann, Ed., Marcel Dekker Inc. , pp. 461, 1993.

[184] W. Steurer, in: *Physical Metallurgy*, Robert W. Cahn and Peter Haasen, Eds., Vol. 1, North-Holland and Elsevier Science B.V., pp. 371, 1996.

[185] S. Ranganathan and A. Inoue, "Application of Pettifor Structure Maps for the Identification of Pseudo-Binary Quasicrystalline Intermetallics", *Acta Mater.* 54, pp. 3647-3656, 2006.

[186] D. Holland-Moritz, T. Schenk, V. Simonet, R. Bellissent, P. Convert, T. Hansen and D. Herlach, "Short-range order in undercooled metallic liquids", *Mater. Sci. Eng.* A (375-377), pp. 98-103, 2004.

[187] J. Basu and S. Ranganathan, "Bulk metallic glasses: A new class of engineering materials", *Sadhana* 28 (3-4), pp. 783-798, 2003.

[188] F. Audebert, in: *Properties and Applications of Nanocrystalline Alloys from Amorphous Precursors*, B. Idzikowski P. Svec, M. Miglierini, Eds., Kluwer Academic Publishers, 301-312, , 2005.

[189] A. Inoue, *Bulk Amorphous Alloys: Preparation and Fundamental Characteristics*, Trans. Tech. Publications, 1998.

[190] A. Inoue, *Bulk Amorphous Alloys: Practical Characteristics and Applications*, Trans. Tech. Publications, 1999.

[191] A. Inoue, "Stabilization of Metallic Supercooled Liquid and Bulk Amorphous Alloys", *Acta Mater. 48*, pp. 279–306, 2000.

[192] J. Saida and A. Inoue, Nanoicosahedral Quasicrystal, in: *Encyclopedia of Nanoscience and Nanotechnology*, H. Nalwa, Ed., Vol. 6, American Scientific Publishers, pp. 795–813, 2004.

[193] A. Inoue and H. Kimura, in: *Aerospace Materials*, B. Cantor, H. Assender and P. Grant, Eds., Institute of Physics Publishing (IoP), London, pp. 150, 2001.

[194] A. Inoue, B. Shen and N. Nishiyama, in: *Bulk Metallic Glasses*: An Overview, M. Miller and P. Liaw, Springer Science+Business Media LLC, pp. 1-25, 2008.

[195] W. Johnson, "Bulk glass-forming metallic alloys: Science and Technology", *MRS Bull.* 24, pp. 42–56, 1999.

[196] W. Johnson, "Bulk amorphous alloys: an emerging engineering material", *J. Mater.* 54, pp. 40-43, 2002.

[197] W. Johnson, A. Inoue and C. Liu, Eds., *Bulk metallic glasses*, Materials Research Society (MRS), Warrendale, PA, 1999.

[198] A. Inoue, A. Yavari, W. Johnson and R. Dauskardt, *Supercooled liquid, bulk glassy and nanocrystalline states of alloy*, Materials Research Society (MRS), Warrendale, PA, 2001.

[199] T. Egami, A. Greer, A. Inoue and S. Ranganathan, Eds., *Supercooled liquids, the glass transition and bulk metallic glasses*, Materials Research Society (MRS), Warrendale, PA, 2003.

[200] A. Inoue, T. Zhang, J. Saida, M. Matsushita, M. Chen and T. Sakurai, "Formation of icosahedral quasicrystalline phase in Zr–Al–Ni–Cu–M (M = Pd, Pt, Au, Ag) systems", *Mater. Trans. Jpn. Inst. Met.* (JIM) 40, pp. 1181-1184, 1999.

[201] A. Inoue, "Amorphous, nanoquasicrystalline and nanocrystalline alloys in Al-based systems", *Prog. Mat. Sci.* 43, pp. 365-520, 1998.

[202] A. Inoue, M. Watanabe, H. Kimura, F. Takahashi, A. Nagata, and T. Masumoto, "High mechanical strength of quasicrystalline phase surrounded by FCC-Al phase in rapidly solidified Al-Mn-Ce alloys", *Mat. Trans. JIM* 33, pp. 723-729, 1992.

[203] A. Inoue and H. Kimura, "High-strength aluminum alloys containing nanoquasicrystalline particles", *Mat. Sci. Eng.* A 286, pp. 1-10, 2000.

[204] M. Galano, F. Audebert, B. Cantor, and I. Stone, "Structural characterization and stability of new nanoquasicrystalline Al-based alloys", *Mater. Sci. Eng.* A (375-377), pp. 1206-1211, 2004.

[205] F. Audebert, F. Prima, M. Galano, M. Tomut, P. Warren, I. Stone, and B. Cantor, "Structural characterization and mechanical properties of nanocomposite Al-based alloys", *Mat. Trans. JIM* (43-8), pp. 2017-2025, 2002.

[206] Y. Liu, G. Yuan, C. Lu, W. Ding and J. Jiang, "Role of nanoquasicrystals on ductility enhancement of as-extruded Mg–Zn–Gd alloy at elevated temperature", *J. Mater. Sci.* 43, pp. 5527-5533, 2008.

[207] A. Singh, M. Watanabe, A. Kato and A. Tsai, "Strengthening Effects of Icosahedral Phase in Magnesium Alloys", *Philos. Mag.* 86 (6-8), 951-956, 2006.

[208] A. Singh and A. Tsai, "A new orientation relationship OR4 of icosahedral phase with magnesium matrix in Mg–Zn–Y alloys", *Scr. Mater.* 53, pp. 1083-1087, 2005.

[209] D. Turnbull, "Kinetics of solidification of supercooled liquid mercury droplets", *J. Chem. Phys.* 20, pp. 411-424, 1952.

[210] D. Holland-Moritz, D. Herlach and K. Urban, "Observation of the undercoolability of quasicrystal-forming alloys by electromagnetic levitation", *Phys. Rev. Lett.* 71, pp. 1196-1199, 1993.

[211] M. Li, K. Nagashio, T. Ishikawa, A. Mizuno, M. Adachi, M. Watanabe, S. Yoda, K. Kuribayashi and Y. Katayama, "Microstructure formation and in situ phase identification from undercooled Co−61.8 at.% Si melts solidified on an electromagnetic levitator and an electrostatic levitator", *Acta Mater.* 56, pp. 2514-2525, 2008.

[212] R. Schaefer, L. Benderski and F. Biancaniello, "Nucleation and growth of aperiodic crystals in aluminum alloys", *J. de Physique* 47 C3, C3-311-C3-320, 1986.

[213] D. Herlach, F. Gillessen, T. Volkmann, M. Wollgarten and K. Urban, "Phase selection in undercooled quasicrystal forming Al-Mn alloy melts", *Phys. Rev.* B 46, pp. 5203-5210, 1992.

[214] J. Lennard-Jones and B. Dent, "The Forces between Atoms and Ions. II", *Proc. Roy. Soc.* (London) A 112, pp. 230-234, 1926.

[215] J. Lennard-Jones and A. Devonshire, "Critical and Co-operative Phenomena. IV. A Theory of Disorder in Solids and Liquids and the Process of Melting", *Proc. R. Soc.* (London) A 170, pp. 464-484, 1939.

[216] C. Kittel, Introduction to Solid State Physics, 7th Ed., John Wiley and Sons, 1996.

[217] K. Kelton, Quasicrystals, in: *Encyclopedia of Materials: Science and Technology*, K. Buschow, R. Cahn, M. Flemings, B. Ilschner, E. Kramer, S. Mahajan, and P. Veyssière, Eds., Elsevier Science Ltd., pp. 7947-7952. 2008.

[218] M. Allen and D. Tildesley, *Computer Simulation of Liquids*, Oxford University Press, 1991.

[219] G. Grest, S. Nagel and A. Rahman, "Longitudinal and Transverse Excitations in a Glass", *Phys. Rev. Lett.* 49, pp. 1271-1274, 1982.

[220] R. Bird, W. Stewart and E. Lightfoot, *Transport Phenomena*, John Wiley and Sons, New York, 1960.

[221] R. Brodkey and H. Hershey, *Transport Phenomena: A Unified Approach*, McGraw-Hill Inc., 1988.

[222] H. Huitema, M. Vlot and J. van der Eerden, "Simulations of crystal growth from Lennard-Jones melt: detailed measurements of the interface structure", *J. Chem. Phys.* 111, pp. 4714-4723, 1999.

[223] D. Nelson, "Order, frustration, and defects in liquids and glasses", *Phys. Rev.* B 28, pp. 5515-5535, 1985.

[224] T. Egami, "Icosahedral order in liquids", *J. Non-Crystalline Solids* 353, pp. 3666-3670, 2007.

[225] I. Fisher, K. Cheon, A. Panchula, P. Canfield, M. Chernikov, H. Ott and K. Dennis, "Magnetic and transport properties of single-grain R-Mg-Zn icosahedral quasicrystals [R = Y, $(Y_{1-x} Gd_x)$, $(Y_{1-x} Tb_x)$, Tb, Dy, Ho, and Er]", *Phys. Rev.* B 59, pp. 308-21, 1999.

[226] A. Inoue, H. Kimura, T. Masumoto, A. Tsai and Y. Bizen, "Al-Ge-(Cr or Mn) and Al-Si-(Cr or Mn) Quasicrystals with High Metalloid Concentration Prepared by Rapid Quenching", *J. Mater. Sci. Lett.* 6, pp. 771-774, 1987.

[227] Y. Grin, U. Schwarz and W. Steurer, in: *Alloy Physics: A Comprehensive Reference*, W. Pfeiler, Ed., WILEY-VCH Verlag GmbH and Co., pp. 19, 2007.

[228] F. Spaepen, "Microscopic mechanism for steady-state inhomogeneous flow in metallic glasses", *Acta Metall.* 25, pp. 407-415, 1977.

[229] D. Turnbull and M. Cohen, "Free-volume model of the amorphous phase: Glass transition", *J. Chem. Phys.* 34, pp. 120-125, 1961.

[230] M. Cohen and G. Grest, "Liquid–glass transition, a free-volume approach", *Phys. Rev.* B 20, pp. 1077-1098, 1979.

[231] R. Aga and J. Morris, in: *Bulk Metallic Glasses: An Overview*, M. Miller and P. Liaw, Eds., Springer Science+Business Media LLC, pp. 57, 2008.

[232] L. Bosio, A. Defrain and I. Epelboin, "Changements de phase du gallium a la pression atmosphérique" (Phase Change in Gallium by Atmospheric Pressure), *J. de Physique* (Paris) 27, pp. 61-71, 1966.

[233] K. Jackson, in: *Growth and Perfection of Crystals*, R. Doremus, B. Roberts and D. Turnbull, Eds., Wiley, New York, pp. 319, 1958.

[234] D. Follstaedt, P. Peercy and J. Perepezko, "Phase selection during pulsed laser annealing of manganese", *Appl. Phys. Lett.* 48, pp.338-340, 1986.

[235] J. Knapp and D. Follstaedt, "Measurements of melting temperatures of quasicrystalline Al-Mn phases", *Phys. Rev. Lett.* 58, pp. 2454-2457, 1987.

[236] D. Follstaedt and J. Knapp, in: *Selected Topics in Electronic Materials*, B. Appleton, D. Biegelsen, W. Brown, and J. Knapp, Eds., Materials Research Society (MRS), Pittsburgh, PA, pp. 263, 1988.

[237] W. Allen, H. Fecht and J. Perepezko, "Melting behavior of Sn-Bi alloy droplets during continuous heating", *Scripta Metall.* 23, pp. 643-648, 1989.

[238] P. Gilgien, A. Zryd and W. Kurz, "Microstructure Selection Maps for Al-Fe Alloys", *Acta Metall. Mater.* 43, pp. 3477-3487, 1995.

[239] W. Boettinger, S. Coriell, A. Greer, A. Karma, W. Kurz, M. Rappaz and R. Trivedi, "Solidification Microstructures: Recent developments, Future Directions", *Acta Mater.* 48, pp. 43-70, 2000.

[240] H. Jones, "Some effects of solidification kinetics on microstructure formation in aluminium-base alloys", *Mater. Sci. Eng.* A (413-414), pp. 165-173, 2005, and references therein.

[241] D. Liang, Y. Bayraktar and H. Jones, "Formation and segregation of primary silicon in Bridgman solidified Al-18.3 wt% Si alloy", *Acta Met. Mater.* 43, pp. 579–585, 1995.

[242] W. Kyffin, W. Rainforth and H. Jones, "Effect of phosphorus additions on the spacing between primary silicon particles in a Bridgman solidified hypereutectic Al-Si alloy", *J. Mater. Sci.* 36, pp. 2667-2672, 2001.

[243] A. Moir and H. Jones, "The effect of temperature gradient and growth rate on spacing between primary silicon particles in rapidly solidified hypereutectic aluminium-silicon alloys", *J. Cryst. Growth* 113, pp. 77-82, 1991.

[244] M. Pierantoni, M. Gremaud, P. Magnin, D. Stoll and W. Kurz, "The coupled zone of rapidly solidified Al-Si alloys in laser treatment", *Acta Met. Mater.* 40, pp. 1637-1644, 1992.

[245] S. Bourban, N. Karapatis, H. Hofmann and W. Kurz, "Solidification Microstructure of Laser Remelted Al_2O_3-ZrO_2 Eutectic", *Acta Mater.* 45, pp. 5069-5075, 1997.

[246] M. Gäumann, C. Bezençon, P. Canalis and W. Kurz, "Single-Crystal Laser Deposition of Superalloys: Processing-Microstructure Maps", *Acta Mater.* 49, pp. 1051-1062, 2001.

[247] M. Burden and J. Hunt, "Cellular and dendritic growth: I and II", *J. Cryst. Growth* 22, pp. 99-116, 1974.

[248] M. Tassa and J. Hunt, "The measurement of Al-Cu dendrite tip and eutectic interface temperatures and their use for predicting the extent of the eutectic range", *J. Cryst. Growth* 34, pp. 38-48, 1976.

[249] W. Kurz, B. Giovanola and R. Trivedi, "Theory of microstructural development during rapid solidification", *Acta Metal.* 34, pp. 823-830, 1986.

[250] J. Hunt and S. Lu, "Numerical modeling of cellular/dendritic array growth: spacing and structure predictions", *Met. Mater. Trans.* A 27, pp. 611-623, 1996.

[251] A. Juarez-Hernandez and H. Jones, "Growth temperatures and microstructure selection during Bridgman solidification of hypereutectic Al–La and Al–Ce alloys", *J. Cryst. Growth* 208, pp. 442-448, 2000.

[252] L. Kaufman and H. Bernstein, *Computer Calculation of Phase Diagrams*, Academic Press, New York, 1970.

[253] J. Ågren, "A thermodynamic analysis of the Fe-C and Fe-N phase diagrams", *Metall. Trans.* A 10, pp.1847-1852, 1979.

[254] Y. Chuang, Y. Chang, and R. Schmid and J. Lin, "Magnetic contributions to the thermodynamic functions of alloys and the phase equilibria of Fe-Ni system below 1200 °K", *Metall. Trans.* A 17, pp. 1361-1372, 1986.

[255] S. Gill and W. Kurz, "Rapidly Solidified Al-Cu Alloys II: Calculation of the Microstructure Selection Map", *Acta Met. Mater.* 43, pp. 139-151, 1995.

[256] W. Boettinger, S. Coriell, A. Greer, A. Karma, W. Kurz, M. Rappaz and R. Trivedi, "Solidification Microstructures: Recent developments", Future Directions, *Acta Mater.* 48, pp. 43-70, 2000.

[257] F. Gärtner, A. Norman, A. Greer, A. Zambon, E. Ramous, K. Eckler and D. Herlach, "Texture analysis of the development of microstructure in Cu-30 at.% Ni alloy droplets solidified at selected undercoolings", *Acta Mater.* 45, pp. 51-66, 1997.

[258] A. Norman, K. Eckler, A. Zambon, F. Gärtner, S. Moir, E. Ramous, D. Herlach and A. Greer, "Application of microstructure-selection maps to droplet solidification: a case study of the Ni-Cu system", *Acta Mater.* 46, pp. 3355-3370, 1998.

[259] W. Boettinger, D. Schechtman, R. Schaefer and F. Biancaniello, "The Effect of Rapid Solidification Velocity on the Microstructure of Ag-Cu Alloys", *Metall. Trans.* A 15, pp. 55-66, 1984.

[260] S. Anderbouhr, S. Gilles, E. Blanquet, C. Bernard and R. Madar, "Thermodynamic Modeling of the Ti-Al-N System and Application to the Simulation of CVD Processes of the (Ti,Al)N Metastable Phase", *Chem. Vap. Deposition* 5, pp. 109-115, 1999.

[261] O. Hunziker and W. Kurz, "Solidification Microstructure Maps in Ni-Al Alloys", *Acta. Mater.* 45, pp. 4981-4992, 1997.

[262] R. Bormann, F. Gartner and K. Zoltzer, "Application of the CALPHAD method for the prediction of amorphous phase formation", *J. Less-Comm. Metals* 145, pp. 19-29, 1988.

[263] R. Bormann, "Thermodynamic and kinetic requirements for inverse melting", *Mater. Sci. Eng.* A (179-180), pp. 31-35, 1994.

[264] P. Turchi, *Phase Diagrams as a Tool for Advanced Materials Design*, The Alloy Phase Diagram International Commission (APDIC) and World Round Robin Seminar (WRRS) series, T. Mohri, Ed., Sapporo, Japan, 2003.

[265] E. Scheil, "Bemerkungen zur Schichtkristallbildung" (Comments on Layered Crystals), *Z. Metallkde.* 34, pp. 70-72, 1942.

[266] W. Bragg and E. Williams, "The Effect of Thermal Agitation on Atomic Arrangement in Alloys", *Proc. Roy. Soc.* A 145, pp. 699-730, 1934.

[267] N. Saunders and A. Miodownik, *CALPHAD (Calculation of Phase Diagrams): A Comprehensive Guide*, Pergamon Materials Series Vol. 1, Elsevier Science Ltd., 1998.

[268] A. Miodownik, in: Proc. Conf. Advances in Materials and Processes, P. Ramakrishnan, Ed., Oxford and IBBH, New Delhi, p. 87, 1993.

[269] T. Mohri, in: *Alloy Physics: A Comprehensive Reference*, W. Pfeiler, Ed., WILEY-VCH Verlag GmbH and Co., pp. 525, 2007.

[270] A. Pelton, in: *Phase Transformations in Materials*, G. Kostorz, Ed., Wiley-VCH Verlag GmbH, pp. 1, 2001.

[271] O. Kubaschewski and C. Alcock, *Metallurgical Thermochemistry*, 5th edn., Pergamon Press, New York, 1979.

[272] A. Pelton, in: *Physical Metallurgy*, Robert W. Cahn and Peter Haasen, Eds., Vol. 1, North-Holland and Elsevier Science B.V., pp. 471, 1996.

[273] G. Raynor, in: *Physical Metallurgy*, 2nd edn., R. W. Cahn, Ed., Amsterdam, North Holland, 1970.

[274] J. MacChesney and P. Rosenberg, in: *Phase Diagrams, Materials Science and Technology*, Vol. 1, A. Alper, Ed., Academic, New York, 1970.

[275] R. Buckley, in: *Techniques of Metals Research*, Vol. IV, Part 1, R. Rapp, Ed., Interscience, New York, 1970.

[276] W. Hume-Rothery, J. Christian, W. Pearson, *Metallurgical Equilibrium Diagrams*, Inst. Phys. (IoP), London 1952.

[277] P. Nash and B. Sundman, Eds., Symp. Applications of Thermodynamics in the Synthesis and Processing of Materials, TMS, Warrendale, PA, 1994.

[278] K. Hack, Ed., *The SGTE Casebook: Thermodynamics at Work*, Institute of Materials, London, 1996.

[279] H. Baker, Ed., *Alloy Phase Diagrams*, ASM Handbook, Vol. 3, ASM International, Materials Park, OH, 1992.

[280] M. Hillert, *Phase Equilibria, Phase Diagrams and Phase Transformations Their Thermodynamic Basis*, 2nd edn., Cambridge University Press, 2007.

[281] C. Thompson and F. Spaepen, "Homogeneous crystal nucleation in binary metallic melts", *Acta Metal.* 31, pp. 2021-2027, 1983.

[282] K. Ishihara, M. Maeda, P. Shingu, "The nucleation of metastable phases from undercooled liquids", *Acta Metal.* 33, pp. 2113-2117, 1985.

[283] J. Richmond, J. Perepezko, S. Lebeau and K. Cooper, in: *Rapid Solidification Processing: Principles and Technologies* III, R. Mehrabian, Ed., NBS, Washington, DC, pp. 90, 1983.

[284] B. Bewlay and B. Cantor, "Modelling of spray deposition: measurements of particle size, gas velocity, particle velocity and spray temperature in gas atomized sprays", *Metal. Trans.* B 21, pp. 899-912, 1990.

[285] B. Bewlay and B. Cantor, "The Relationship Between Thermal History and Microstructure in Spray-Deposited Tin-Lead Alloys", *J. Mater. Res.* 6, pp. 1433-54, 1991.

[286] B. Cantor, "Microstructure Development During Rapid Solidification", Proc. 22nd Risø International Sym. Mater. Sci. (Science of Metastable and Nanocrystalline Alloys), A. Dinesen, M. Eldrup, D. Jensen, S. Linderoth, T. Pedersen, N. Pryds, A. Pedersen and J. Wert, Eds., Risø National Laboratory, Roskilde, Denmark, pp. 483-493, 2001.

[287] D. Turnbull, "Kinetics and metastable structure formation in rapid solidification processing", Proceedings of Second Israel Materials Engineering Conference, Beersheva, Ben Gurion University, *Israel*, pp. 1-10, 1984.

[288] I. Maxwell and A. Hellawell, "A simple model for grain refinement during solidification", *Acta Metal.* 23, pp. 229-237, 1975.

[289] P. Boswel and G. Chadwick, "The grain size of splat-quenched alloys", *Scripta Metall.* 11, pp. 459-65, 1977.

[290] C. Levi, "The evolution of microcrystalline structures in supercooled metal powders", *Met. Trans.* A 19, pp. 699-708, 1988.

[291] K. Suslick, S. Choe, A. Cichowlas and M. Grinstaff, "Sonochemical Synthesis of Amorphous Iron", *Nature* 353, pp. 414-416, 1991.

[292] L. Bendersky, F. Biancaniello, W. Boettinger, J. Perepezko, "Microstructural characterization of rapidly solidified Nb-Si alloys", *Mater. Sci. Eng.* 89, , pp. 151-159, 1987.

[293] N. Al-Aqeeli, G. Mendoza-Suarez, C. Suryanarayana and R. Drew, "Development of new Al-based nanocomposites by mechanical alloying", *Mater. Sci. Eng.* A 480, pp. 392-396, 2008.

[294] L. Katgerman and F. Dom, "Rapidly solidified aluminium alloys by melt spinning", *Mater. Sci. Eng.* A 375-377, pp. 1212-1216, 2004.

[295] Powder Metal Technologies and Applications, ASM Handbook, Vol. 7, W. Eisen, B. Ferguson, R. German, R. Iaccoca, P. Lee, D. Madan, K. Moyer, H. Sanderow and Y. Trudel, Eds., ASM International, 1998.

[296] P. Shingu, K. Kobayashi., R. Suzuki and K. Takeshita, "Cooling Process of the Single Roller Chill Block Casting", Proceedings of Rapidly Quenched Metals, Vol. IV, T. Masumoto and K. Suzuki, Eds., Sendai, Japan Institute of Metals, pp. 57-60, 1982.

[297] L. Katgerman, P. Van Den Brink, in: Rapidly Quenched Metals, Vol. IV, T. Masumoto and K. Suzuki, Eds., Sendai, Japan Institute of Metals, 1982, pp. 61.

[298] K. Takeshita, P. Shingu, "An analysis of the ribbon formation process by the Single Roller Chill Block Casting", *Trans. Jpn. Inst. Metals* (*Trans. JIM*) 24, pp. 529-530, 1983.

[299] H. Davies, in: Rapidly Quenched Metals, S. Steeb and H. Warlimont, Eds., Vol. V, Elsevier Science, 1985, pp. 101.

[300] L. Katgerman, "Theoretical analysis of ribbon thickness formation during melt spinning", *Scripta Metall.* 14, , pp. 861-864, 1980.

[301] L. Katgerman, in: Rapidly Quenched Metals, S. Steeb and H. Warlimont, Eds., Vol. V, Elsevier, Amsterdam, 1985, pp. 819.

[302] E. Matthys, Ed., Melt-spinning and Strip Casting; Research and Implementation, TMS, Warrendale, 1993.

[303] E. Matthys, W. Truckner, Eds., Melt-spinning, Strip Casting and Slab Casting, TMS, Warrendale, 1996.

[304] N. El-Mahallawy and M. Taha, "Melt Spinning of Al-Cu Alloys: Modelling of Heat Transfer", *J. Mater. Sci. Lett.* 6, pp. 885-889, 1987.

[305] H. Liebermann, "The dependence of the Geometry of glassy alloy ribbons on the chill block melt-spinning process parameters", *Mater. Sci. Eng.* 43, pp. 203-210, 1980.

[306] H. Liebermann, "Coaxial jet melt-spinning of glassy alloy ribbons", *J. Mater. Sci.* 15, pp. 2771-2776, 1980.

[307] P. Shingu and K. Ishihara, in: Rapidly Solidified Alloys: Processes, Structures, Properties and Applications, H. Liebermann, Ed., Marcel Dekker Inc., 1993, pp. 103, and references therein.

[308] K. Nagashio and K. Kuribayashi, "Experimental Verification of Ribbon Formation Process in Chill-Block Melt Spinning", *Acta Mater.* 54, pp. 2353-2360, 2006.

[309] R. Valiev, R. Islamgaliev and I. Alexandrov, "Bulk nanostructured materials from severe plastic deformation", *Prog. Mater. Sci.* 45, pp. 103-189. 2000, and references therein.

[310] R. Valiev, In: Synthesis and Processing of Nanocrystalline Powder, The Minerals, Metals and Materials Society (TMS), D. Bourell, Ed., 1996, p.153.

[311] I. Alexandrov, Y. Zhu, T. Lowe, R. Islamgaliev and R. Z. Valiev, "Consolidation of nanometer sized powders using severe plastic torsional straining", *Nanostr. Mater.* 10, pp. 45-54, 1998.

[312] I. Alexandrov, R. K. Islamgaliev, R. Valiev, Y. Zhu and T. Lowe, "Microstructures and properties of nanocomposites obtained through SPTS consolidation of powders", *Metall. Mater. Trans.* A 29, 2253-2260, 1998.

[313] J. Groza, in: Non-equilibrium Processing of Materials, C. Suryanarayana, Ed., Pergamon Materials Series, Vol. 2, Elsevier Science Ltd., 1999, pp. 347.

[314] H. Liebermann and C. Graham, Jr., "Production of amorphous alloy ribbons and effects of apparatus parameters on ribbon dimensions", *IEEE Trans. Mag.* MAG-12, pp. 921-23, 1976.

[315] R. Maringer and C. Mobley, in: Rapidly Quenched Metals III, Vol. 1, B. Cantor, Ed., The Metals Society, London, 1978, pp. 49.

[316] T. Anthony and H. Cline, "Dimensional Variations in Newtonian-Quenched Metal Ribbons Formed by Melt Spinning and Melt Extraction, *J. Appl. Phys.* 50, 245-254, 1979.

[317] S. Charter, D. Mooney, R. Cheese and B. Cantor, "Melt spinning of crystalline alloys", *J. Mater. Sci.* 15, pp. 2658-2661, 1980.

[318] H. Hillmann and H. Hilzinger, in: Rapidly Quenched Metals III, Vol. 1, B. Cantor, Ed., The Metals Society, London, 1978, p-22.

[319] S. Huang and H. Fiedler, "Amorphous ribbon formation and the effects of casting velocity", *Mater. Sci. Eng.* 51, pp. 39-46, 1981.

[320] S. Huang and H. Fiedler, "Effects of Wheel Surface Conditions on the Casting of Amorphous Metal Ribbons", *Metal. Trans.* A 12, pp. 1107-1112, 1981.

[321] J. Vincent, H. Davies and J. Herbertson, in: Continuous Casting of Small Sections, Y. Murty and F. Mollard, Eds., *Metals Society of the American Institute of Mining, Metallurgical and Petroleum Engineers* (*Met. Soc. AIME*), 1981, pp. 103.

[322] J. Vincent, J. Herbertson and H. A. Davies, in: Rapidly Quenched Metals IV, Vol. 1, T. Masumoto and K. Suzuki, Eds., The Japan Institute of Metals, Sendai, 1982, pp. 77.

[323] J. Vincent and H. Davies, in: Solidification Technology in the Foundry and Casthouse, K. Beecroft, Ed., Metals Society, London, 1983, pp. 153.

[324] J. Vincent, J. Herbertson and H. Davies, Comments on "The geometry of melt spun ribbon", *J. Mater. Sci. Lett.* 2, pp. 88-90, 1983.

[325] F. Luborsky and H. Liebermann, "Effect of melt temperature on some properties of $Fe_{80.5}B_{15}Si_4C_{0.5}$ and $Fe_{40}Ni_{40}B_{20}$ amorphous alloys", *Mater. Sci. Eng.* 49, pp. 257-261, 1981.

[326] S. Huang, R. LaForce, A. Ritter and R. Goehner, "Rapid Solidification Characteristics in Melt Spinning a Ni-Base Superalloy", *Metall. Trans.* A 16, pp. 1773-1779, 1985.

[327] M. Kramer, H. Mecco, K. Dennis, E. Vargonova, R. McCallum and R. Napolitano, "Rapid Solidification and Metallic Glass Formation: Experimental and Theoretical Limits", *J. Non-Crystalline Solids* 353, pp. 3633-3639, 2007.

[328] Y. N. Chen, University of Pennsylvania, by personal communication cited in Ref. 305.

[329] K. Nagashio and K. Kuribayashi, "Direct observation of the melt/substrate interface in melt spinning", *Mater. Sci. Eng.* A 449-451, pp. 1033-1035, 2007.

[330] K. Nagashio and K. Kuribayashi, "Direct observation of the melt/substrate interface in chill-block melt spinning", *Space Utiliz. Res.* 22, pp. 22-25, 2006.

[331] H. Liebermann and R. Bye, in: Rapid Solidified Crystalline Alloys, S. Das, B. Kear and C. Adam, Eds., *TMS-AIME*,1985, p. 61.

[332] H. Schlichting, Boundary-Layer Theory, (trans. J. Kestin), McGraw-Hill Book Company, New York, 6th edn., 1968.

[333] J. Lienhard IV and J. Lienhard V, A Heat Transfer Textbook, Phlogiston Press, Cambridge, Massachusetts, U.S.A., 2006.

[334] K. Takeshita and P. Shingu, "An Analysis of the Melt Puddle Formation in the Single Roller Chill Block Casting", *Trans. Japan Inst. Metall.* (*Trans. JIM*) 27, pp. 141-148, 1986.

[335] A. Bejan and A. D. Kraus, Heat Transfer Handbook, John Wiley and Sons Inc, 2003.

[336] H. Vogel, "Das temperatur-abhängigkeitsgesetz der viskosität von flüssigkeiten" (The temperature dependence law of viscosity of fluids), *Z. Phys.*, 22, 645-646, 1921.

[337] G. Fulcher, "Analysis of recent measurements of viscosity of glasses", *J. Amer. Ceram. Soc.* 6, 339-355, 1925.

[338] H. Liebermann, "Rapidly Solidified Alloys Made by Chill Block Melt-Spinning Processes", *J. Crystal Growth* 70, 497, 1984.

[339] S. Takayama and T. Oi, "The analysis of casting conditions of amorphous alloys", *J. Appl. Phys.* 50, pp. 1595-7 and pp. 4962-5, 1979.

[340] G. Stephani, H. Mühlbach, H. Fiedler, and G. Richter, "Infrared measurements of the melt puddle in planar flow casting", *Mater. Sci. Eng.* 98, pp. 29-32, 1988.

[341] J. Carpenter and P. Steen, "Heat transfer and solidification in planar-flow melt-spinning: high wheel speeds", *Int. J. Heat Mass Transfer* 40, pp. 1993-2007, 1997.

[342] V. Tkatch, S. Denisenko and B. Selyakov, "Computer Simulation of $Fe_{80}B_{20}$ Alloy Solidification in the Melt Spinning Process", *Acta Metall. Mater.* 43, pp. 2485-2491, 1995.

[343] V. Tkatch, S. Denisenko and B. Selyakov, "Direct Measurements of the Cooling Rates in the Single Roller Rapid Solidification Technique", *Acta Mater.* 45, pp. 2821-2826, 1997.

[344] J. Mullin, Crystallization, 4th edn., Reed Educational and Professional Pub. Ltd., Butterworth-Heinemann, Oxford, 2001.

[345] D. Kirwan and R. Pigford, Crystallization kinetics of pure and binary melts, AIChE J. 15, pp. 442-449, 1969.

[346] R. Strickland-Constable, Kinetics and Mechanics of Crystallization, Academic Press, London, 1968.

[347] E. Vogt and G. Frommeyer, in: Rapidly Solidified Materials, P. Lee and R. Carbonara, ASM, Metals Park, Ohio, 1986, p. 291.

[348] G. Frommeyer and W. Frech, "Continuous casting and rapid solidification of wires produced by a newly developed shape flow casting technique", *Mater. Sci. Eng.* A 226-228, pp. 1019-1024, 1997.

[349] W. Kim and B. Cantor, "The variation of grain size with cooling rate during melt spinning", *Scripta Met. Mat.* 24, pp.633-637, 1990.

[350] K. Takeshita, P. Shingu, "Thermal Contact during the Cooling by the Single Roller Chill Block Casting", *Trans. JIM* 27, pp.454-462, 1986.

[351] A. Greer, K. Rutherford and I. Hutchings, "Wear resistance of amorphous alloys and related materials", *Int. Mater. Rev.* 47, pp. 87-112, 2002.

[352] L. Katgerman, F. Kievits, H. Kleinjan and W. Zalm, Proceedings 4th International Yugoslav Symposium on Aluminium, Titograd, 1982, cited in Ref. 304.

[353] H. Kuiken, "Solidification of a liquid on a moving sheet", *Inter. J. Heat Mass Transfer* 20, pp. 309-314, 1977.

[354] S. Patankar, Numerical Heat Transfer and Fluid Flow, Hemisphere Pub. Corp., New York, 1980.

[355] G. Korn, Advanced Dynamic-system Simulation: Model-Replication Techniques and Monte Carlo Simulation, John Wiley and Sons Inc., 2007.

[356] S. Chapra, Applied Numerical Methods with MATLAB for Engineers and Scientists, 2nd edn., McGraw-Hill Science, 2006.

[357] A. Shabana, Computational Dynamics, 2nd Edn., John Wiley and Sons, 2001.

[358] D. Raabe, Computational Materials Science: The Simulation of Materials, Microstmctures and Properties, WILEY-VCH Verlag GmbH, 1998.

[359] I. Dimov, Monte Carlo Methods For Applied Scientists, World Scientific Publishing Co. Pte. Ltd., 2008.

[360] M. Shimono, in: Springer Handbook of Materials Measurement Methods, H. Czichos, T. Saito and L. Smith, Eds., Springer Science+Business Media, 2007, pp. 915, and references therein.

[361] L. Nastac, Modeling and Simulation of Microstructure Evolution in Solidifying Alloys, Kluwer Academic Publishers, Boston, 2004.

[362] B. Karpe, B. Kosec, T. Kolenko and M. Bizjak, "Effect of the chill wheel cooling during continuous free jet melt-spinning", *RMZ-Materials and Geoenvironment* 56, pp. 401-414, 2009.

[363] Y. Joo, J. Sun, M. Smith, R. Armstrong, R. Brown and R. Ross, "Two-Dimensional Numerical Analysis of Non-Isothermal Melt Spinning with and without Phase Transition", *J. Non-Newtonian Fluid Mech.* 102, pp. 37-70, 2002.

[364] G. Wang and E. Matthys, "Numerical Modelling of Phase Change and Heat Transfer During Rapid Solidification Processes: Use of Control Volume Integrals with Element Subdivision", *Inter. J. Heat Mass Transfer* 35, 141-153, 1992.

[365] G. Wang, V. Prasad and S. Sampath, "An Integrated Model for Dendritic and Planar Interface Growth and Morphological Transition in Rapid Solidification", *Metal. Mater. Trans.* A 31, pp. 735-746, 2000.

[366] G. Wang and E. Matthys, in: Melt-Spinning and Strip Casting: Research and Implementation, E. Matthys, Ed., The Minerals, Metals and Materials Society (TMS), Warrendale, PA, 1992, pp. 263.

[367] P. Wilde and E. Matthys, "Experimental investigation of the planar flow casting process: development and free surface characteristics of the solidification puddle", *Mater. Sci. Eng.* A 150, pp. 237-247, 1992.

[368] T. Blase, Z. Guo, Z. Shi, K. Long and W. Hopkins, "A 3D conjugate heat transfer model for continuous wire casting", *Mater. Sci. Eng.* A 365, pp. 318–324, 2004.

[369] A. Bichi, W. Smith and J. Wissink, "Solidification and downstream meniscus prediction in the planar-flow spin casting process", *Chem. Eng. Sci.* 63 pp. 685-695, 2008.

[370] M. Bussmann, J. Mostaghimi, D. Kirk and J. Graydon, "A numerical study of steady flow and temperature fields within a melt spinning puddle", *Inter. J. Heat Mass Transfer* 45, pp. 3997-4010, 2002.

[371] J. Chen, W. Frazier and A. Tseng, "Heat transfer in Melt Spinning of Intermetallic Materials", First International Conference on Transport Phenomena in Processing, S. Güçeri, Ed., Technomic Pub. Co. Inc., Lancaster, Pennsylvania, USA, 1992, pp. 239-249.

[372] H. Liu, W. Chen, S. Qiu and G. Liu, "Numerical simulation of initial development of fluid flow and heat transfer in planar flow casting process", *Metall. Mater. Trans.* B 40, pp. 411-429, 2009.

[373] E. Theisen, M. Davis, S. Weinstein and P. Steen, "Transient behavior of the planar-flow melt spinning process", *Chem. Eng. Sci.* 65, pp. 3249-3259, 2010.

[374] D. Warrington, H. Davies and N. Shohoji, in: Rapidly Quenched Metals, Vol. IV, T. Masumoto and K. Suzuki, Eds., Japan Institute of Metals, Sendai, 1982, pp. 69.

[375] M. Tenwick, H. Davies, in: Rapidly Quenched Metals, S. Steeb and H. Warlimont, Eds., Vol. V, Elsevier Science, 1985, pp. 67.

[376] H. Mühlbach, G. Stephani, R. Sellger, H. Fiedler, "Cooling Rate and Heat Transfer Coefficient during PFC of Microcrystalline Steel Ribbons", *Int. J. Rapid Solid.* (Now; International Journal of Non-Equilibrium Processing, from 1997 till now) 3, pp.83-94, 1987.

[377] H. Fiedler, H. Mühlbach, G. Stephani, "The Effect of main processing parameters on the geometry of amorphous metal ribbons during planar flow casting (PFC)", *J. Mater. Sci.* 19, 3229-3235, 1984.

[378] S. Coriell and R. Sekerka, in: *Rapid Solidification Processing: Principles and Technologies* II, R. Mehrabian, B. Kear, and M. Cohen, Eds., Claitor's Publishing, Baton Rouge, LA, 1980, pp. 35.

[379] C. Byrne, S. Weinstein and P. Steen, "Capillary stability limits for liquid metal in melt spinning, Chem. Eng. Sci. 61, pp. 8004-8009, 2006.

[380] E. Schubert and H. Bergmann, in: Rapidly Solidified Alloys: Processes, Structures, Properties and Applications, H. Liebermann, Ed., Marcel Dekker Inc., 1993, pp. 195.

[381] S. David, S. Babu and J. Vitek, "Welding: Solidification and Microstructure", *JOM*, pp. 14-20, June 2003.

[382] H. Liebermann, in: Rapidly Quenched Metals III, Vol. 1, B. Cantor, Ed., Metals Society, London, 1978, pp. 34.

[383] T. Duffar, in: Alloy Physics: A Comprehensive Reference, W. Pfeiler, Ed., Wiley-VCH Verlag GmbH and Co. KGaA, Weinheim, 2007, pp. 63.

[384] R. Sekerka, "The Theory of Morphological Stability", *Am. Assoc. Cryst. Growth Newslett.* 16, 2-4, 1986.

[385] W. Mullins and R. Sekerka, "Stability of a Planar Interface During Solidification of a Dilute Binary Alloy", *J. Appl. Phys.* 35, pp. 444-451, 1964.

[386] S. Coriell, G. McFadden and R. Sekerka, "Cellular Growth during Directional Solidification", *Ann. Rev. Mat. Sci.* 15, pp. 119-145, 1985.

[387] S. Coriell and R. Sekerka, "Oscillatory morphological instabilities due to non-equilibrium segregation", *J. Cryst. Growth* 61, pp. 499-508, 1983.

[388] U. Mohammed, "Effect of Rapid Solidification on Structural, Mechanical and Physical Properties of [Lead-Tin-Antimony] Ternary Alloy", M.Sc. Thesis, Physics Dep., Faculty of Science in Damietta, Al-Mansoura University, Egypt, 2009.

[389] I. Ohnaka, "Microsegregation and Macrosegregation", in: Casting, J. Davis, Ed., ASM Handbook, Vol. 15, ASM International, 1998, pp. 136-141.

[390] M. Aziz, "Model for solute redistribution during rapid solidification", *J. Appl. Phys.* 53, pp. 1158-1168, 1982.

[391] R. Napolitano, H. Meco, "The role of melt pool behavior in free-jet melt spinning", *Metall. Mater. Trans.* A 35, pp. 1539-1553, 2004.

[392] M. Kramer, Y. Tang, K. Dennis and R. McCallum, "Effects of Partial pressure on the microstructure of melt spun Nd-Fe-B", *Mater. Res. Soc. Symp. Proc.* 577 (Advanced Hard and Soft Magnetic Materials), pp. 57-62, 1999.

[393] H. Cline and T. Anthony, "The effect of harmonics on the capillary instability of liquid jets", *J. Appl. Phys.* 49, pp. 3203-8, 1978.

[394] H. Liebermann, "Manufacture of amorphous alloy ribbons", *IEEE Trans. Magnetics* MAG-15, pp. 1393-1397, 1979.

[395] M. Matsurra, M. Kikuchi, M. Yagi and K.Suzuki, "Effects of Ambient Gases on Surface Profile and Related Properties of Amorphous Alloy Ribbons Fabricated by Melt-Spinning", *Japanese J. Appl. Phys.* 19, pp. 1781-7, 1980.

[396] C. Mobley, R. Maringer and L. Dillinger, in: *Rapid Solidification Processing: Principles and Technologies* I, R. Mehrabian, B. Kear and M. Cohen, Eds., Claitor's, Baton Rouge, LA, 1978, pp. 223.

[397] G. Hewitt, in: Handbook of Heat Transfer, 3rd edn., W. Rohsenow, J. Hartnett and Y. Cho, Eds., McGraw-Hill Companies, Inc., 1998, pp. 15.1.

[398] M. Wu, A. Ludwig and L. Ratke, in: Solidification and Crystallization, D. Herlach, Ed., Wiley-VCH Verlag GmbH and Co. KGaA, Weinheim, 2004, pp. 34.

[399] S. Marashi, A. Abedi, S. Kaviani, S. Aboutalebi, M. Rainforth and H. Davies, "Effect of melt-spinning roll speed on the nanostructure and magnetic properties of stoichiometric and near stoichiometric Nd–Fe–B alloy ribbons", J. Phys. D: Appl. Phys. 42, pp. 115410-17, 2009.

[400] S. Huang, in: Rapidly Quenched Metals IV, Vol. 1, T. Masumoto and K. Suzuki, Eds., The Japan Institute of Metals, Sendai, 1982, pp. 65.

[401] G. Wang and E. Matthys, "Experimental determination of the interfacial heat transfer during cooling and solidification of molten metal droplets impacting on a metallic substrate: effect of roughness and superheat", *Inter. J. Heat and Mass Transfer* 45, pp. 4967-4981, 2002.

[402] S. Huang and A. Ritter, in: Rapidly Solidified Materials, B. Berkowitz and R. Scattergood, Eds., TMS-AIME, Warrendale, PA, 1983, pp. 25.

[403] H. Davies, N. Shohoji, and D. Warrington, in: *Rapid Solidification Processing: Principles and Technologies* II, R. Mehrabian, B. H. Kear, and M. Cohen, Eds., Claitor's Publishing, Baton Rouge, LA, 1980, pp. 153.

[404] B. Cantor, W. Kim, B. Bewlay and A. Gillen, "Microstructure-cooling rate correlations in melt-spun alloys", *J. Mater. Sci.* 26, pp. 1266-1276, 1991.

[405] E. Miksch, "Solidification of Ice Dendrites in Flowing Supercooled Water", *Trans. AIME* 245, pp. 2069-2072, 1969.

[406] K. Murakami, T. Fujiyama, A. Koike, and T. Okamoto, "Influence of melt flow on the growth directions of columnar grains and columnar dendrites", *Acta Metall.* 31, pp. 1425-1432, 1983.

[407] K. Murakami, H. Aihara, and T. Okamoto, "Growth direction of columnar crystals solidified in flowing melt", *Acta Metall.* 32, pp. 933-939, 1984.

[408] P. Siutsova, E. Neumerzhitskaya and V. Shepelevich, "Comparison of structure and properties of binary and ternary rapidly solidified alloys of the Al–Ni–Zr system", *J. Alloys Compounds* 479, pp. 161-165, 2009.

[409] M. Wu, A. Ludwig and A. Bührig-Polaczek, in: Solidification and Crystallization, D. Herlach, Ed., Wiley-VCH Verlag GmbH and Co. KGaA, Weinheim, 2004, pp. 204.

[410] C. Suryanarayana and T. Anantharaman, "Solidification of aluminium-germanium alloys at high cooling rates", *J. Mater. Sci.* 5, pp. 992-1004, 1970.

[411] I. Donald and H. Davies, "The influence of composition on the formation and stability of Ni-Si-B metallic glasses", *J. Mater. Sci.* 15, pp. 2754-2760, 1980.

[412] R. Trivedi and W. Kurz, "Solidification of Single-Phase Alloys", in: *Casting*, J. Davis, Ed., ASM Handbook, Vol. 15, ASM International, 1988, pp. 114-119.

[413] J. Hunt, in: Solidification and Casting of Metals, J. Hunt, Ed., The Metals Society, London, 1979, pp. 3.

[414] W. Kurz, D.J. Fisher, "Dendrite growth at the limit of stability: tip radius and spacing", *Acta Metall.* 29, pp. 11-20, 1981.

[415] S. Das and F. Froes, in: Rapidly Solidified Alloys: Processes, Structures, Properties and Applications, H. Liebermann, Ed., Marcel Dekker Inc., 1993, pp. 339.

[416] C. Chang, S. Das, and D. Raybould, in: Rapidly Solidified Materials, P. Lee and R. Carbonara, Eds., ASM, Metals Park, Ohio, 1986, pp. 129.

[417] S. Isserow F. J. Rizzitano, "Microquenched Magnesium ZK60A Alloy", *Int. J. Powder Metall. Powder Tech.* 10, pp. 217-227, 1974.

[418] F. Hehmann, F. Sommera and B. Predel, "Extension of solid solubility in magnesium by rapid solidification", *Mater. Sci. Eng.* A 125, pp. 249-265, 1990.

[419] F. Hehmann and H. Jones, in: Rapidly Solidified Alloys and Their Mechanical and Magnetic Properties, B. C. Giessen, D. Polk, and A. Taub, Eds., Materials Research Society, Pittsburgh, PA, 1986, pp. 259.

[420] K. Schemme, "Rapid Solidification of Mg-Li-Y Alloys", in: Magnesium Alloys and Their Applications, B. Mordike, F. Hehmann and Garmisch-Partenkirchen, Eds., DGM-Informationsges, Oberursel, Germany, 1992, pp. 519.

[421] S. Das, L. Davis, "High performance aerospace alloys via rapid solidification processing", *Mater. Sci. Eng.* 98, pp. 8-12, 1998.

[422] F. Hehmann, "Developments in magnesium alloys by rapid solidification processing: an update", In: Advanced aluminium and magnesium alloys, Proc. Int. Conf. on Light Metals, Cahn and Effenberg, Eds., ASM International, Ohio, 1990, pp. 781.

[423] P. Jardim, I. Solórzano, J. Vander Sande, B. You and W. Park, in: Magnesium Alloys and their Applications, K. Kainer, Ed., Wiley-VCH Verlag Gmbh, 2000, pp. 584.

[424] T. Sanders, Jr., and E. Starke, Jr., Eds., Aluminum-Lithium Alloys I, Proc. 1st Int. Conf. Aluminum-Lithium Alloys, *AIME*, NY, 1981.

[425] Aluminum-Lithium Alloys II, Proc. 2nd Int. Conf. on Aluminum-Lithium Alloys, T. Sanders, Jr., and E. Starke, Jr., Eds., AIME, NY, 1984.

[426] P. Meschter, R. Lederick, and J. O'Neal, in: Proc. Int. Conf. PM Aerospace Materials, A Metal Powder Report Conf., Luzern, Nov. 1987, MPIF, *APMI*, Princeton, NJ, USA, 1988, pp. 85.

[427] N. Kim, D. Skinner, K. Okazaki, and C. Adam, in: Aluminum-Lithium Alloys III, Proc. 3rd Inst. Conf. Aluminum-Lithium Alloys, C. Baker, P. Gregson, S. Harris and C. Peel, Eds., Institute of Metals, London, 1986, pp. 78.

[428] D. Sinner, in: Dispersion Strengthened Aluminum Alloys, Y. Kim and W. Griffith, Eds., The Minerals, Metals and Materials Society (TMS), Warrendale, PA, 1988, pp. 181.

[429] M. Zedalis, D. Raybould, D. Skinner and S. Das, in: Processing of Structural Metals by Rapid Solidification, F. Froes and S. Savage, Eds., ASM, Metals Park, OH, 1987, pp. 347.

[430] P. Gilman and S. Das, in: Proc. Int. Conf. PM Aerospace Materials, A Metal Powder Report Conf., Luzern, Nov. 1987, MPIF, *APMI*, Princeton, NJ, USA, 1988, pp. 27.

[431] G. Hildenman and M. Koczak, in: New Light Alloys, AGARD (Advisory Group for Aerospace Research and Development, NATO) Lecture Series No. 174, IGA P. Costa, Ed., 1990, pp. (5-1)-(5-13).

[432] I. Pontikakas, and H. Jones, *Metal Science* 16, pp. 27, 1982; cited in Ref. 431.

[433] D. Vojtech, J. Verner, B. Bártová and K. Saksl, "Rapid solids hold hope for strong aluminium alloys", *MPR* (Metal Powder Report) 61, pp. 32-35, 2006.

[434] J. Kim, D. Kim, K. Shin and N. Kim, "Modification of Mg2Si morphology in squeeze cast Mg-Al-Zn-Si alloys by Ca or P addition", *Scripta Mater.* 41, pp. 333-340. 1999.

[435] G. Yaun, Z. Liu, Q. Wang and W. Ding, "Microstructure refinement of Mg-Al-Zn-Si alloys", *Mater. Lett.* 56, pp. 53-58, 2002.

[436] F. Hehmann and M. Weidemann, "Selected Processing for Nonequilibrium Light Alloys and Products", US Patent No. 6,908,516 B2, June 21, 2005.

[437] H. Gjestland, G. Nussbaum, G. Regazzoni, O. Lohne and O. Bauger, "Stress-relaxation and creep behaviour of some rapidly solidified magnesium alloys", *Mat. Sci. Eng.* A 134, pp. 1197-1200, 1991.

[438] G. Nussbaum, P. Sainfort, G. Regazzoni, H. Gjestland, "Strengthening mechanisms in the rapidly solidified AZ 91 magnesium alloy", *Scripta Metall.* 23, pp. 1079-1084, 1989.

[439] B. Bronfin, E. Aghion, S. Schumann, P. Bohling and K. Kainer, "Magnesium Alloy for High Temperature Applications", US Patent No. 6,139,651, Oct. 31, 2000.

[440] B. Bronfin, E. Aghion, F. von Buch, S. Schumann and H. Friedrich, in: Magnesium: Proc. 6th Intern. Conf. Magnesium Alloys, K. Kainer, Ed., WILEY-VCH Verlag GmbH and Co. KGaA, Weinheim, 2004, pp. 55.

[441] J. Cizek, I. Prochazka, I. Stulikova, B. Smola, R. Kuzel, V. Cherkaska, R. Islamgaliev and O. Kulyasova, in: Magnesium: Proc. 6th Intern. Conf. Magnesium Alloys, K. Kainer, Ed., Wiley-VCH Verlag GmbH and Co. KGaA, Weinheim, 2004, pp. 202.

[442] S. Sastry, T. Peng, P. Meschter and J. O'Neil, "Rapid Solidification Processing of Titanium Alloys", *J. Metals* 35, pp. 21-28, 1983.

[443] S. Whang, "Rapidly Solidified Ti Alloys Containing Novel Additives", *J. Metals* 36, 34-40, 1984.

[444] S. Sastry, T. Peng and J. O'Neil, in: Titanium; Science and Technology, Vol. 1, G. Lutjering, U. Zwicker and W. Bunk, Eds., DGM, Oberursel, West Germany, 1985, pp. 397.

[445] J. Immarigeon, R. Holt, A. Koul, L. Zhao, W. Wallace and J. Beddoes, "Lightweight Materials for Aircraft Applications", *Mater. Charact.* 35, pp. 41-67, 1995.

[446] G. Lütjering and J. Williams, Titanium, 2nd ed., Engineering Materials and Processes series, B. Derby, Ed., Springer 2007.

[447] I. Anderson and B. Rath, in: Rapid Solidified Crystalline Alloys, S. Das, B. Kear, and C. Adam, Eds., *TMS-AIME*,1985, pp. 219.

[448] S. Fujiwara and S. Miwa, "Shape memory effects in rapidly quenched Cu-12%Al and Cu-12%Al-1%Si alloys", *Mater. Sci. Eng.* 98, pp. 509-513, 1988.

[449] D. Morris and M. Morris, "Rapid solidification and mechanical alloying techniques applied to Cu-Cr alloys", *Mater. Sci. Eng.* A 104, pp. 201-213, 1988.

[450] D. Morris, M. Morris, J. Joye, "High strength Cu-Zr alloys prepared by rapid solidification techniques", *Mater. Sci. Eng.* A 158, pp. 111-117, 1992.

[451] M. Kamal and El-Said Gouda, "Electrical and mechanical properties of liquid rapidly quenched Cu-Al-Ni Shape memory alloys", *Radiation Effects and Defects in Solids* 163, pp. 273-240, 2008.

[452] R. Patterson, A. Cox, and E. Van Reuth, in: Rapid Solidification Technology Source Book, R. Ashbrook, Ed., ASM, Metals Park, OH, 1983, pp. 414.

[453] R. Patterson, A. Cox, and E. Van Reuth, "Rapid solidification rate processing and application to turbine engine materials", *J. Metals* 32, pp. 34-39, 1980.

[454] H. Chin and A. Adair, in: Superalloys, M. Gell, C. Kortovich, R. Bricknell, W. Kent, and J. Radavich, Eds., ASM, Metals Park, OH, 1984, pp. 335.

[455] E. Schulson and D. Barker, "A Brittle to Ductile Transition in Ni-Al of a Critical Grain Size", *Scripta Metall.* 17, pp. 519-522, 1983.

[456] S. Huang, K. Chang and A. Taub, in: Rapidly Solidified Materials, P. Lee and R. Carbonara, Eds., ASM, Metals Park, OH, 1986, pp. 255.

[457] R. Adler and S. Hsu, "Alloying effects on the processing and properties of melt spun nickel-metalloid ribbons", *Mater. Sci. Eng.* 97, pp. 207-213, 1988.

[458] G. Yurek, D. Eisen, and A. Garrett-Reed, "Oxidation behavior of a fine-grained rapidly solidified 18-8 stainless steel", *Metall. Trans.* A 13, pp. 473-485, 1982.

[459] G. Olson, in: Rapid Solidification Processing: Principles and Technologies IV, R. Mehrabian and P. Parrish, Eds., Claitor's, Baton Rouge, LA, 1988, pp. 82.

[460] E. Slaughter and S. Das, in: Rapid Solidification Processing: Principles and Technologies II, R. Mehrabian, B. Kear, and M. Cohen, Eds., Claitor's Publishing Division, Baton Rouge, LA, 1980, pp. 354.

[461] D. Branagan and Y. Tang, "Developing extreme hardness (>15 GPa) in iron based nanocomposites", *Composites Part A: Applied Science and Manufacturing* 33, pp. 855-859, 2002.

[462] J. Wadsworth, C. Roberts and E. Rennhack, "Creep behaviour of hot isostatically pressed niobium alloy powder compacts", *J. Mater. Sci.* 17, pp. 2539-2546, 1982.

[463] R. Perkins, K. Chiang and G. Meier, "Formation of alumina on Nb-Al alloys", Scripta Metall. 22, pp. 419-424, 1988.

[464] M. Hebsur and I. Locci, "Microstructures in rapidly solidified niobium aluminides", *NASA Tech. Memo* 19890030822, NASA Glenn Research Center, 1988. Also: I. Locci, T. Bloom and M. Hebsur, "Rapid Solidification Research at the NASA Lewis Research Center", in: Rapidly solidified materials: Properties and processing, P. Lee, Ed., Metals Park OH, ASM International, 1988, pp. 207.

[465] D. Bowden, Rapid Solidification Processing of Niobium-Based Alloys, *Adv. Mater. Manufacture Process* 3, pp. 79-89, 1988.

[466] D. Vujic, S. Whang and S. Cytron, in: Science and Technology of Rapidly Quenched Alloys, M. Tenhover, L. Tanner and W. Johnson, Eds., Materials Research Society, Pittsburgh, PA, 1987, pp. 307.

[467] J. Lasalle, D. Raybould, E. Limoncelli, S. Das and S. Cytron, "Rapidly solidified tungsten-based alloys", *Mater. Sci. Eng.* 98, pp. 165-168, 1988.

[468] T. Kanomata, Y. Sato, Y. Sugawara, H. Kimura, T. Kaneko and A. Inoue, "Specific heat of Zr-based metallic glasses", *J. Alloys Comp.* 461, pp. 39-41, 2008.

[469] K. Soda, K. Shimba, S. Yagi, M. Kato, T. Takeuchi, U. Mizutani, T. Zhang, M. Hasegawa, A. Inoue, T. Ito and S. Kimura, "Electronic structure of bulk metallic glass $Zr_{55}Al_{10}Cu_{30}Ni_5$", *J. Electron Spectroscopy and Related Phenomena* 144-147, pp. 585-587, 2005.

[470] R. Busch, S. Schneider, A. Peker, and W. Johnson, "Decomposition and primary crystallization in undercooled Zr41.2Ti13.8Cu12.5Ni10.0Be22.5 melts", *Appl. Phys. Lett.* 67, pp. 1544-1546, 1995.

[471] S. Scudino, U. Kuhn, L. Schultz, D. Nagahama, K. Hono, and J. Eckert, "Microstructure evolution upon devitrification and crystallization kinetics of $Zr_{57}Ti_8Nb_{2.5}Cu_{13.9}Ni_{11.1}Al_{7.5}$ melt-spun glassy ribbon", *J. Appl. Phys.* 95, pp. 3397-3403, 2004.

[472] S. Das, in: Intermetallic Compounds, Vol. 3: Structural Applications of Intermetallic Compounds, J. Westbrook and R. Fleischer, Eds., John Wiley and Sons Ltd., 2000, pp. 179.

[473] H. Stadelmaier and B. Reinsch, in: Intermetallic Compounds, Vol. 4: Magnetic, Electrical and Optical Properties and Applications of Intermetallic Compounds, J. Westbrook and R. Fleischer, Eds., John Wiley and Sons Ltd., 2000, pp. 31.

[474] K. Yoshimi, S. Hanada and A. Inoue, "Intermetallic Compound and Method of Manufacturing the Same", US Patent No. 2003/0056863 A1, Mar. 27, 2003.

[475] Aerospace Materials, B. Cantor, H. Assender and P. Grant, IoP Publishing Ltd., 2001.

[476] J. Williams and E. Starke, Jr., "Progress in structural materials for aerospace systems", *Acta Mater.* 51, pp. 5775-5799, 2003.

[477] E. Starke, Jr. and J. Staleyt, "Application of Modern Aluminum Alloys to Aircraft", *Prog. Aerospace Sci.* 32, pp. 131-172, 1996.

[478] T. Srivatsan, W. Soboyejo and R. Lederich, "The Cyclic Fatigue and Fracture Behavior of a Titanium Alloy Metal Matrix Composite", *Eng. Fracture Mechanics* 52, pp. 467-491, 1995.

[479] M. Kamal, A. Shaban, M. El-Kady and R. Shalaby, "Irradiation, mechanical and structural behavior of Al-Zn-Based alloys rapidly quenched from melt", *Radiation Effects and Defects in Solids* 138, pp. 307-318, 1996.

[480] A. Shaban, S. Hammad, A. Daoud and M. Kamal, "Thermal and ionizing radiation treatment of the rapidly solidified Al-Zn-Si melt spun alloys", *Radiation Effects and Defects in Solids* 143, pp. 179-191, 1997.

[481] M. Kamal, R. Shalaby and A. Issa, "X- Ray and Microhardness investigations of rapidly solidified Aluminium- Silicon eutectic alloys from molten state", Proc. 8th Arab Intern. Conf. Material Science, Alex., Egypt, 2004, pp. 205-214.

[482] New Light Alloys, AGARD (Advisory Group for Aerospace Research and Development, NATO) Lecture Series No. 174, IGA P. Costa, Ed., 1990.

[483] I. Polmear, Light Alloys: From Traditional Alloys to Nanocrystals, 4th edn., Elsevier Butterworth-Heinemann Pub., 2006.

[484] A. Rosochowski, L. Olejnik and M. Richert, in: Advanced Methods in Material Forming, D. Banabic, Ed., Springer-Verlag Berlin Heidelberg, 2007, pp. 215.

[485] M. Fine and E. Starke, Jr., Eds., Rapidly Solidified Powder Aluminum Alloys: A Symposium, ASTM STP (Special Technical Publication) 890, Philadelphia, PA, 1986.

[486] M. Galano, F. Audebert, A. Escorial, I. Stone and B. Cantor, "Nanoquasicrystalline Al-Fe-Cr-based alloys with high strength at elevated temperature", *J. Alloys Compounds* 495, pp. 372-376, 2010.

[487] N. Stoloff, "Intermetallics: Mechanical Properties", in: *Encyclopedia of Materials: Science and Technology*, K. Buschow, R. Cahn, M. Flemings, B. Ilschner, E. Kramer, S. Mahajan, and P. Veyssière, Eds., Elsevier Science Ltd., pp. 4213-4225, 2008.

[488] S. Huang, "Titanium Aluminum Alloys Modified by Chromium and Niobium and Method of Preparation", U.S. Patent No. 4,879,092, Nov. 7, 1989.

[489] S. Kampe, J. Brupbacher and D. Nagle, "Process for Rapid Solidification of Intermetallic-Second Phase Composites", US Patent No. 4,915,905, Apr. 10, 1990.

[490] S. Huang and J. Chesnutt, in: Intermetallic Compounds, Vol. 3: Structural Applications of Intermetallic Compounds, J. Westbrook and R. Fleischer, Eds., John Wiley and Sons Ltd., 2000, pp. 75, and references therein.

[491] M. Loretto, in: Aerospace Materials, B. Cantor, H. Assender and P. Grant, IoP Publishing Ltd., 2001, pp. 229.

[492] F. Appel and R. Wagner, "Intermetallics: Titanium Aluminides", in: *Encyclopedia of Materials: Science and Technology*, K. Buschow, R. Cahn, M. Flemings, B. Ilschner, E. Kramer, S. Mahajan, and P. Veyssière, Eds., Elsevier Science Ltd., pp. 4246-4265, 2008, and references therein.

[493] H. Jones, in: Non-equilibrium Processing of Materials, C. Suryanarayana, Ed., Pergamon Materials Series, Vol. 2, Elsevier Science Ltd., 1999, pp. 23.

[494] K. Hashimoto, in: Rapidly Solidified Alloys: Processes, Structures, Properties and Applications, H. Liebermann, Ed., Marcel Dekker Inc., 1993, pp. 591; and references therein.

[495] A. Mitsuhashi, K. Asami, A. Kawashima, and K. Hashimoto, "The corrosion behavior of amorphous nickel base alloys in a hot concentrated phosphoric acid", *Corrosion Sci.* 27, pp. 957-970, 1987.

[496] R. Spear and G. Gardner, "Dendrite Cell Size", *Trans. AFS* 71, pp. 209-215, 1963.

[497] H. Jones, "Developments in aluminum alloys by solidification at higher cooling rates", *Aluminium* 54, pp. 274-281, 1978.

[498] H. Jones, "The effect on lattice parameter and hardness of manganese in extended solid solution in aluminium", *J. Mater. Sci. Lett.* 1, pp. 405-406, 1982.

[499] G. Thursfield and M. Stowell, "Mechanical properties of Al-8 wt% Fe-based alloys prepared by rapid quenching from the liquid state", *J. Mater. Sci.* 9, pp. 1644-1660, 1974. *See Also*; M. Lebo and N. Grant, *Metal. Trans.* 5, pp. 1547, 1974.

[500] E. Collings, C. Mobley, R. Maringer and H. Gegel, in: Rapidly Quenched Metals III, Vol. 1, B. Cantor, Ed., The Metals Society, London, 1978., pp. 188.

[501] O. Ruano, L. Eiselstein and O. Sherby, "Superplasticity in Rapidly Solidified White Cast Irons", *Met. Trans.* A 13, pp. 1785-1792, 1982.

[502] T. Tietz and I. Palmers, in: Advances in Powder Technology, G. Chin, Ed., ASM, Metals Park, Ohio, 1982, pp. 189.

[503] R. Thomson, in: Physical Metallurgy, R. Cahn and P. Haasen, Eds., Vol. 3, North-Holland and Elsevier Science B.V., pp. 2207, 1996.

[504] Y. Soifer and N. Kobelev, "Anomalous elastic behaviour of nanostructured materials", *Nanostruct. Mater.* 6, 647-650, 1995.

[505] H. Gleiter, "Nanocrystalline materials", *Prog. Mats. Sci.* 33, pp. 223-315, 1989.

[506] N. Akhmadeev, N. Kobelev, R. Mulyukov, Y. Shaefer and R. Valiev, "The effect of heat treatment on the elastic and dissipative properties of copper with the submicrocrystalline structure", *Acta Metall. Mater.* 41, pp. 1041-1046, 1993.

[507] J. Weertman, in: Nanostructured Materials: Processing, Properties and Potential Applications, Carl C. Koch, Ed., Noyes Publications, 2002, pp. 397.

[508] A. Bassi and C. Bottani, Mechanical Behavior of Nanomaterials, *Encyclopedia of Nanoscience and Nanotechnology*, H. Nalwa, Ed., American Scientific Publishers, Vol. 5, 2004, pp. 53-71, and references therein.

[509] K. Ramesh, Nanomaterials: Mechanics and Mechanisms, Springer Science+Business Media, LLC, 2009.

[510] C. Suryanarayana and F. Froes, "Mechanical, Chemical, and Electrical Applications of Rapidly Solidified Alloys", in: Rapidly Solidified Alloys: Processes, Structures, Properties and Applications, H. Liebermann, Ed., Marcel Dekker Inc., 1993, pp. 737; and references therein.

[511] K. Kendig and D. Miracle, "Strengthening mechanisms of an Al-Mg-Sc-Zr alloy", *Acta Mater.* 50, pp. 4165-4175, 2002.

[512] J. Li, Plastic Deformation of Noncrystalline Materials, in: *Encyclopedia of Materials: Science and Technology*, K. Buschow, R. Cahn, M. Flemings, B. Ilschner, E. Kramer, S. Mahajan, and P. Veyssière, Eds., Elsevier Science Ltd., 2008, pp. 7071-7079.

[513] K. Mawella, R. Honycombe and P. Howell, "A microstructural study of a melt-spun ultra high-strength alloy steel", *J. Mater. Sci.* 17, pp. 2850-2854, 1982.

[514] C. Eiselstein, O. Ruano and O. Sherby, "Structural characterization of rapidly solidified white cast iron powders", *J. Mater. Sci.* 18, pp. 483-492, 1983.

[515] M. Kamal, A. Shaban, M. El- Kady, and R. Shalaby, "Determination of structure-property of rapidly quenched Aluminium-based bearing alloys before and after gamma irradiation", 2nd Intern. Conf. Engineering Mathematics and Physics (ICEMP- 94), Vol. 2, pp. 107-121, 1994.

[516] M. Kamal, A. El-Bediwi and T. El-Ashram, "Structural, Thermal and Mechanical properties of rapidly solidified Sn-Sb-Cu-Al bearing alloys", *Tratamente Termice Si Ingineria Suprafetelor* (Heat Treatment and Surface Engineering), Romania, Vol. IV, pp. 36-50, 2004.

[517] M. Kamal, A. Shaban, M. El-Kady, A. Daoud and R. Alarashi, "Rapidly solidified of Sn-Sb-Ag ternary bearing alloys", *U. Scientist Phyl. Sciences* 8, pp. 166-172, 1996.

[518] B. Duflos and B. Cantor, in: Rapidly Quenched Metals III, B. Cantor, Ed., Institute of Metals, London, Vol. 1, 1978, pp. 110.

[519] R. Smallman and R. Bishop, Modern Physical Metallurgy and Materials Engineering, 6th edn., Butterworth-Heinemann, Reed Educational and Professional Publishing Ltd., Oxford, 1999.

[520] E. Hall, "The deformation and ageing of mild steel: III; Discussion of results", *Proc. Phys. Soc. London.* 64, pp. 747-753, 1951.

[521] N. Petch, "The cleavage strength of polycrystals", *J. Iron Steel Inst.* 174, pp. 25-28, 1953.

[522] G. E. Dieter and D. Bacon, Mechanical Metallurgy, McGraw-Hill, 1988.

[523] P. Haasen, in: Physical Metallurgy, R. Cahn and P. Haasen, Eds., Vol. 3, North-Holland and Elsevier Science B.V., pp. 2009, 1996.

[524] W. Hosford, Mechanical Behavior of Materials, Cambridge University Press, 2005.

[525] C. Zener, "A theoretical criterion for the initiation of slip bands", *Phys. Rev.* 69, 128-129, 1946.

[526] J. Hirth and J. Lothe, Theory of Dislocations, 2nd edn., Malabar, FL, Krieger, 1992.

[527] N. Stoloff, Intermetallics: Mechanical Properties, *Encyclopedia of Materials: Science and Technology*, K. Buschow, R. Cahn, M. Flemings, B. Ilschner, E. Kramer, S. Mahajan, and P. Veyssière, Eds., Elsevier Science Ltd., 2008, pp. 4213-4225.

[528] A. Chokshi, A. Rosen, J. Karch and H. Gleiter, "On the validity of the Hall-Petch relationship in nanocrystalline materials", *Scripta Mater.* 23, pp. 1679-1683, 1989.

[529] W. Oliver and G. Pharr, "An improved technique for determining hardness and elastic modulus using load and displacement sensing indentation experiments", *J. Mater. Res.* 7, 1564-1583, 1992.

[530] B. Bhushan, in: Springer Handbook of Nanotechnology, 2nd edn., B. Bhushan, Ed., 2007, pp.1305-1338.

[531] E. Gdoutos, Fracture Mechanics: An Introduction, 2nd edn., Springer, Dordrecht, The Netherlands, 2005.

[532] C. Lepienski and C. Foerster, Nanomechanical Properties by Nanoindentation, in: *Encyclopedia of Nanoscience and Nanotechnology*, American Scientific Publishers, H. Nalwa, Ed., Vol. 7, 2004, pp. 1-20.

[533] A. Minor, S. Asif, Z. Shan, E. Stach, E. Cyrankowski, T. Wyrobek and O. Warren, "A new view of the onset of plasticity during the nanoindentation of aluminium", *Nature Materials* 5, pp. 697-702, 2006.

[534] S. Takeuchi, "The mechanism of the inverse Hall-Petch relation of nanocrystals", *Scripta Mater.* 44, pp. 1483-1487, 2001.

[535] M. Zhao, J. Li and Q. Jiang, "Hall-Petch relationship in nanometer size range", *J. Alloys Compounds* 361, pp. 160-154, 2003.

[536] T. Nieh and J. Wang, "Hall-Petch relationship in nanocrystalline Ni and Be-B alloys", *Intermetallics* 13, pp. 377-385, 2005.

[537] M. Cordill, N. Moody and W. Gerberich, "The role of dislocation walls for nanoindentation to shallow depths", *Intern. J. Plasticity* 25, pp. 281-301, 2009.

[538] M. Cordill, M. Lund, J. Parker, C. Leighton, A. Nair, D. Farkas, N. Moody and W. Gerberich, "The Nano-Jackhammer effect in probing near-surface mechanical properties", *Intern. J. Plasticity* 25, pp. 2045-2058, 2009.

[539] C. Krenn, D. Roundy, M. Cohen, D. Chrzan and J. Morris, "Connecting atomistic and experimental estimates of ideal strength", *Phys. Rev.* B 65, pp. 13411-16, 2002.

[540] R. Siegel, "Mechanical Behavior of Nanophase Materials", *Mater. Sci. Forum* 235-238, pp. 851-860, 1997.

[541] R. Siegel and G. Fougere, "Mechanical properties of nanophase metals", *Nanostr. Mater.* 6, pp. 205-216, 1995.

[542] J. Weertman and R. Averback, in: Nanomaterials: Synthesis, Properties, and Applications, A. Edlestein and R. Cammarata, Eds., Institute of Physics Publishers, Bristol, 1996, pp. 323-345.

[543] J. Li, "Petch relations and grain boundary sources", *Trans. Metall. Soc. AIME* 277, pp. 239-247, 1963.

[544] M. Meyers and E. Ashworth, "A model for the effect of grain size on the yield stress of metals", *Philos. Mag.* (Philosophical Magazine) A 46, pp. 737-759, 1982.

[545]   M. Ashby, "Deformation of plastically non-homogeneous materials", *Philos. Mag.* 21, pp. 399-424, 1970.

[546]   R. Armstrong, G. Hughes, In: Advanced materials for the twenty-first century, Y. Chubng, Ed., The Julia Weertman Symp., *TMS*, 1999, pp. 409.

[547]   J. Shu and N. Fleck, "Strain gradient crystal plasticity: size-dependent deformation of bicrystals", *J. Mech. Phys. Solids* 47, pp. 297-324, 1999.

[548]   D. Hughes, N. Hansen and D. Bammann, "Geometrically necessary boundaries, incidental dislocation boundaries and geometrically necessary dislocations", *Scripta Mater.* 48, pp. 147-153, 2003.

[549]   N. Hansen, "Hall-Petch relation and boundary strengthening", *Scripta Mater.* 51, pp. 801-806, 2004.

[550]   M. Myers, A. Mishra and D. Benson, "Mechanical properties of nanocrystalline materials", *Prog. Mater. Sci.* 51, pp. 427-556, 2006.

[551]   J. Strudel, in: Physical Metallurgy, R. Cahn and P. Haasen, Eds., Vol. 3, North-Holland and Elsevier Science B.V., 1996, pp. 2105.

[552]   W. Park, "Alloy designing and characterization of rapidly solidified Al-Zr(-V) base alloys", *Materials and Design* 17, pp. 85-88, 1996.

[553]   A. Inoue, H. Kimura, "Fabrications and mechanical properties of bulk amorphous, nanocrystalline, nanoquasicrystalline alloys in aluminum-based system", *J. Light Met.* 1, pp. 31-41, 2001.

[554]   N. Adkins and P. Tsakiropoulos, "Rapid solidification of Al-Cr-Zr-Mn alloys", *Mater. Sci. Eng.* A 134, pp. 1158-1161, 1991.

[555]   U. Prakash, T. Raghu, A. Gokhale and S. Kamat, "Microstructure and mechanical properties of RSP/M Al-Fe-V-Si and Al-Fe-Ce alloys", *J. Mater. Sci.* 34, pp. 5061-5065, 1999.

[556]   A. Inoue, H. Kimura and Y. Horio, "High Strength and High Rigidity Aluminum-Based Alloy and Production Method Therefore", US Patent No. 5,858,131, Jan. 12, 1999.

[557]   I. Mathy and G. Scharf, "High Temperature-Resistant Aluminum Alloy and Process for its Production", US Patent No. 4,832,737, May 23, 1989.

[558]   E. Vogt, G. Frommeyer, J. Wittig and H. Sassik, "The solidification structures of planar flow cast Fe-(20-47) at.% Al ribbons", *Mater. Sci. Eng.* 98, pp. 295-299, 1988.

[559]   W. Park, "Alloy designing and characterization of rapidly solidified Al-Zr-V base alloys", *Materials and Design* 17, pp. 85-88, 1996.

[560]   C. Wayman, Introduction to the Crystallography of Martensitic Transformations, McMillan Series in Material Science, New York, 1964.

[561]   Smithells Metals Reference Book, 8th edn., W. Gale and T. Totemeier, Eds., Elsevier Inc., 2004.

[562]   T. Nieh, J. Wadsworth and O. Sherby, Superplasticity in Metals and Ceramics, Cambridge University Press, New York, 1997.

[563]   R. Mishra, R. Valiev and A. Mukherjee, "The observation of tensile superplasticity in nanocrystalline materials", *Nanostruct. Mater.* 9, pp. 473-476, 1997.

[564]   N. Ridley, "Superplastic Forming", in: Handbook of Aluminum, Vol. 1: Physical Metallurgy and Processes, G. Totten and D. MacKenzie, Ed., Marcel Dekker Inc., 2003, pp. 1105.

[565]   R. Valiev and T. Langdon, "Principles of equal-channel angular pressing as a processing tool for grain refinement", *Prog. Mater. Sci.* 51, pp. 881-981, 2006.

[566]   S. Kalpakjian, "Mechanical Testing for Metalworking Processes", in: ASM Handbook, Vol. 8: *Mechanical Testing and Evaluation*, H. Kuhn and D. Medlin, Eds., 2000, pp. 70-78.

[567]   L. Lu, M. L. Sui and K. Lu, "Superplastic Extensibility of Nanocrystalline Copper at Room Temperature", *Science* 287, pp. 1463-1466, 2000.

[568]   S. Hannula, J. Koskinen, E. Haimi and R. Nowak, "Mechanical Properties of Nanostructured Materials", *Encyclopedia of Nanoscience and Nanotechnology*, H. Nalwa, Ed., American Scientific Publishers, Vol. 5, 2004, pp. 131-162.

[569]   R. Figueiredo and T. Langdon, "Achieving Microstructural Refinement in Magnesium Alloys through Severe Plastic Deformation", *Mater. Trans. JIM* (Materials Transactions of the Japan Institute of Metals) 50, pp. 111-116, 2009.

[570]   A. Argon, in: Physical Metallurgy, Vol. 3, R. Cahn and P. Haasen, Eds., Elsevier Science, 1996, pp. 1957.

[571]   Farghalli A. Mohamed, "Superplastic Deformation at Elevated Temperatures", in: ASM Handbook, Vol. 8: *Mechanical Testing and Evaluation*, H. Kuhn and D. Medlin, Eds., 2000, pp. 412-424.

[572]   G. Rai and N. Grant, "On the measurements of superplasticity in an Al-Cu alloy", *Metall. Trans.* A 6, pp. 385-390, 1975.

[573]   E. Taleff, G. Henshall, D. Lesuer, T. Nieh and J. Wadsworth, in: Superplasticity and Superplastic Forming, A. Ghosh and T. Bieler, Eds., The Minerals, Metals and Materials Society (*TMS*), Warrendale, PA, 1995, pp. 3.

[574] K. Higashi and N. Ridley, in: Superplasticity and Superplastic Forming, C. Hamilton and N. Paton, Eds., The Minerals, Metals, and Materials Society (*TMS*), Warrendale, PA, Blaine, Washington, 1988, p. 447.

[575] C. Hamilton and L. Ascani, "Method of Making a Metallic Structure by Combined Superplastic Forming and Forging", U. S. Patent No. 4,113,522, Sep. 12, 1978.

[576] C. Humphries and N. Ridley, "Cavitation during the superplastic deformation of an α/β brass", *J. Mater. Sci.* 13, pp. 2477-2482, 1978.

[577] J. Wadsworth, J. Lin and O. Sherby, "Superplasticity in a tool steel", *Metals Technol.* 8, pp. 190-193, 1981.

[578] J. Wittenauer, P. Schepp and B. Walser, in: Superplasticity and Superplastic Forming, C. Hamilton and N. Paton, Eds., The Minerals, Metals and Materials Science Society, Warrendale, PA, Blaine, WA, 1988, p. 507.

[579] A. Ghosh, in: Superplastic Forming of Structural Alloys, N. Paton and C. Hamilton, Ed., TMS-AIME, Warrendale, PA, 1982, pp. 85.

[580] R. Menzies, J. Edington and G. Davies, "Superplastic Behavior of Powder Consolidated Nickel-Base Superalloy IN-100", *Metal Sci.* 15, pp. 210-216, 1981.

[581] V. Valitov, G. Salishchev and S. Mukhtarov, "Superplasticity of Nickel-based Alloys with submicrocrystalline structure", *Mater. Sci. Forum* 243-5, pp. 557-562, 1997.

[582] M. Mahoney and R. Crooks, in: Superplasticity and Superplastic Forming, C. Hamilton and N. Paton, Eds., The Minerals, Metals, and Materials Society (TMS), Warrendale, PA, Blaine, Washington, 1988, p. 73.

[583] M. Mahoney and R. Crooks, in: Superplasticity in Aerospace, C. Heikkenen and T. McNelly, Eds., *TMS-AIME*, Warrendale, PA. 1988, p. 331.

[584] N. Paton and C. Hamilton, "Microstructural influences on superplasticity in Ti-6Al-4V", *Metall. Trans.* A 10, pp. 241-250, 1979.

[585] A. Ghosh and C. Hamilton, "Mechanical behavior and hardening characteristics of a superplastic Ti-6Al-4V alloy", *Metall. Trans.* A 10, pp. 699-706, 1979.

[586] C. Koch, D. Morris, K. Lu, and A. Inoue, "Ductility of nanostructured materials", *MRS Bull.* 24, pp. 54-58, 1999.

[587] K. Kitazono and E. Sato, "Internal stress superplasticity in Al-Be eutectic alloy during triangular temperature profile", *Acta Mater.* 46, pp. 207-213, 1998.

[588] S. Pickard and B. Derby, "The deformation of particle reinforced metal matrix composites during temperature cycling", *Acta Metall. Mater.* 38, pp. 2537-2552, 1990.

[589] S. Pickard and B. Derby, in: 9th Risø International Symposium on Metallurgy and Materials Science, S. Andersen, H. Lilholt and O. Pedersen, Eds., Risø National Laboratory, Roskilde, Denmark, 1988, p. 447.

[590] M. Lozinsky, "Grain boundary adsorption and superhigh plasticity", *Acta Metall.* 9, pp. 689-694, 1961. *See Also*; M. Lozinsky and I. Simeonova, "Superhigh plasticity of commercial iron under cyclic fluctuations of temperature", *Acta Metall.* 7, pp. 709-715, 1959.

[591] P. Zwigl and D. Dunand, "Transformation superplasticity of iron and Fe/TiC metal matrix composites", *Metall. Mater. Trans.* A 29, pp. 565-575, 1998.

[592] T. Massalski, S. Bhattacharyya and J. Perepezko, "Enhancement of plastic flow during massive transformations", *Metall. Trans.* A 9, pp. 53-56, 1978.

[593] D. Oelschlagel and V. Weiss, "Superplasticity of Steels During the Ferrite-Austenite Transformation", *Trans. Am. Soc. Metals* (Trans. ASM) 59, pp. 143-154, 1966.

[594] Y. Takayama, N. Furushiro and S. Hori, in: Titanium Science and Technology, G. Lutjering, U. Zwicker and W. Bunk, Eds., Deutsche Gesellschaft fur Metallkunde (German Society for Metallurgy), Munich, 1985, pp. 753.

[595] N. Furushiro, H. Kuramoto, Y. Takayama and S. Hori, "Fundamental Characteristics of the Transformation Superplasticity in a Commercially-Pure Titanium", *Trans. ISIJ* 27, pp. 725-729, 1987.

[596] C. Schuh and D. Dunand, "Load transfer during transformation superplasticity of Ti-6Al-4V/TiB whisker-reinforced composites", *Scripta Mater.* 45, pp. 631-638, 2001.

[597] R. Kot, G. Krause and V. Weiss, in: The Science, Technology and Applications of Titanium, R. Jaffe and N. Promisel, Eds., Pergamon, Oxford. 1970, p. 597.

[598] K. Nuttall and D. McCooeye, in: Mechanical Behavior of Materials, The Society of Materials Science, Japan, 1974, p. 129.

[599] M. Mahoney and C. Bampton, "Fundamentals of Diffusion Bonding", in: ASM Handbook, Vol. 6; *Welding, Brazing, and Soldering*, D. Olson, T. Siewert, S. Liu, and G. Edwards, Eds., 1993, pp. 156-159.

[600] C. Paez, R. Messler, Design and Fabrication of Advanced Titanium Structures, American Institute of Aeronautics and Astronautics, AIAA-79 0757, 1979.

[601] F. Campbell, Manufacturing Technology for Aerospace Structural Materials, 2006 Elsevier Ltd.

[602] J. Gerken and W. Owzarski, A Review of Diffusion Welding, *Welding Research Council Bulletin* No. 109, Oct 1965.

[603] D. Hull and T. Clyne, in: An Introduction to Composite Materials, 2nd edn., Cambridge University Press, 1992, pp. 280.

[604] O. Buck, R. Thompson, D. Rehbein, D. Palmer and L. Brasche, "Contacting Surfaces: A Problem in Fatigue and Diffusion Bonding", *Metall. Trans.* A 20, pp. 627-635, 1989.

[605] M. Donachie, Jr., Titanium: A Technical Guide, 2nd edn., ASM International, 2000.

[606] A. Ogawa, H. Fukai and C. Ouchi, in: Beta Titanium Alloys in the 1990s, D. Eylon, R. Boyer and P. Koss, Eds., TMS, 1993, pp. 513.

[607] T. Imamura, Advanced materials and process technologies for aerospace structures, in: Aerospace Materials, B. Cantor, H. Assender and P. Grant, IoP Publishing Ltd., 2001, pp. 15-27.

[608] M. Ashby and R. Verrall, "Diffusion-Accommodated Flow and Superplasticity", *Acta Metall.* 21, pp. 149-163, 1973.

[609] U. Kocks, A. Argon, and M. Ashby, Thermodynamics of Slip, Pergamon Press, New York, 1975.

[610] B. Burton, in: Diffusion and Defect Monograph Series #3, Y. Adda, A. LeClaire, L. Slifkin, and F. Wohlbier, Eds., Trans. Tech. Pub., Ohio, 1977.

[611] M. Ashby, G. Edward, J. Davenport and R. Verrall, "Application of bound theorems for creeping solids and their application to large strain diffusional flow", *Acta Metall.* 26, pp. 1379-1388, 1978.

[612] J. Edington, "Microstructural aspects of superplasticity", *Metall. Trans.* A 13, pp. 703-715, 1982.

[613] A. Mukherjee, in: Plastic Deformation and Fracture of Materials, R. Cahn, P. Haasen, and E. Kramer, Eds., Materials Science and Technology, Vol. 6, H. Mughrabi, Ed., VCH Publishers, Weinheim, 1993, p. 407.

[614] J. Karch, R. Birringer and H. Gleiter, "Ceramics ductile at low-temperature", *Nature* 330, pp. 556-558, 1987.

[615] P. Sanders, M. Rittner, E. Kiedaisch, J. Weertman, H. Kung, and Y. Lu, "Creep of nanocrystalline Cu, Pd, and Al-Zr", *Nanostr. Mater.* 9, pp. 433-440, 1997.

[616] J. Bird, A. Mukherjee and J. Dorn, "Correlations between High-Temperature Creep Behavior and Structure", Proc. Symp. Quantitative Relation Between Properties and Microstructure, D. Brandon and A. Rosen, Eds., Israel University Press, 1969, pp. 22.

[617] A. Mukherjee, J. Bird and J. Dorn, "Experimental correlation for high-temperature creep", *Trans. ASM* 62, pp. 155-179, 1969.

[618] A. Sergueeva, V. Stolyarov, R. Valiev and A. Mukherjee, "Superplastic behaviour of ultrafine-grained Ti-6Al-4V alloys", *Mater. Sci. Eng.* A 323, pp. 318-325, 2002.

[619] K. Padmanabhan and H. Gleiter, "Optimal structural superplasticity in metals and ceramics of microcrystalline- and nanocrystalline-grain sizes", *Mater. Sci. Eng.* A 381, pp. 28-38, 2004.

[620] J. Markmann, P. Bunzel, H. Rösner, K. Liu, K. Padmanabhan, R. Birringer, H. Gleiter and J. Weissmüller, "Microstructure evolution during rolling of inert-gas condensed palladium", *Scripta Mater.* 49, pp. 637-644, 2003.

[621] Gouthama and K. Padmanabhan, "Transmission electron microscopic evidence for cavity nucleation during superplastic flow", *Scr. Mater.* 49, pp. 761-766, 2003.

[622] K. Padmanabhan and J. Schlipf, "A model for grain boundary sliding and its relevance to optimal structural superplasticity - I. Theory", *Mater. Sci. Technol.* 12, pp. 391-399, 1996.

[623] K. Kannan, C. Johnson, and C. Hamilton, "The Role of Flow Properties and Damage Accumulation in Superplastic Ductility of Al-Mg-Mn Alloys", *Mater. Sci. Forum* 243-245, pp. 125-130, 1997.

[624] K. Padmanabhana and M. Basariya, "Mesoscopic grain boundary sliding as the rate controlling process for high strain rate superplastic deformation", *Mater. Sci. Eng.* A 527, pp. 225-234, 2009.

[625] K. Padmanabhan, "Mechanical properties of nanostructured materials", *Mater. Sci. Eng.* A 304-306, pp. 200-205, 2001.

[626] C. Zener, Elasticity and Anelasticity of Metals, University of Chicago Press, Chicago, 1948.

[627] F. McClintock and A. Argon, Mechanical Behavior of Materials, Addison-Wesley, 1966.

[628] L. Goodman, "Material Damping and Slip Damping", in: Harris Shock and Vibration Handbook, C. Harris and A. Piersol, Eds., 5th Ed., McGraw-Hill, 2002.

[629] T. Kê, "Internal Friction and Precipitation from the Solid Solution of N in Tantalum", *Phys. Rev.* 74, pp. 914-916, 1948.

[630] T. Kê, "Micro-mechanism of grain boundary relaxation in metals", *Scripta Metall. Mater.* 24, pp. 347-352, 1990.

[631] X. Guan and T. Kê, "Non-linear mechanical relaxation associated with the viscous sliding of grain boundaries in aluminium bicrystals", *J. Alloys Comp.* 211-212, pp. 480-483, 1994.

[632] T. Kê, R. Mehl and A. Medalist, "Fifty-year study of grain-boundary relaxation", *Met. Mat. Trans.* A 30, pp. 2267-2295, 1999.

[633] J. Ide, "Some dynamic methods of determination of Young's modulus", *Rev. Sci. Instr.* 6, pp. 296-298, 1935.

[634] M. Kamal, M. El-Tonsy, I. Fouda, M. Radwan and H. Hosny, "Structure changes and mechanical properties of rapidly solidified $Fe_{85}B_{15}$ melt spun alloy", *The Bulletin of the Faculty of Science Mansoura University* 20, pp. 1-18, 1993.

[635] S. Wiederhorn, R. Fields, S. Low, G. Bahng, A. Wehrstedt, J. Hahn, Y. Tomota, T. Miyata, H. Lin and B. Freeman, in: Springer Handbook of Materials Measurement Methods, H. Czichos, T. Saito and L. Smith, Eds., Springer Science+Business Media, Inc., 2006, pp. 283.

[636] M. Blanter, I. Golovin, H. Neuhäuser and H. Sinning, Internal Friction in Metallic Materials: A Handbook, Springer-Verlag Berlin Heidelberg, 2007.

[637] J. Woirgard, J. Amirault, J. DeFouquet, in: Proc. 5th Intern. Conf. Internal Friction and Ultrasonic Attenuation in Crystalline Solids (ICIFUAS-5), D. Lenz, K. Lücke, Eds., Vol. 1, Springer, Berlin, 1975, pp. 392.

[638] A. Rivière, "High temperature relaxation in single crystals", *Scr. Mater.* 43, pp. 991-995, 2000.

[639] J. Li, in: Rapidly Solidified Alloys: Processes, Structures, Properties and Applications, H. Liebermann, Ed., Marcel Dekker Inc., 1993, pp. 379, and references therein.

[640] NASA Fact Sheets (NASA-FS); FS-2004-08-011-KSC, FS-2004-03-005-KSC, FS-2004-03-005-KSC, and FS-2001-03-60-MSFC, www.nasa.gov.

[641] J. Groza, in: Nanostructured Materials: Processing, Properties and Potential Applications, C. Koch, Ed., Noyes Publications 2002, pp. 115.

[642] A. Manthiram, D. Bourell and H. Marcus, "Nanophase Materials in Solid Freeform Fabrication", *J. Metals* (JOM) 45, pp.66-70, 1993.

[643] M. Mayo and D. Hague, "Porosity-grain growth relationships in the sintering of nanocrystalline ceramics", *Nanostr. Mater.* 3, pp. 43-52, 1993.

[644] M. Mayo, "Processing of Nanocrystalline Ceramics from Ultrafine Particles", *Int. Mater. Rev.* 41, pp. 85-115, 1996.

[645] H. Gleiter, in: Physical Metallurgy, R. Cahn and P. Haasen, Vol. 1, North-Holland and Elsevier Science B.V., 1996, pp. 843.

[646] N. Anoshkin and G. Demchenkov, "Material science and technological aspects of rapidly solidified titanium alloy production", *Mater. Sci. Eng.* A 243, pp. 263-268, 1998.

[647] W. Schlump and J. Willbrand, *VDI Nachrichten* 917, pp. 23, 1992; cited in Ref. 645.

[648] J. Newkirkn, in: Handbook of Aluminum, Vol. 1: Physical Metallurgy and Processes, G. Totten and D. MacKenzie, Eds., Marcel Dekker, Inc., 2003, pp. 1251.

[649] S. Kang, Sintering: Densification, Grain Growth and Microstructure, Elsevier Butterworth-Heinemann Pub., 2005.

[650] J. Hansen, R. Rusin, M. Teng and D. Johnson, "Combined-Stage Sintering Model", *J. Am. Ceram. Soc.* 75, pp. 1129-1135, 1992.

[651] R. Porat, S. Berger and R. Rosen, "Dilatometric Study of the Sintering Mechanism of Nanocrystalline Cemented Carbides", *Nanostr. Mater.* 7, pp. 429-436, 1996.

[652] H. Sanderow, in: *Powder Metal Technologies and Applications*, ASM Handbook, Vol. 7, W. Eisen, B. Ferguson, R. German, R. Iaccoca, P. Lee, D. Madan, K. Moyer, H. Sanderow and Y. Trudel, Eds., 1998, pp. 9-15.

[653] H. Exner and E. Aut, in: Physical Metallurgy, R. Cahn and P. Haasen, Eds., Vol. 3, North-Holland and Elsevier Science B.V., 1996, pp. 2627.

[654] H. Hahn, J. Logas and R. Averbach, "Sintering Characteristics of Nanocrystalline TiO_2", *J. Mater. Res.* 5, pp. 609-614, 1990.

[655] G. Thuenissen, A. Winnubst and A. Burggraaf, "Sintering Kinetics and Microstructure Development of Nanoscale Y-TZP Ceramics", *J. Europ. Ceram. Soc.* 11, pp. 319-324, 1993.

[656] J. Rankin and B. Sheldon, "In Situ TEM Sintering of Nano-Sized ZrO_2 Particles", *Mater. Sci. Eng.* A 204, pp. 48-53, 1995.

[657] J. Bonevics and L. Marks, "The Sintering Behavior of Ultrafine Alumina Particles", *J. Mater Res.* 7, pp. 1489-1500, 1992.

[658] R. Averback, H. Zhu, R. Tao and H. Hofler, in: Synthesis and Processing of Nanocrystalline Powder, D. Bourell, Ed., TMS, Warrendale, PA, 1996, pp. 203.

[659] G. Messing and M. Kumagai, "Low-Temperature Sintering of α-Alumina Seeded Bohemite Gels", *Am. Ceram. Bull.* 73, pp. 88-91, 1994.

[660] G. Shaik and W. Milligan, "Consolidation of Nanostructured Metal Powders by Rapid Forging: Processing, Modeling, and Subsequent Mechanical Behavior", *Met. Mater. Trans.* A 28, pp. 895-904, 1997.

[661] E. Gutmanas, "Materials with Fine Microstructures by Advanced Powder Metallurgy", *Prog. Mater. Sci.* 34, pp. 261-366, 1990.

[662] H. Zhu and R. Averback, "Sintering Process of Two Nanoparticles: A Study by Molecular Dynamics Simulations", *Phil. Mag. Lett.* 73, pp.27-33, 1996.

[663] H. Zhu and R. Averback, "Sintering of Nano-Particle Powders: Simulations and Experiments", *Mater. Manuf. Processes* 11, pp. 905-923, 1996.

[664] L. Trusov, V. Lapovok and V. Novikov, in: Science of Sintering, D. Uskokovic, H. Plamour III, and R. Spriggs, Eds., Plenum Press, New York, 1989, pp. 185.

[665] R. Valiev, "Approach to Nanostructured Solids through the Studies of Submicron Grained Polycrystals", *Nanostr. Mater.* 6, pp. 73-82, 1995.

[666] J. Horváth, "Diffusion in Nanocrystalline Materials", *Def. Diff. Forum* (Defect and Diffusion Forum) 66-69, pp. 207-228, 1989.

[667] S. Schumacher, R Birringer, R Strauß and H Gleiter, "Diffusion of silver in nanocrystalline copper between 303 and 373 °K", *Acta Metall.* 37, pp. 2485-2488, 1989.

[668] B. Bokstein, H. Bröse, L. Trusov and T. Khvostantseva, "Diffusion in nanocrystalline nickel", *Nanostr. Mater.* 6, pp. 873-876, 1995.

[669] E. Ivanov, "Preparation of Cu_xHg_{1-x} Solid Solutions by Mechanical Alloying", *Mater. Sci. Forum* 88-90, pp. 475-480, 1992.

[670] R. Siegel and H. Hahn, in: Current Trends in the Physics of Materials, M. Youssouff, Eds., World Sci. Pub. Co., Singapore, 1987, p. 403.

[671] C. Suryanarayana and C. Koch, "Nanocrystalline materials - Current research and future directions", *Hyperfine Interactions* 130, pp. 5-44, 2000.

[672] H. Gleiter, "Our thoughts are ours, their ends none of our own: Are there ways to synthesize materials beyond the limitations of today?", *Acta Mater.* 56, pp. 5875-5893, 2008.

[673] D. Şopu, K. Albe, Y. Ritter and H. Gleiter, "From nanoglasses to bulk massive glasses", *Appl. Phys. Lett.* 94, pp. 191911-3, 2009.

[674] H. Exner, "Principles of Single Phase Sintering", *Rev. Powder Metallurg. Phys. Ceram.* 1, pp. 7-251, 1979.

[675] M. Yeadon, J. Yang, M. Ghaly, D. Olynick, R. Averback and J. Gibson, in: Nanophase and Nanocomposite Materials, S. Komarneni, J. Parker and H. Wollenberger, Eds., MRS, Pittsburgh, 1997, pp. 179.

[676] A. Frisch, W. Kaysser and G. Petzow, in: Science of Sintering: New Directions for Materials Processing and Control, D. Uskokovic, H. Palmour, R. Spriggs, Eds., Plenum Press, New York, 1989, pp. 311.

[677] J. Adlerborn H. Larker, J. Nilsson and B. Mattsson, "Method of Manufacturing an Object of a Powdered Material by Isostatic Pressing", US Patent No. 4,568,516, Feb. 4, 1986.

[678] V. Kodash, J. Groza, K. Cho, B. Klotz and R. Dowding, "Field-assisted sintering of Ni nanopowders", *Mater. Sci. Eng.* A 385, pp. 367-371, 2004.

[679] J. Ahn, M. Huh and J. Park, "Effect of Green Density on Subsequent Densification and Grain Growth of Nanophase SnO_2 Powder during Isothermal Sintering", *Nanostr. Mater.* 8, pp. 637-643, 1997.

[680] C. Suryanarayana, in: *Powder Metal Technologies and Applications*, ASM Handbook, Vol. 7, W. Eisen, B. Ferguson, R. German, R. Iaccoca, P. Lee, D. Madan, K. Moyer, H. Sanderow and Y. Trudel, Eds., 1998, pp. 80.

[681] M. Mayo, "Grain Growth and the Processing of Nanocrystalline Ceramics", *Mater. Sci. Forum* 204-206, pp. 389-398, 1996.

[682] A. Khoei, Computational Plasticity in Powder Forming Processes, Elsevier Ltd., 2005.

[683] G. Skandan, "Processing of Nanostructured Zirconia Ceramics", Nanostr. Mater. 5, pp. 111-126, 1995.

[684] G. Skandan, H. Hahn, B. Kear, M. Roddy and R. Cannon, "The Effect of Applied Stress on Densification of Nanostructured Zirconia during Sinter-Forging", *Matt. Lett.* 20, pp. 305-309, 1994.

[685] J. Jamnik and R. Raj, "Space-Charge-Controlled Diffusional Creep: Volume Diffusion Case", *J. Amer. Ceram. Soc.* 79, pp. 193-198, 1996.

[686] Z. Munir, U. Anselmi-Tamburini and M. Ohyanagi, "The effect of electric field and pressure on the synthesis and consolidation of materials: A review of the spark plasma sintering method", *J. Mater. Sci.* 41, pp. 763-777, 2006.

[687] R. Raj, "Analysis of the sintering pressure", *J. Am. Ceram. Soc.* 70, C-210-C-211, 1987.

[688] A. Pannikkat and R. Raj, "Measurement of an electrical potential induced by normal stress applied to the interface of an ionic material at elevated temperatures", *Acta Mater.* 47, pp. 3423-3431, 1999.

[689] J. Carsley, J. Ning, W. Milligan, S. Hackney and E. Aifantis, "A simple, mixtures-based model for the grain size dependence of strength in nanophase metals", *Nanostr. Mater.* 5, pp. 441-448, 1995.

[690] J. Groza and R. Dowding, "Nanoparticulate Materials Densification", *Nanostr. Mater.* 1, pp. 749-768, 1996.

[691] T. Allen, Powder Sampling and Particle Size Determination, Elsevier B. V., 2003.

[692] A. Latapie and D. Farkas, "Effect of grain size on the elastic properties of nanocrystalline alpha-iron", *Scripta Mater.* 48, pp. 611-615, 2003.

[693] A. Guinier, X Ray Diffraction in Crystals, Imperfect Crystals and Amorphous Bodies, W. H. Freeman and Company, USA, 1963.

[694] T. Allen, Particle Size Measurement, 5th edn. , 2 Volumes, Chapman and Hall, Springer, 1996 and 1997.

[695] H. Merkus, Particle Size Measurements: Fundamentals, Practice, Quality, Springer Science+Business Media B.V., 2009.

[696] D. Jia, K. Ramesh, and E. Ma, "Effects of nanocrystalline and ultrafine grain sizes on constitutive behavior and shear bands in iron", *Acta Mater.* 51, pp. 3495-3509, 2003.

[697] M. Rittner, J. Weertman and J. Eastman, "Structure-Property Correlations in Nanocrystalline Al-Zr Alloy Composites", *Acta Mater.* 44, pp. 1271-1286, 1996.

[698] G. Thomas, R. Siegel, J. Eastman, "Grain boundaries in nanophase palladium: High resolution electron microscopy and image simulation", *Scripta Metall. Mater.* 24, pp. 201-206, 1990.

[699] U. Herr, J. Jing, R. Birringer, U. Gonser and H. Gleiter, "Investigation of nanocrystalline iron materials by Mössbauer spectroscopy", *Appl. Phys. Lett.* 50, pp. 472-474, 1987.

[700] M. Trudeau, A. Tschöpe and J. Ying, "XPS Investigation of Surface Oxidation and Reduction in Nanocrystalline $Ce_xLa_{1-x}O_{2-y}$", *Surface Interf. Analysis* 23, pp. 219-226, 1995.

[701] M. El-Eskandarany, Mechanical Alloying for Fabrication of Advanced Engineering Materials, Noyes Publications New York, USA, 2001.

[702] C. Suryanarayana, Mechanical Alloying and Milling, Marcel Dekker New York, 2004.

[703] D. Olynick, J. Gibson, and R. Averbach, "*In Situ* Ultra-High Vacuum Transmission Electron Microscopy Studies of Nanocrystalline Copper", *Mat. Sci. Eng.* A 204, pp. 54-58, 1995.

[704] D. L. Olynick, J. Gibson and R. Averbach, "Trace Oxygen Effects on Copper Nanoparticle Size and Morphology", *Appl. Phys. Lett.* 68, pp. 343-345, 1996.

[705] M. Payne, in: Mechanical Properties and Deformation Behavior of Materials Having Ultra-Fine Microstructures, M. Nastasi, D. Markin, and H. Gleiter, Eds., Kluwer Academic Pub., Dordrecht, Netherlands, 1993, pp. 37.

[706] T. Rabe and R. Wasche, "Sintering Behavior of Nanocrystalline Titanium Nitride Powders", *Nanostr. Mater.* 6, pp. 357-360, 1995.

[707] M. Eldrup, M. Bentzon, A. Pedersen, S. Linderoth, N. Pedersen and B. Larsen, in: Mechanical Properties and Deformation Behavior of Materials Having Ultra-Fine Microstructures, M. Nastasi, D. Parkin and H. Gleiter, Eds., Kluwer Academic Pub., Dordrecht, Netherlands, 1993, pp. 571.

[708] G. Chen, C. Suryanarayana, and F. Froes, "Structure of mechanically alloyed Ti-Al-Nb powders", *Metall. Mater. Trans.* A 26, pp. 1379-1387, 1995.

[709] T. Klassen, M. Oehring and R. Bormann, "Microscopic mechanisms of metastable phase formation during ball milling of intermetallic TiAl phases", *Acta Mater.* 45, pp. 3935-3948, 1997.

[710] J. Estrada, J. Duszczyk and B. Korevaar, "Heating Sequences and Hydrogen Evolution in Alloyed Aluminum Powders", *J. Mater. Sci.* 26, pp. 1631-1634, 1991.

[711] A. Ziani and S. Pelletier, "Supersolidus Liquid-Phase Sintering Behavior of Degassed 6061 Al Powder", *Int. J. Powder Metall.* 35, pp. 49-58, 1999.

[712] W. Kawamura, A. Inoue and T. Masumoto, "Mechanical properties of amorphous alloy compacts prepared by a closed processing system", *Scripta Met. Mater.* 29, pp. 25-30, 1993.

[713] P. Beiss, "Forming", in: Group VIII: Advanced Materials and Technologies, Vol. 2: Materials, Subvolume A: Powder Metallurgy Data Part 1: Metals and Magnets, P. Beiss, R. Ruthardt, H. Warlimont, Eds., Landolt-Börnstein Numerical Data and Functional Relationships in Science and Technology New Series, Springer-Verlag Berlin, Heidelberg, Germany, 2003, pp. (3-1)-(3-36).

[714] V. Viswanathan, T. Laha, K. Balani, A. Agarwal and S. Seal, "Challenges and advances in nanocomposite processing techniques", *Mater. Sci. Eng.* R (*Reports*) 54, pp. 121-285, 2006.

[715] B. Murphy and T. Courtney, "Synthesis of Cu-NbC Nanocomposites by Mechanical Alloying", *Nanostr. Mat.* 4, pp. 365-369, 1994.

[716] R. German, Powder Metallurgy of Iron and Steel, John Wiley and Sons Inc., New York, 1998.

[717] J. Nilsson and B. Mattsson, "Method for Isostatic or Pseudo-Isostatic Pressing Employing a Surrounding Casing of Glass", U.S. Patent No. 5,284,616, Feb. 8, 1994.

[718] T. Yamamoto, A. Otsuki, K. Ishihara and P. Shingu, "Synthesis of near net shape high density TiB-Ti composite", *Mater. Sci. Eng.* A 239-240, pp. 647-651, 1997.

[719] A. Hostatter, W. Lichti and J. Papp, "Powder Metal Consolidation of Multiple Preforms", US Patent No. 4,594,219, Jun. 10, 1986.

[720] P. Feng, W. Xiong, L. Yu, Y. Zheng and Y. Xia "Phase evolution and microstructure characteristics of ultrafine Ti(C,N)-based cermet by spark plasma sintering", *Intern. J. Refractory Metals and Hard Materials* 22, pp. 133-138, 2004.

[721] U. Anselmi-Tamburini, Y. Kodera, M. Gasch, C. Unuvar, Z. Munir, M. Ohyanagi and S. Johnson, "Synthesis and characterization of dense ultra-high temperature thermal protection materials produced by field activation through spark plasma sintering (SPS): I. Hafnium Diboride, *J. Mater. Sci.* 41, pp. 3097-3104, 2006.

[722] E. Gaffet and G. Le Caër, "Mechanical Processing for Nanomaterials", in: *Encyclopedia of Nanoscience and Nanotechnology*, H. Nalwa, Ed., American Scientific Publishers, Vol. 5, pp. 91-129, 2004.

[723] R. Dobedoe, G. West and M. Lewis, "Spark plasma sintering of ceramics, *Bull. Eu. Ceram. Soc.* 1, pp. 19-24, 2003.

[724] W. Chen, U. Anselmi-Tamburini, J. Garay, J. Groza and Z. Munir, "Fundamental investigations on the spark plasma sintering/synthesis process I. Effect of dc pulsing on reactivity", *Mater. Sci. Eng.* A 394, pp. 132-138, 2005.

[725] ASTM standards: ASTM A 989-98 R02 (2002): Standard Specification for Hot Isostatically-Pressed Alloy Steel Flanges, Fittings, Valves, and Parts for High Temperature Service.

[726] ASTM standards: ASTM A 988-98 R02 (2002): Standard Specification for Hot Isostatically-Pressed Stainless Steel Flanges, Fittings, Valves, and Parts for High Temperature Service.

[727] ASTM standards: ASTM B 834-95 R03 (2003): Standard Specification for Pressure Consolidated Powder Metallurgy Iron-Nickel-Chromium-Molybdenum (UNS N08367) and Nickel-Chromium-Molybdenum-Columbium (Nb) (UNS N06625) Alloy Pipe Flanges, Fittings, Valves, and Parts.

[728] R. Averback, H. Hofler, H. Hahn and J. Logas, "Sintering and Grain Growth in Nanocrystalline Ceramics", *Nanostr. Mater.* 1, pp. 173-178, 1992.

[729] D. Hague and M. Mayo, "Modeling densification during sinter-forging of yttria-partially-stabilized zirconia", *Mater. Sci. Eng.* A 204, pp. 83-89, 1995.

[730] D. Bourell and J. Groza, in: *Powder Metal Technologies and Applications*, ASM Handbook, Vol. 7, W. Eisen, B. Ferguson, R. German, R. Iaccoca, P. Lee, D. Madan, K. Moyer, H. Sanderow and Y. Trudel, Eds., 1998, pp. 583.

[731] S. Singh, A. Jha and Kumara, "Dynamic effects during sinter forging of axi-symmetric hollow disc preforms", *Inter. J. Machine Tools and Manuf.* 47, pp. 1101-1113, 2007.

[732] J. Duva and P. Crow, "The densification of powders by power-law creep during hot isostatic pressing", *Acta Metall. Mater.* 40, pp. 31-35, 1992.

[733] P. Chandrasekhar and S. Singh, "Investigation of dynamic effects during cold upset-forging of sintered aluminium truncated conical preforms", *J. Mater. Proc. Tech.* 211, pp. 1285-1295, 2011.

[734] N. Boucharat, R. Hebert, H. Rösner, R. Valiev and G. Wilde, "Nanocrystallization of amorphous $Al_{88}Y_7Fe_5$ alloy induced by plastic deformation", *Scripta Mater.* 53, pp. 823-828, 2005.

[735] N. Boucharat, R. Hebert, H. Rösner, R. Valiev and G. Wilde, "Synthesis routes for controlling the microstructure in nanostructured $Al_{88}Y_7Fe_5$ alloys", *J. Alloys Comp.* 434-435, pp. 252-254, 2007.

[736] G. Wilde, "Review: Nanostructures and nanocrystalline composite materials - synthesis, stability and phase transformations", *Surf. Interface Anal.* 38, pp. 1047-1062, 2006.

[737] O. Senkov, S. Senkova, J. Scott and D. Miracle, "Compaction of amorphous aluminum alloy powder by direct extrusion and equal channel angular extrusion", *Mat. Sci. Eng.* A 393, pp. 12-21, 2005.

[738] V. Stolyarov, Y. Zhu, T. Lowe, R. Islamgaliev and R. Valiev, "Processing nanocrystalline Ti and its nanocomposites from micrometer-sized Ti powder using high pressure torsion", *Mater. Sci. Eng.* A 282, pp. 78-85, 2000.

[739] T. Shimokawa, T. Hiramoto, T. Kinari and S. Shintaku, "Effect of Extrinsic Grain Boundary Dislocations on Mechanical Properties of Ultrafine-Grained Metals by Molecular Dynamics Simulations", *Mater. Trans. JIM* 50, pp. 2-10, 2009.

[740] C. Xu, M. Furukawa, Z. Horita and T. Langdon, "Achieving a superplastic forming capability through severe plastic deformation", *Adv. Eng. Mater.* 5, pp. 359-364, 2003.

[741] S. Tjong and H. Chen, "Nanocrystalline materials and coatings", *Mater. Sci. Eng.* R 45, pp. 1-88, 2004.

[742] A. Shpak, V. Varyukhin, V. Tkatch, V. Maslov, Y. Beygelzimer, S. Synkov, V. Nosenko and S. Rassolov, "Nanostructured $Al_{86}Gd_6Ni_6Co_2$ bulk alloy produced by twist extrusion of amorphous melt-spun ribbons", *Mater. Sci. Eng.* A 425, pp. 172-177, 2006.

[743] V. Stolyarov, Y. Beygel'zimer, D. Orlov and R. Valiev, "Refinement of Microstructure and Mechanical Properties of Titanium Processed by Twist Extrusion and Subsequent Rolling", *Phys. Met. Metallogr.* (The Physics of Metals and Metallography) 99, pp. 204-211, 2005.

[744] D. Orlov, Y. Beygelzimer, S. Synkov, V. Varyukhin, N. Tsuji and Z. Horita, "Microstructure Evolution in Pure Al Processed with Twist Extrusion", *Mater. Trans. JIM* 50, pp. 96-100, 2009.

[745] Y. Beygelzimer, V. Varyukhin, S. Synkov and D. Orlov, "Useful properties of twist extrusion", *Mater. Sci. Eng.* A 503, pp. 14-17, 2009.

[746] C. Suryanarayana, G. Korth and F. Froes, "Compaction and Characterization of Mechanically Alloyed Nanocrystalline Titanium Aluminides", *Met. Mater. Trans.* A 28, pp. 293-302, 1997.

[747] S. Glade and N. Thadhani, "Shock Consolidation of Mechanically Alloyed Amorphous Ti-Si Powders", *Met. Mater. Trans.* A 26, pp. 2565-2569, 1995.

[748] T. Chen, J. Hampikian and N. Thadhani, "Synthesis and Characterization of Mechanically Alloyed and Shock-Consolidated Nanocrystalline NiAl Intermetallic", *Acta Mater.* 47, pp. 2567-2579, 1999.

[749] G. Korth and R. Williamson, "Dynamic Consolidation of Metastable Nanocrystalline Powders", *Metal. Mater. Trans.* A 26, pp. 2571-2578, 1995.

[750] K. Hokamoto, S. Tanaka, M. Fujita, S. Itoh, M. Meyers and H. Chen, "High temperature shock consolidation of hard ceramic powders", *Physica* B 239, pp. 1-5, 1997.

[751] S. Takeda and H. Shishiba, "Method of Producing Sintered Ceramic Materials", US Patent No. 5,139,720, Aug. 18, 1992.

[752] J. Flinn and G. Korth, "Die-Target for Dynamic Powder Consolidation", US Patent No. 4,599,060, Jul. 8, 1986.

[753] A. Niiler, G. Moss and R. Eichelberger, "Fabrication of Ceramics by Shock Compaction of Materials Prepared by Combustion Synthesis", US Patent No. 5,114,645, May 19, 1992.

[754] M. Meyers, D. Benson and E. Olevsky, "Shock Consolidation: Microstructurally-Based Analysis and Computational Modeling", *Acta Mater.* 47, pp. 2089-2108, 1999.

[755] S. Ando, Y. Mine, K. Takashima, S. Itoh and H. Tonda, "Explosive compaction of Nd-Fe-B powder", *J. Mater. Proc. Tech.* 85, pp. 142-147, 1999.

[756] W. Gourdin, "Energy Deposition and Microstructural Modification in Dynamically Consolidated Metal Powders", *J. Appl. Phys.* 55, pp. 172-181, 1984.

[757] A. Mamalis, I. Vottea and D. Manolakos, "On the modelling of the compaction mechanism of shock compacted powders", *J. Mater. Proc. Tech.* 108, pp. 165-178, 2001.

[758] N. Thadhani, R. Graham, T. Royal, E. Dunbar, M. Anderson and G. Holman, "Shock-induced chemical reactions in titanium-silicon powder mixtures of different morphologies: Time-resolved pressure measurements and materials analysis", *J. Appl. Phys.* 82, pp. 1113-28, 1997.

[759] D. Eakins and N. Thadhani, "Mesoscale simulation of the configuration-dependent shock compression response of Ni + Al powder mixtures", *Acta Mater.* 56, pp. 1496-1510, 2008.

[760] S. B. Bhaduri and S. Bhaduri, in: Non-equilibrium Processing of Materials, C. Suryanarayana, Ed., Pergamon Materials Series Vol. 2, Elsevier Science Ltd., 1999, pp. 289-313.

[761] L. Kecskes, Hot Explosive Consolidation of W-Ti Alloys: Microstructural Effects, *Army Research Laboratory* (ARL) TR-2615, Aberdeen Proving Ground, MD 21005-5069, USA, November 2001.

[762] J. Li, S. Meng, J. Han and B. Wang, "Energy and Deformation during Explosive Compaction of ZrB_2-SiC Ultrahigh Temperature Ceramics", *Scholarly Research Exchange* 2008, Article ID: 754838.

[763] A. Mamalis, "Near net-shape manufacturing of bulk ceramic high-T_c superconductors for application in electricity and transportation", *J. Mater. Proc. Tech.* 108, pp. 126-140, 2001.

[764] A. Mamalis, "Technological aspects of high-T_c superconductors", *J. Mater. Proc. Tech.*, 99, pp. 1-31, 2000.

[765] A. Mamalis, A. Szalay, D. Pantelis, G. Pantazopoulos, I. Kotsis and M. Enisz, "Fabrication of multi-layered steel/superconductive ceramic (Y-Ba-K-Cu-O)/silver rods by explosive powder compaction and extrusion", *J. Mater. Proc. Tech.* 57, pp. 155-163, 1996.

[766] A. Zurek and M. Meyers, in: High-Pressure Shock Compression of Solids II: Dynamic Fracture and Fragmentation, L. Davison, D. Grady and M. Shahinpoor, Eds., 1996 Springer-Verlag New York, Inc., pp. 25.

[767] S. Chengwei, Z. Shiming, W. Yanping and L. Cangli, in: High-Pressure Shock Compression of Solids II: Dynamic Fracture and Fragmentation, L. Davison, D. Grady and M. Shahinpoor, Eds., Springer-Verlag New York, Inc., 1996, pp. 71.

[768] M. Lewis and H. Schreyer, in: High-Pressure Shock Compression of Solids II: Dynamic Fracture and Fragmentation, L. Davison, D. Grady and M. Shahinpoor, Eds., 1996 Springer-Verlag New York, Inc., pp. 452.

[769] A. Ivnnov, in: High-Pressure Shock Compression of Solids VII: Shock Waves and Extreme States of Matter, V. Fortov, R. Trunin, L. Al'tshuler and A. Funtikov, Eds., Springer-Verlag New York, Inc., 2004, pp. 491.

[770] A. Farinha, R. Mendes, J. Baranda, R. Calinas and M. Vieira, "Behavior of explosive compacted/consolidated of nanometric copper powders", *J. Alloys Compounds* 483, pp. 235-238, 2009.

[771] G. Venz, P. Killen and N. Page, "Hot shock compaction of nanocrystalline alumina", *J. Mater. Sci.* 38, pp. 2935-2944, 2003.

[772] Z. Jin, C. Rockett, J. Liu, K. Hokamoto and N. Thadhani, "Underwater Explosive Shock Consolidation of Nanocomposite $Pr_2Fe_{14}B/\alpha$-Fe Magnetic Powders", *Mater. Trans. JIM* 46, pp. 1-4, 2005.

[773] M. Zohoor and A. Mehdipoor, "Explosive compaction of tungsten powder using a converging underwater shock wave", *J. Mater. Proc. Tech.* 209, pp. 4201-4206, 2009.

[774] K. Hokamoto, S. Shang, L. Yu and M. Meyers, in: Proc. Shock Wave and High Strain Rate Phenomena in Material, M. A. Meyers, L. Murr and K. Staudhammer, Eds., Marcel Dekker Inc., pp. 453-462, 1992.

[775] K. Hokamoto, S. Tanaka, M. Fujita, R. Zhang, T. Kodama, T. Awano and Y. Ujimoto, "An improved high temperature shock compression and recovery system using underwater shock wave for dynamic compaction of powders", *J. Mater. Process. Technol.* 85, pp. 153-157, 1999.

[776] L.J. Kecskes, "Hot explosive compaction of tungsten-titanium and molybdenum-titanium alloys", *J. Metal Powder Rep.* 52, pp. 38, 1997.

[777] K. Hokamoto, Y. Ujimoto and M. Fujita, "Possibility of Industrial Application by the Use of Underwater Shock Wave for Explosive Welding of Various Plates", *Mater. Sci. Forum* 426-432, pp. 4039-4044, 2003.

[778] K. Raghukandan, K. Hokamoto, J. Lee, A. Chiba and B. Pai, "An investigation on underwater shock consolidated carbon fiber reinforced Al composites", *J. Mater. Process. Technol.* 134, pp. 329-337, 2003.

[779] N. Page, in: Shock Waves and Shock Tubes XV, D. Bershader and R. Hanson, Eds., Stanford University Press, 1986, p. 878.

[780] M. Baer and W. Trott, in: Shock Compression of Condensed Matter, M. Furnish, N. Thadhani and Y. Horie, Eds., American Institute of Physics, Woodbury, NY, 2001, pp. 713-716.

[781] R. Graham, Solids under High-Pressure Shock Compression: Mechanics, Physics, and Chemistry, Springer-Verlag New York, 1993.

[782] T. Vogler, M. Lee and D. Grady, "Static and dynamic compaction of ceramic powders", *Intern. J. Solids Structures* 44, pp. 636-658, 2007.

[783] J. Millett, N. Bourne, G. Gray III and I. Jones, "The response of TiAl based alloys to one dimensional shock Loading", *Acta Mater.* 50, pp. 4801-4811, 2002.

[784] T. Richmond and H. Güntherodt, in: Rapidly Solidified Alloys: Processes, Structures, Properties and Applications, H. Liebermann, Ed., Marcel Dekker Inc., 1993, pp. 431.

[785] M. Howson, "Incipient localisation and electron-electron correlation effects in metallic glass alloys", *J. Phys. F: Metal Phys.* 14, pp. L25-L31, 1984.

[786] M. Howson and D. Greig, "Temperature dependence of conductivity arising from electron-electron interaction effects in amorphous metals", *Phys. Rev.* B 30, pp. 4805-4806, 1984.

[787] U. Mizutani and T. Matsuda, "Resistivity, hall coefficient and thermopower in $Mg_{70}Cu_{30-x}Zn_x$ ($0 \leq x \leq 30$) metallic glasses", *J. Non-Cryst. Solids* 94, pp. 345-352, 1987.

[788] R. Cochrane and J. Ström-Olsen, "Scaling behavior in amorphous and disordered metals", *Phys. Rev.* B 29, pp. 1088-1090, 1984.

[789] G. Hertel, D. Bishop, E. Spencer, J. Rowell and R. Dynes, "Tunneling and Transport Measurements at the Metal-Insulator Transition of Amorphous Nb: Si", *Phys. Rev. Lett.* 50, pp. 743-746, 1983.

[790] T. Rosenbaum, K. Andres, G. Thomas, and R. Bhatt, "Sharp Metal-Insulator Transition in a Random Solid", *Phys. Rev. Lett.* 45, pp. 1723-26, 1980.

[791] R. Dynes and J. Garno, "Metal-Insulator Transition in Granular Aluminum", *Phys. Rev. Lett.* 46, pp. 137-140, 1981.

[792] R. Morel, Y. Huai and R. Cochrane, "Resistivity and Hall effect in sputtered NiZr metallic glasses", *J. App. Phys.* 64, pp. 5462-5464, 1988.

[793] J. Mooij, "Electrical conduction in concentrated disordered transition-metal alloys", *Phys. Status Solidi.* A 17, pp. 521-530, 1973.

[794] P. Häussler, "Evidence for a new Hume-Rothery phase with an amorphous structure", *Z. Phys.* B (Zeitschrift für Physik B: Condensed Matter) 53, pp. 15-26, 1983.

[795] G. Tangonan, "Electrical resistivity of amorphous Pd-Cu-P alloys", *Physics Letters* A 54, pp. 307-308, 1975.

[796] Z˘. Marohnić, E. Babić and D. Pavuna, "Galvanomagnetic effects in $Fe_{40}Ni_{40}B_{20}$", *Phys. Lett.* A 63, pp. 348-350, 1977.

[797] K. Rao, R. Malmhäll, G. Bäckström, and S. Bhagat, "Hall effect in amorphous magnetic materials METGLAS 2826 and 2826B", *Solid State Commun.* 19, pp. 193-195, 1976.

[798] J. Gaebner and H. Chen, "Specific Heat of an Icosahedral Superconductor $Mg_3Zn_3Al_2$", *Phys. Rev. Lett.* 58, pp. 1945-48, 1987.

[799] J. Chevrier, D. Pavuna and F. Cyrot-Lackmann, "Electronic properties and superconductivity of rapidly quenched Al-Si alloys", *Phys. Rev.* B 36, pp. 9115-21, 1987.

[800] A. Inoue, T. Masumoto and H. Chen, "Correlation between superconducting characteristics and normal electrical resistivity for Zr-Si and Zr-Ge alloys in metastable BCC and amorphous phases", J. Phys. F: *Metal Physics*, 13, pp. 2603-2614, 1983.

[801] J. Bednoz and K. Müller, "Possible high-T_c superconductivity in the Ba-La-Cu-O system", *Z. Phys.* B 64, pp. 189-193, 1986.

[802] M. Wu, J. Ashburn and C. Torng, "Superconductivity at 93 °K in a new mixed-phase Y-Ba-Cu-O compound system at ambient pressure", *Phys. Rev. Lett.* 58, pp. 908-910, 1987.

[803] T. Komatsu, K. Imai, K. Matusita, M. Takata, Y. Iwai, A. Kawakami, Y. Kaneko and T. Yamashita, "Liquid Quenched Superconductor Ba-Y-Cu-O with $T_{c,zero}$ = 88 °K and AC Josephson Effect at 77 °K", *Jpn. J. Appl. Phys.* 26, pp. L1148-L1149, 1987. *See also*; T. Komatsu, K. Imai, K. Matusita, M. Ishii, M. Takata and T. Yamashita, "Crystalline Phases in Superconductor Ba-Y-Cu-O with High-T_c Prepared by Melting Method", *Jpn. J. Appl. Phys.* 26, L1272-L1273, 1987.

[804] J. McKittrick, M. McHenry, C. Heremans, P. Standley, T. Prasanna, G. Kalonji, R. O'Handley and M. Foldaeki, "Rapid solidification processing of high-T_c superconductors: Microstructural features and phase relationships", *Physica* C 153-155, pp. 369-370, 1988.

[805] W. Johnson, in: Glassy Metals I, H. Günterodt and H. Beck, Eds., Springer-Verlag, Berlin, 1981, pp. 191.

[806] A. Inoue, A. Hoshi, C. Suryanarayana and T. Masumoto, "Ductile superconducting Ti-Nb-Si-B alloys with a duplex structure of amorphous and crystalline phases", *Scripta Metall.* 14, pp. 1077-1082, 1980.

[807] A. Inoue, Y. Takahashi, A. Hoshi, C. Suryanarayana and T. Masumoto, "Superconductivity in amorphous + crystalline Ti-(Nb or V)-Si-B ductile alloys obtained by rapid quenching from the melt", *J. Appl. Phys.* 52, pp. 4711-9, 1981.

[808] A. Inoue, S. Okamoto, N. Toyota, T. Fukase, K. Matsuzaki and T. Masumoto, "Effect of annealing on the superconducting properties of two amorphous alloys: $Nb_{70}Zr_{15}Si_{15}$ and $Zr_{85}Si_{15}$", *J. Mater. Sci.* 19, pp. 2719-2730, 1984. *See Also*; A. Inoue *et al*, in: Rapidly Quenched Metals IV, Vol. 1, T. Masumoto and K. Suzuki, Eds., The Japan Institute of Metals, Sendai, 1982, pp. 1249.

[809] H. v. Löhneysen, in: Rapidly Solidified Alloys: Processes, Structures, Properties and Applications, H. Liebermann, Ed., Marcel Dekker Inc., 1993, pp. 461.

[810] T. Egami, in: Glassy Metals I, H. Beck and H.-J. Günterodt, Eds., Springer-Verlag, Berlin, New York, 1981, pp. 25.

[811] H. Hsieh, T. Egami, Y. He, S. Poon and G. Shiflet, "Short range ordering in amorphous $Al_{90}Fe_xCe_{10-x}$", *J. Non-Cryst. Solids* 135, pp. 248-254, 1991.

[812] T. Egami, in: Amorphous Materials: Modeling of Structure and Properties, V. Vitek, Ed., TMS-AIME, Warrendale, 1983, pp. 127.

[813] T. Egami, in: Rapidly Solidified Alloys: Processes, Structures, Properties and Applications, H. Liebermann, Ed., Marcel Dekker Inc., 1993, pp. 231.

[814] P. Duwez and S. Lin, "Amorphous Ferromagnetic Phase in Iron-Carbon-Phosphorus Alloys", *J. Appl. Phys.* 38, pp. 4096-4097, 1967.

[815] K. Buschow and F. de Boer, Physics of Magnetism and Magnetic Materials, Kluwer Academic Publishers, 2004.

[816] H. Heinrich, T. Haag and J. Geiger, "A metallic glass tip: a promising field electron emission source", *J. Phys.* D: *Appl. Phys.* 11, 2439-2442, 1978.

[817] H. Warlimont, "Metallic glasses", *Physics in Technology* 11, pp. 28-31, 1980.

[818] A. Hernando and M. Vázquez, in: Rapidly Solidified Alloys: Processes, Structures, Properties and Applications, H. Liebermann, Ed., Marcel Dekker Inc., 1993, pp. 553.

[819] R. O'Handley and C. Chou, "Magnetoelastic effects in metallic glasses", *J. Appl. Phys.* 49, pp. 1653-1658, 1978.

[820] G. Herzer, "Grain structure and magnetism of nanocrystalline ferromagnets", *IEEE Trans. Mag.* 25, pp. 3327-3329, 1989.

[821] G. Herzer, in: Magnetic Materials, K. Buschow, Ed., North-Holland Pub. Co., Amsterdam, Vol. 10, 1996, pp. 415.

[822] E. Machlin, Materials Science in Microelectronics, Vol. 2: The effects of structure on properties in thin films, 2nd edn., Elsevier Ltd, 2006.

[823] M. Getzlaff, Fundamentals of Magnetism, Springer-Verlag, Berlin, 2008.

[824] G. Herzer, in: Rapidly Quenched Metals, H. Frederiksson and S. Savage, Eds., Elsevier Science Ltd, Oxford, London, 1991, pp. 1.

[825] P. Slick, in: Ferromagnetic materials, E. Wohlfarth, Ed., North-Holland Pub. Co., Amsterdam, Vol. 2, 1980, pp. 189.

[826] H. Fujimori, H. Yoshimoto, T. Masumoto and T. Mitera, "Anomalous eddy current loss and amorphous magnetic materials with low core loss", *J. Appl. Phys.* 52, pp. 1893-1898, 1981.

[827] F. Michelini, L. Ressier, J. Degauque, P. Baulès, A. Fert, J. Peyrade and J. Bobo, "Permalloy thin films on MgO(001): Epitaxial growth and physical properties", *J. Appl. Phys.* 92, pp. 7337-7340, 2002.

[828] P. Andricacos and N. Robertson, "Future directions in electroplated materials for thin-film recording heads", *IBM J. Res. Dev.* (IBM Journal of Research and Development) 42, pp. 671-680, 1998.

[829] S. Ohnuma, H. Fujimori, T. Masumoto, X. Xiong, D. Ping and K. Hono, "Fe-Co-Zr-O nanogranular soft-magnetic thin films with a high magnetic flux density", *Appl. Phys. Lett.* 82, pp. 946-948, 2003.

[830] H. Okumura, D. Twisselmann, R. McMichael, M. Huang, Y. Hsu, D. Laughlin and M. McHenry, "Magnetic and structural characterization and ferromagnetic resonance study of thin film HITPERM soft magnetic materials for data storage applications", *J. Appl. Phys.* 93, pp. 6528-6530, 2003.

[831] H. Zijlstra, in: Ferromagnetic Materials, Vol. 3, E. Wohlfarth, Ed., North-Holland, Amsterdam, 1983, pp. 37.

[832] E. Luborsky, J. Livingston and G. Chin, in: Physical Metallurgy, R. Cahn and P. Haasen, Vol. 3, North-Holland and Elsevier Science B.V., 1996, pp. 2501.

[833] K. Kronmüller, in: Supermagnets, Hard Magnetic Materials, G. Long and F. Grandjean, Eds., NATO ASI Series E, Vol. 331, Kluwer Academic Pub., Dordrecht, Netherlands, 1991, p. 461.

[834] R. Anderson, A. Gangulee and L. Romankiw, "Annealing Behaviour of Electroplated Permalloy Thin Films", *J. Elect. Mater.* (*TMS-AIME*) 2, pp. 161-176, 1973.

[835] T. Jagielinski, "Materials for Future High Performance Magnetic Recording Heads", *MRS Bull.* 15, pp. 36-44, 1990.

[836] H. Träuble, in: Moderne Probleme der Metallphysik (Modern Problems of Metal Physics), Vol. 2, A. Seeger, Ed., Springer- Verlag, Berlin, 1966, p. 372.

[837] E. Adler and H. Pfeiffer, "The influence of grain size and impurities on the magnetic properties of the soft magnetic alloy 47.5% NiFe", *IEEE Trans. Mag.* MAG-10, pp. 172-174, 1974.

[838] C. Bean and J. Livingston, "Superparamagnetism", *J. Appl. Phys.* 30, pp. S120-S129, 1959. *See also*; J. Livingston and C. Bean, *J. Appl. Phys.* 30, S318, 1959.

[839] K. Ahn, "Magnetic Properties Observed During Vacuum Deposition of Permalloy Films", *J. Appl. Phys.* 37, pp. 1481-1482, 1966.

[840] L. Neel, "Remarques sur la théorie des propriétés magnétiques des couches minces et des grains fins" (Notes on the theory of magnetic properties of thin films and fine grains), *J. Phys. Radium* 17, pp. 250-255, 1956.

[841] I. Hashim and H. Atwater, "In Situ Analysis of Magnetic and Structural Properties of Epitaxial and Polycrystalline $Ni_{80}Fe_{20}$ Thin Films", *MRS Symp. Proc.* 313, pp. 363-368, 1993.

[842] S. Cheng, M. Kryder and M. Mathur, "Stress related anisotropy studies in d.c.-magnetron sputtered TbCo and Tb Films", *IEEE Trans. Magn.* 25, pp. 4018-20, 1989.

[843] C. Prados, E. Marinero and A. Hernando, "Magnetic interactions and anisotropy in amorphous TbFe Films", *J. Magn. Magn. Mater.* (Journal of Magnetism and Magnetic Materials) 165, pp. 414-16, 1997.

[844] A. Clark and K. Hathaway, in: Handbook of Giant Magnetostrictive Materials, G. Engdahl, Ed., Academic Press, 2000, pp. 1.

[845] M. Aus, B. Szpunar, U. Erb, G. Palumbo and K. Aust, "Electrical, Magnetic and Mechanical Properties of Nanocrystalline Nickel", *MRS Symp. Proc.* 318, pp. 39-44 1994.

[846] W. Kinner, *Mater. Eng.* 89, pp. 32, 1979; cited in Ref. 19.

[847] L. Mendelsohn, E. Nesbitt and G. Bretts, "Glassy metal fabric: A unique magnetic shield", *IEEE Trans. Mag.* Mag-12, pp. 924-926, 1976.

[848] L. Mendelsohn and E. Nesbitt, "Flexible Electromagnetic Shield Comprising Interlaced Glassy Alloy Filaments", US Patent No. 4,030,892, June 21, 1977.

[849] L. Mendelsohn and E. Nesbitt, "Flexible Electromagnetic Shield Comprising Interlaced Glassy Alloy Filaments", US Patent No. 4,126,287, Nov. 21, 1978.

[850] C. Smith, in: Rapidly Solidified Alloys: Processes, Structures, Properties and Applications, H. Liebermann, Ed., Marcel Dekker Inc., 1993, pp. 617.

[851] T. Rikitake, Magnetic and Electromagnetic Shielding, Terra Scientific Pub. Co. (TERRAPUB), Tokyo, 1987.

[852] R. Fitzpatrick, Advanced Classical Electromagnetism, a graduate level course of lectures at The University of Texas at Austin, 1996.

[853] E. Kurgan, in: Computational science - ICCS 2004, 4th Intern. Conf. 2004 Proc. (Lecture Notes in Computer Science), Part II, M. Bubak, G. van Albada, P. Sloot and J. Dongarra, Eds., Kraków, Poland, Springer-Verlag, Berlin, 2004, pp. 252-259.

[854] J. Fraden, Handbook of Modern Sensors, Springer-Verlag New York, Inc., 2004.

[855] J. Bain, "Magnetic Recording Devices: Inductive Heads Properties", in: *Encyclopedia of Materials: Science and Technology*, K. Buschow, R. Cahn, M. Flemings, B. Ilschner, E. Kramer, S. Mahajan, and P. Veyssière, Eds., Elsevier Science Ltd., 2008, pp. 4868-4879.

[856] L. Ranno, in: Alloy Physics: A Comprehensive Reference, W. Pfeiler, Ed., Wiley-VCH Verlag GmbH and Co., 2007, pp. 911.

[857] A. Inoue, in: Non-equilibrium Processing of Materials, C. Suryanarayana, Ed., Pergamon Materials Series Vol. 2, Elsevier Science Ltd., 1999, pp. 375.

[858] S. Elliott, "Amorphous Materials: Medium-range Order", in: *Encyclopedia of Materials: Science and Technology*, K. Buschow, R. Cahn, M. Flemings, B. Ilschner, E. Kramer, S. Mahajan, and P. Veyssière, Eds., Elsevier Science Ltd., 2008, pp. 215-220.

[859] R. Cywinski and S. Kilcoyne, "Amorphous Intermetallic Alloys: Resistivity", in: *Encyclopedia of Materials: Science and Technology*, K. Buschow, R. Cahn, M. Flemings, B. Ilschner, E. Kramer, S. Mahajan, and P. Veyssière, Eds., Elsevier Science Ltd., 2008, pp. 160-166.

[860] M. Howson and B. Gallagher, "Electron transport properties of metallic glasses", *Phys. Rep.* 170, pp. 267-324, 1988.

[861] R. Stephens, "Low-Temperature Specific Heat and Thermal Conductivity of Noncrystalline Dielectric Solids", *Phys. Rev.* B 8, pp. 2896-2905, 1973.

[862] N. Kovalenko, Y. Krasny and U. Krey, Physics of Amorphous Metals, Wiley-VCH Verlag GmbH, Weinheim, 2001.

[863] S. Hasegawa and R. Ray, "Iron-boron metallic glasses", *J. Appl. Phys.* 49, pp. 4174-4179, 1978.

[864] A. Makino, A. Inoue, and T. Masumoto, "Compositional Effect on Soft Magnetic Properties of Co-Fe-Si-B Amorphous Alloys with Zero Magnetostriction", *Mater. Trans. Japan Inst. Met.* (JIM) 31, pp. 884-890, 1990.

[865] W. Dörner, H. Mehrer, P. Pokela, E. Kowala and M. Nicolet, "Diffusion of [195]Au in amorphous W-N diffusion barriers", *Mater. Sci. Eng.* B10, pp. 165-169, 1991.

[866] C. Suryanarayana, F. Froes and R. Rowe, "Rapid solidification processing of titanium alloys", *Int. Mater. Rev.* 36, pp. 85-123, 1991.

[867] / Y. Yokoyama, K. Fukaura, and A. Inoue, "Effect of Ni addition on fatigue properties of bulk glassy $Zr_{50}Cu_{40}Al_{10}$ alloys", *Mater. Trans. JIM* 45, pp. 1672-1678, 2004.

[868] T. Hufnagel, C. Fana, R. Ott, J. Li and S. Brennan, "Controlling shear band behavior metallic glasses through microstructural design", *Intermetallics* 10, pp. 1163-1166, 2002.

[869] A. Inoue, T. Zhang and T. Masumoto, "Zr-Al-Ni amorphous-alloys with high glass transition temperature and significant supercooled liquid region", *Mater. Trans. JIM* 31, pp. 177-183, 1990.

[870] E. Matsubara, T. Tamura, Y. Waseda, T. Zhang, A. Inoue and T. Masumoto, "An anomalous X-ray structural study of an amorphous $La_{55}Al_{25}Ni_{20}$ alloy with a wide supercooled liquid region", *J. Non-Cryst. Solids* 150, pp. 380-385, 1992.

[871] R. Ruhl, B. Giessen, M. Cohen and N. Grant, "New microcrystalline phases in the Nb-Ni and Ta-Ni systems", *Acta Metall.* 15, pp. 1693-1702, 1967.

[872] A. Inoue, K. Ohtera and T. Masumoto, "New Amorphous Al-Y, Al-La and Al-Ce Alloys Prepared by Melt Spinning", *Japan J. Appl. Phys.* 27, pp. L736-L739 1988.

[873] A. Inoue, N. Matsumoto and T. Masumoto, "Al-Ni-Y-Co Amorphous Alloys with High Mechanical Strengths, Wide Supercooled Liquid Region and Large Glass-Forming Capacity", *Mater. Trans. Japan Inst. Met.* (JIM) 31, pp. 493-500, 1990.

[874] A. Inoue, T. Zhang and T. Masumoto, "Al-La-Ni Amorphous Alloys with a Wide Supercooled Liquid Region", *Mater. Trans. JIM* 30, pp. 965-972, 1989.

[875] K. Agyeman, R. Müller and C. Tsuei, "Alloying effect on superconductivity in amorphous lanthanum-based alloys", *Phys. Rev.* B 19, 193-198, 1979.

[876] L. Xia, M. Tang, M. Pan, W. Wang, and Y. Dong, "Glass forming ability and magnetic properties of $Nd_{48}Al_{20}Fe_{27}Co_5$ bulk metallic glass with distinct glass transition", *J. Phys. D: Appl. Phys.* 37, pp. 1706-1709, 2004.

[877] A. Calka, M. Madhava, D. Polk, B. Giessen, H. Matyja and J. Vander-Sande, "A transition-metal-free amorphous alloy: $Mg_{70}Zn_{30}$", *Scripta Metall.* 11, pp. 65-70, 1977.

[878] A. Gebert, R. Subba Rao, U. Wolff, S. Baunack, J. Eckert, and L. Schultz, "Corrosion behaviour of the $Mg_{65}Y_{10}Cu_{15}Ag_{10}$ bulk metallic glass", *Mater. Sci. Eng.* A 375-377, pp. 280-284, 2004.

[879] R. RAY and E. Musso, "Amorphous Alloys in the U-Cr-V System", US Patent No. 3,981,722, Sep. 21, 1976.

[880] K. Wong, A. Drehman and S. Poon, "Properties of superconducting uranium based metallic glasses", *Physica* B+C 135, pp. 299-301, 1985.

[881] S. Elliott, Physics of Amorphous Materials, Longman Inc., New York, 1984.

[882] A. Taub, in: Rapidly Quenched Metals V, S. Steeb and H. Warlimont, Eds., Elsevier, Amsterdam, 1985, p. 1611.

[883] M. Hagiwara, A. Inoue and T. Masumoto, in: Rapidly Quenched Metals V, S. Steeb and H. Warlimont, Eds., Elsevier, Amsterdam, 1985, p. 1779.

[884] A. Argon, G. Hawkins and H. Kuo, "Reinforcement of mortar with Metglas fibres", *J. Mater. Sci.* 14, pp. 1707-1716, 1979.

[885] A. Fels, E. Hornbogen, and K. Friedrich, "A study of reinforcement by prestressed metallic glass ribbons", *J. Mater. Sci. Lett.* 3, pp.639-642, 1984.

[886] W. Kadir, C. Hayzelden and B. Cantor, "Epoxy resin-metallic glass composites", *J. Mater. Sci.* 15, pp. 2663-2664, 1980.

[887] S. Cytron, "A metallic glass-metal matrix composite", *J. Mater. Sci. Lett.* 1, pp. 211-213, 1982.

[888] A. Holbrook, in: Progress in Powder Metallurgy, Vol. 41, H. Sanderow, W. Giebelhausen, and K. Kulkarni, Eds., Metal Powder Industries Federation, Princeton, NJ, 1986, , pp. 679.

[889] K. Hashimoto, in: Rapidly Quenched Metals V, S. Steeb and H. Warlimont, Eds., Elsevier, Amsterdam, 1985, p. 1449.

[890] H. Yoshioka, K. Asami, A. Kawashima and K. Hashimoto, "Laser processed corrosion-resistant amorphous Ni-Cr-P-B surface alloys on a mild steel", *Corrosion Sci.* 27, pp. 981-995, 1987.

[891] K. Asami, A. Kawashima and K. Hashimoto, "Chemical properties and applications of some amorphous alloys", *Mater. Sci. Eng.* 99, pp. 475-481, 1988.

[892] M. Yatsuzuka, T. Yamasaki, H. Uchida and Y. Hashimoto, "Amorphous layer formation on a $Ni_{65}Cr_{15}P_{16}B_4$ alloy by irradiation of an intense pulsed ion beam", *App. Phys. Lett.* 67, pp. 206-207, 1995.

[893] T. Nieh, in: Bulk Metallic Glasses: An Overview, M. Miller and P. Liaw, Eds., Springer Science+Business Media LLC, 2008, pp. 147.

[894] J. Lewandowski and P. Lowhaphandu, "Effects of hydrostatic pressure on the flow and fracture of a bulk amorphous metal", *Phil. Mag.* A 82, pp. 3427-3441, 2002.

[895] K. Mondal and K. Hono, "Geometry Constrained Plasticity of Bulk Metallic Glass", *Mater. Trans. JIM* 50, pp. 152-157, 2009.

[896] H. Choi-Yim, R. Busch, U. Kosster and W. L. Johnson, "Synthesis and characterization of particulate reinforced $Zr_{57}Nb_5Al_{10}Cu_{15.4}Ni_{12.6}$ bulk metallic glass composites", *Acta Mater.* 47, pp. 2455-2462, 1999.

[897] C. Hays, C. Kim and W. Johnson, "Microstructure controlled shear band pattern formation and enhanced plasticity of bulk metallic glasses containing in situ formed ductile phase dendrite dispersions", *Phys. Rev. Lett.* 84, pp. 2901-2904, 2000.

[898] J. Lewandowski, M. Shazly and A. Nouri, "Intrinsic and extrinsic toughening of metallic glasses", *Scripta Mater.* 54, pp. 337-341, 2006.

[899] J. Brennan and K. Prewo, "Silicon carbide fiber reinforced glass-ceramic matrix composites exhibiting high strength and toughness", *J. Mater. Sci.* 17, pp. 2371-2383, 1982.

[900] R. Conner, R. Dandliker and W. Johnson, "Mechanical properties of tungsten and steel fiber reinforced $Zr_{41.25}Ti_{13.75}Cu_{12.5}Ni_{10}Be_{22.5}$ metallic glass matrix composites", *Acta Mater.* 46, pp. 6089-6102, 1998.

[901] G. Wang and P. Liaw, in: Bulk Metallic Glasses: An Overview, M. Miller and P. Liaw, Eds., Springer Science+Business Media, LLC, 2008, pp. 169.

[902] Y. Kawamura, Y. Ohno, "Spark welding of $Zr_{55}Al_{10}Ni_5Cu_{30}$ bulk metallic glasses", *Scripta Mater.* 45, pp. 127-132, 2001.

[903] Y. Kawamura and Y. Ohno, "Superplastic bonding of bulk metallic glasses using friction", *Scripta Mater.* 45, pp. 279-285, 2001.

[904] C. Wong, C. Shek, "Friction welding of $Zr_{41}Ti_{14}Cu_{12.5}Ni_{10}Be_{22.5}$ bulk metallic glass", *Scripta Mater.* 49, pp. 393-397, 2003.

[905] Y. Kawamura, T. Shoji and Y. Ohno, "Welding technologies of bulk metallic glasses", *J. Non-Cryst. Solids* 317, pp. 152-157, 2003.

[906] W. Wanga, C. Dongb and C. Shek, "Bulk metallic glasses", *Mater. Sci. Eng.* R 44, pp. 45-89, 2004.

[907] K. Miyoshi and D. Buckley, "Mechanical-contact-induced transformation from the amorphous to the partially crystalline state in metallic glass", *Thin Solid Films* 118, pp. 363-373, 1984.

[908] T. Masumoto, "Recent progress in amorphous metallic materials in Japan", *Mater. Sci. Eng.* A 179-180, pp. 8-16, 1994.

[909] T. Waniuk, R. Busch, A. Masuhr and W. Johnson, "Equilibrium viscosity of the $Zr_{41.2}Ti_{13.8}Cu_{12.5}Ni_{10}Be_{22.5}$ bulk metallic glass-forming liquid and viscous flow during relaxation, phase separation, and primary crystallization", *Acta Mater.* 46, pp. 5229-5236, 1998.

[910] W. Myung, S. Ryu, I. Hwang, H. Kim, T. Zhang, A. Inoue and A. Greer, "Viscous flow behavior of bulk amorphous $Zr_{55}Al_{10}Ni_5Cu_{30}$ alloys", *Mater. Sci. Eng.* A 304-306, pp. 691-695, 2001.

[911] Y. Kawamura, T. Shibata, A. Inoue and T. Masumoto, "Superplastic deformation of $Zr_{65}Al_{10}Ni_{10}Cu_{15}$ metallic glass", *Scripta Mater.* 37, pp. 431-436, 1997.

[912] Y. Saotome, T. Zhang, A. Inoue, "Microforming of MEMS Parts With Amorphous Alloys", in: Bulk Metallic Glasses; *Mater. Res. Soc.* (MRS) *Symp. Proc.* 554, pp. 385-390, 1999.

[913] Y. Kawamura, T. Shibata, A. Inoue, and T. Masumoto, "Deformation behavior of $Zr_{65}Al_{10}Ni_{10}Cu_{15}$ glassy alloy with wide supercooled liquid region", *Appl. Phys. Lett.* 69, pp. 1208-1210, 1996.

[914] M. Bletry, P. Guyot, Y. Brechet, J. Blandin, and J. Soubeyroux, "Homogeneous deformation of bulk metallic glasses in the super-cooled liquid state", *Mater. Sci. Eng.* A 387-389, pp. 1005-1011, 2004.

[915] C. Chiang, J. Chu, C. Lo, Z. Wang, W. Wang, J. Wang and T. Nieh, "Homogeneous plastic deformation in bulk amorphous $Cu_{60}Zr_{20}Hf_{10}Ti_{10}$ alloy", *Intermetallics* 12, pp. 1057-1061, 2004.

[916] D. Bae, J. Park, J. Na, D. Kim, Y. Kim and J. Lee, "Deformation behavior of Ti-Zr-Ni-Cu-Be metallic glass and composite in the supercooled liquid region", *J. Mater. Res.* 19, pp. 937-942, 2004.

[917] Y. Yang, Y. Song, W. Wu and M. Wang, "Multi-pass overlapping laser glazing of FeCrPC and CoNiSiB alloys", *Thin Solid Films* 323, pp. 199-202, 1998.

[918] H. Hillenbrand, E. Hornbogen, and U. Köster, in: Rapidly Quenched Metals IV, Vol. 2, T. Masumoto and K. Suzuki, Eds., Japan Institute of Metals, Sendai, 1982, pp. 1369-1372.

[919] M. Knobel, J. Schoenmaker, J. Sinnecker, T. Sato, R. Grossinger and W. Hofstetter, "Giant magneto-impedance in nanocrystalline $Fe_{73.5}Cu_1Nb_3Si_{13.5}B_9$ and $Fe_{86}Zi_7B_6Cu_1$ ribbons", *Mater. Sci. Eng.* A 226-228, pp. 546-549, 1997.

[920] R. Grössinger, in: Alloy Physics: A Comprehensive Reference, W. Pfeiler, Ed., Wiley-VCH Verlag GmbH and Co., 2007, pp. 911.

[921] J. Lenz, "A review of magnetic sensors", *Proc. IEEE* 78, pp. 973-989, 1990.

[922] K. Ong and C. Grimes, "Magnetostrictive Nanomaterials for Sensors", in: *Encyclopedia of Nanoscience and Nanotechnology*, H. Nalwa, Ed., American Scientific Publishers, Vol. 5, 2004, pp. 1-28.

[923] J. Tang, "Magnetotransport in Nanogranular Materials", in: *Encyclopedia of Nanoscience and Nanotechnology*, H. Nalwa, Ed., American Scientific Publishers, Vol. 5, 2004, pp. 29-40.

[924] R. Hasegawa, Glassy Metals: Magnetic, Chemical and Structural Properties, CRC Press, Boca Ratou, 1982.

[925] F. Luborsky, in: Amorphous Metallic Alloys, F. Luborsky, Ed., Butterworth, London, 1983, pp. 360.

[926] T. Egami, P. Flanders and C. Graham, "Amorphous Alloys as Soft Magnetic Materials", *AIP* (American Institute of Physics) *Conf. Proc.* 24, pp. 697-702, 1975.

[927] K. Mohri, "Review on recent advances in the field of amorphous-metal sensors and transducers ", *IEEE Trans. Magn.* MAG-20, pp. 942-947, 1984.

[928] K. Mohri, in: Rapidly Quenched Metals, Vol. II, S. Steeb and H. Warlimont, Eds., North-Holland, Amsterdam, 1985, p. 1687.

[929] H. Warlimont, "The impact of amorphous metals on the field of soft magnetic materials", *Mat. Sci. Eng.* 99, 1-10, 1988.

[930] G. Fish, "Soft magnetic materials", *Proc. IEEE,* 78, pp. 947-972, 1990.

[931] K. Mohri, S. Takeuchi and T. Fujimoto, "Sensitive magnetic sensors using amorphous Wiegand-type ribbons", *IEEE Trans. Magn.* MAG-17, pp. 3370-3372, 1981.

[932] D. Webb, D. Forester, A. Ganguly and C. Vittoria, "Applications of amorphous magnetic-layers in surface-acoustic-wave devices", *IEEE Trans. Magn.* MAG-15, pp. 1410-1415, 1979.

[933] O. Nielsen, B. Hernando, J. Petersen and F. Primdahl, "Miniaturisation of low-cost metallic glass flux-gate sensors", *J. Magn. Magn. Mater.* 83, pp. 405-406, 1990.

[934] D. Son, "A new type of fluxgate magnetometer using apparent coercive field strength measurement", *IEEE Trans. Magn.* 25, pp. 3420-3422, 1989.

[935] K. Mohri, T. Kondo and J. Yamasaki, in: Rapidly Quenched Metals, Vol. II, S. Steeb and H. Warlimont, Eds., North-Holland, Amsterdam, 1985, p. 1659.

[936] K. Koo and G. Sigel, Jr., "Characteristics of fiber-optic magnetic-field sensors employing metallic glasses", *Opt. Lett.* 7 (Optical Society of America), pp. 334-336, 1982.

[937] K. Koo and G. Sigel, Jr., "Detection scheme in a fiber-optic magnetic-field sensor free from ambiguity due to material magnetic hysteresis", *Opt. Lett.* 9 (Optical Society of America), pp. 257-259, 1984.

[938] K. Koo, F. Bucholt, A. Dandridge and A. Tveten, "Stability of a fiber-optic magnetometer", *IEEE Trans. Magn.* MAG-22, pp. 141-144, 1986.

[939] A. Maeland, in: Hydrides for Energy Storage, A. Andresen and A. Maeland, Eds., Pergamon, Oxford, UK, 1978, p. 447.

[940] F. Spit, J. Drijver and S. Radelaar, "Hydrogen sorption by the metallic glass $Ni_{64}Zr_{36}$ and by related crystalline compounds", *Scripta Metall.* 14, pp. 1071-1076, 1980.

[941] A. Maeland, L. Tanner and G. Libowitz, "Hydrides of Metallic Glass Alloys", *J. Less-Common Metals* 74, pp. 279-285, 1980.

[942] S. Harris, O. Hartmann, R. Wäppling, R. Hempelmann, H. Fell, C. Scott and A. Maeland, "Diffusion in amorphous and crystalline Cu-Ti studied by the muon spin relaxation technique", *J. Alloys Comp.* 185, pp. 35-44, 1992.

[943] A. Maeland, in: Rapidly Quenched Metals V, S. Steeb and H. Warlimont, Eds., Elsevier, Amsterdam, 1985, p. 1507.

[944] G. Libowitz and A. Maeland, "Interactions of hydrogen with metallic glass alloys", *J. Less Common Metals* 101, pp. 131-143, 1984.

[945] H. Liu, "Magnesium-Nickel Nanocrystalline and Amorphous Alloys for Batteries", in: Encyclopedia of Nanoscience and Nanotechnology, H. Nalwa, Ed., American Scientific Publishers, Vol. 4, 2004, pp. 775-789.

[946] I. Konstanchuk, E. Ivanov, V. Boldyrev, "Interaction of alloys and intermetallic compounds obtained by mechanochemical methods with hydrogen", *Russian Chem. Rev.* 67, pp. 69-79, 1998.

[947] E. Ivanov, I. Konstanchuk, A. Stepanov, V. Boldyrev, "Magnesium mechanical alloys for hydrogen storage", *J. Less Common Metals* 131, pp. 25-29, 1987.

[948] L. Huang, G. Liang and Z. Sun, "Hydrogen-storage properties of amorphous Mg-Ni-Nd alloys", *J. Alloys Comp.* 421, pp. 279-282, 2006.

[949] X. Xiao, L. Chen, Z. Hang, X. Wang, S. Li, C. Chen, Y. Lei and Q. Wang, "Microstructures and electrochemical hydrogen storage properties of novel Mg-Al-Ni amorphous composites", *Electrochemistry Communications* 11, pp. 515-518, 2009.

[950] Y. Zhang, D. Zhao, B. Li, Y. Qi, S. Guo and X. Wang, "Hydrogen storage behaviours of nanocrystalline and amorphous $Mg_{20}Ni_{10-x}Co_x$ ($x = 0$-4) alloys prepared by melt spinning", *Trans. Nonferrous Met. Soc. China* 20, pp. 405-411, 2010.

[951] Y. Zhang, B. Li, H. Ren, S. Guo, Z. Wu and X. Wang, "An investigation on the hydrogen storage characteristics of the melt-spun nanocrystalline and amorphous $Mg_{20-x}La_xNi_{10}$ ($x = 0$, 2) hydrogen storage alloys", *Mater. Chem. Phys.* 115, pp. 328-333, 2009.

[952] W. Oelerich, T. Klassen and R. Bormann, "Hydrogen sorption of nanocrystalline Mg at reduced temperatures by metal-oxide catalysts", *Adv. Eng. Mater.* 3, pp. 487-490, 2001.

[953] / W. Oelerich., T. Klassen and R. Bormann, "Metal oxides as catalysts for improved hydrogen sorption in nanocrystalline Mg-based materials", *J. Alloys Comp.* 315, pp. 237-242, 2001.

[954] K. Zeng, T. Klassen, W. Oelerich and R. Bormann, "Thermodynamic analysis of the hydriding process of Mg-Ni alloys", *J. Alloys Comp.* 283, pp. 213-224, 1999.

[955] T. Klassen, W. Oelerich and R. Bormann, "Nanocrystalline Mg-based hydrides: Hydrogen storage for the zero-emission vehicle", in: Metastable, Mechanically Alloyed and Nanocrystalline Materials, Ismanam-2000 (Proc. Intern. Symp. Metastable, Mechanical Alloyed and Nanocrystalline Materials. Oxford, UK, July 9-14, 2000), *Mater. Sci. Forum* 360-362, pp. 603-608, 2001.

[956] F. Spit, K. Blok, E. Hendriks, G. Winkels, W. Turkenberg, J. Drijver and S. Radelaar, in: Rapidly Quenched Metals IV, T. Masumoto and K. Suzuki, Eds., The Japan Inst. Metals, Sendai, 1982, p. 1635.

[957] C. Janot, Quasicrystals: A Primer, 2nd edn., Oxford University Press, Oxford, 1996.

[958] D. Sordelet and J. Dubois, Eds., Quasicrystals, *MRS Bull.* 22, No. 11, Nov. 1997.

[959] C. Suryanarayana and H. Jones, "Formation and Characteristics of Quasicrystalline Phases: A Review", *Intern. J. Rapid Solidification* 3, pp. 253-293, 1988.

[960] H. Trebin, Ed., Quasicrystals: Structure and Physical Properties, Wiley-VCH GmbH and Co. KGaA, Weinheim, 2003.

[961] J. Dubois, P. Thiel, A. Tsai and K. Urban, Eds., Quasicrystals: Preparation, Properties and Applications, *MRS Symp. Proc.* 553, Warrendale, PA, 1999.

[962] J. Suck, M. Schreiber and P. Häussler, Quasicrystals: An Introduction to the Structure, Physical Properties, and Applications, Springer Series in Materials Science 55, Springer-Verlag, New York, 2002.

[963] K. Kelton, "Quasicrystals: structure and stability", *Int. Mater. Rev.* 38, pp. 105-137, 1993.

[964] J. Dubois, "Quasicrystals", *J. Phys. C: Condens. Matter* 13, pp. 7753-7762, 2001.

[965] T. Fujiwara and Y. Ishii, Eds., Quasicrystals, 1st edn., Handbook of Metal Physics, Elsevier B. V., 2008.

[966] C. Suryanarayana, in: Non-equilibrium Processing of Materials, C. Suryanarayana, Ed., Pergamon Materials Series, Vol. 2, Elsevier Science Ltd., 1999, pp. 49.

[967] A. Tsai, A. Inoue and T. Masumoto, "A Stable Quasicrystal in Al-Cu-Fe System", *Jpn. J. Appl. Phys.* 26. pp. L1505-L1507, 1987.

[968] A. Tsai, A. Inoue, Y. Yokoyama and T. Masumoto, "New icosahedral alloys with superlattice order in the Al-Pd-Mn system prepared by rapid solidification", *Phil. Mag. Lett.* 61, pp. 9-14, 1990.

[969] A. Niikura, A. Tsai, A. Inoue and T. Masumoto, "Stable Zn-Mg-Rare-Earth face-centred icosahedral alloys with pentagonal dodecahedral solidification morphology", *Phil. Mag. Lett.* 69, pp. 351-355, 1994. *See also*: R. Sterzel, E. Dahlmann, A. Langsdorf, and W. Assmus, *Mat. Sci. Eng.* A 294-296, pp. 124-126, 2000.

[970] T. Ishimasa, in: Quasicrystals, T. Fujiwara and Y. Ishii, Eds., Handbook of Metal Physics, Elsevier 2008, pp. 49.

[971] P. Canfield, M. Caudle, C. Ho, A. Kreyssig, S. Nandi, M. Kim, X. Lin, A. Kracher, K. Dennis, R. McCallum and A. Goldman, "Solution growth of a binary icosahedral quasicrystal of $Sc_{12}Zn_{88}$", *Phys. Rev. B* 81, pp. 20201-4, 2010.

[972] P. Thiel and J. Dubois, "Quasicrystals: Reaching Maturity for Technological Applications", *Materials Today* 2, pp. 3-7, 1999.

[973] E. Vedmedenko, "Quasicrystals: Magnetism", in: *Encyclopedia of Materials: Science and Technology*, K. Buschow, R. Cahn, M. Flemings, B. Ilschner, E. Kramer, S. Mahajan, and P. Veyssière, Eds., Elsevier Science Ltd., pp. 1-5, 2008.

[974] J. Dubois, Useful Quasicrystals, World Scientific Publishing Co., Singapore, London, 2005.

[975] M. Feuerbacher, H. Klein, P. Schall, M. Bartsch, U. Messerschmidt and K. Urban, "Plasticity of Quasicrystals", in: Quasicrystals: Preparation, Properties and Applications, J. Dubois, P. Thiel, A. Tsai and K. Urban, Eds., *MRS Symp. Proc.* 553, Warrendale, PA, 1999.

[976] M. Wollgarten, M. Beyss, K. Urban, H. Liebertz and U. Koster, "Direct evidence for plastic deformation of quasicrystals by means of a dislocation mechanism", *Phys. Rev. Lett.* 71, pp. 549-552, 1993.

[977] D. Caillard, in: Alloy Physics; A Comprehensive Reference, W. Pfeiler, Ed., Wiley-VCH Verlag GmbH and Co., 2007, pp. 281.

[978] S. Takeuchi and K. Edagawa, in: Quasicrystals, T. Fujiwara and Y. Ishii, Eds., Handbook of Metal Physics, Elsevier B. V. 2008, pp. 267.

[979] T. Fujiwara and Y. Ishii, in: Quasicrystals, T. Fujiwara and Y. Ishii, Eds., Handbook of Metal Physics, Elsevier 2008, pp. 1.

[980] K. Edagawa, A. Waseda, K. Kimura, H. Ino, "Annealing effects on X-ray diffraction peaks of melt-spun Al-Cu-Fe quasicrystals", *Mater. Sci. Eng.* A 134, pp. 939-942, 1991.

[981] A. Waseda, K. Edagawa and H. Ino, "Effect of heat treatment on the X-ray diffraction of Al-Cu-Fe quasicrystals", *Phil. Mag. Lett.* 62, pp. 183-186, 1990.

[982] S. Takeuchi, "Bulk Mechanical Properties of Quasicrystals", *MRS Symp. Proc.* 553, pp. 283-294, 1999.

[983] D. Caillard, G Vanderschaeve, L. Bresson and D. Gratias, "Transmission electron microscopy study of dislocations and extended defects in as-grown icosahedral Al-Pd-Mn single grains", *Phil. Mag.* A 80, pp. 237-253, 2000.

[984] F. Mompiou, L. Bresson, P. Cordier and D. Caillard, "Dislocation climb and low-temperature plasticity of an Al-Pd-Mn quasicrystal", *Phil. Mag.* 83, pp. 3133-3157, 2003.

[985] M. de Boissieu, R. Currat and S. Francoual, in: Quasicrystals, T. Fujiwara and Y. Ishii, Eds., Handbook of Metal Physics, Elsevier B. V. 2008, pp. 107.

[986] P. Steinhardt and S. Ostlund, The Physics of Quasicrystals, World Scientific, Singapore, 1987.

[987] J. Socolar, T. Lubensky and P. Steinhardt, "Phonons, phasons, and dislocations in quasicrystals", *Phys. Rev.* B 34, pp. 3345-3360, 1986.

[988] T. Lubensky, S. Ramaswamy and J. Toner, "Hydrodynamics of icosahedral quasicrystals", *Phys. Rev.* B 32, pp. 7444-7452, 1985.

[989] G. Coddens and W. Steurer, "Time-of-flight neutron-scattering study of phason hopping in decagonal Al-Co-Ni quasicrystals", *Phys. Rev.* B 60, pp. 270-276, 1999.

[990] G. Coddens, S. Lyonnard, B. Hennion, and Y. Calvayrac, "Triple-axis neutron-scattering study of phason dynamics in Al-Mn-Pd quasicrystals", *Phys. Rev.* B 62, pp. 6268-6295, 2000.

[991] J. Dolinsek, B. Ambrosini, P. Vonlanthen, J. Gavilano, M. Chernikov and H. Ott, "Atomic Motion in Quasicrystalline $Al_{70}Re_{8.6}Pd_{21.4}$: A Two-Dimensional Exchange NMR Study", *Phys. Rev. Lett.* 81, pp. 3671-3674, 1998.

[992] K. Edagawa and K. Kajiyama, "High temperature specific heat of Al-Pd-Mn and Al-Cu-Co quasicrystals", *Mater. Sci. Eng.* A 294-296, pp. 646-649, 2000.

[993] K. Edagawa, K. Kajiyama, R. Tamura and S. Takeuchi, "High-temperature specific heat of quasicrystals and a crystal approximant", *Mater. Sci. Eng.* A 312, pp. 293-298, 2001.

[994] K. Edagawa, P. Mandal, K. Hosono, K. Suzuki and S. Takeuchi, "*In situ* high-resolution transmission electron microscopy observation of the phason-strain relaxation process in an Al-Cu-Co-Si decagonal quasicrystal", *Phys. Rev.* B 70, pp. 184202-9, 2004.

[995] T. Srivatsan, T. Sudarshant and E. Laverniaj, "Processing of Discontinuously-Reinforced Metal Matrix Composites by Rapid Solidification", *Prog. Mater. Sci.* 39, pp. 317-409, 1995.

[996] D. Shaïtura and A. Enaleeva, "Fabrication of Quasicrystalline Coatings: A Review", *Crystallography Reports* 52, pp. 945-952, 2007.

[997] L. Bresson and D. Gratias, "Plastic deformation in AlCuFe icosahedral phase", *J. Non-Cryst. Solids* 153-154, pp. 468-472, 1993.

[998] J. Dubois, "The applied physics of quasicrystals", *Physica Scripta* T49A, pp. 17-23, 1993.

[999] J. Dubois, S. Kang, P. Archambault and B. Colleret, "Thermal diffusivity of quasicrystalline and related crystalline alloys", *J. Mat. Res.* 8, pp. 38-43, 1993.

[1000] P. Archambault, Ph. Plaindoux, E. Belin-Ferré and J. Dubois, in: Quasicrystals: Preparation, Properties and Applications, J. Dubois, P. Thiel, A. Tsai and K. Urban, Eds., *MRS Symp. Proc.* 553, Warrendale, PA, 1999, pp. 409.

[1001] Z. Minevski, C. Tennakoon, K. Anderson, C. Nelson, F. Burns, D. Sordelet, C. Haering and D. Pickard, "Electrocodeposited Quasicrystalline Coatings for Non-Stick Wear Resistant Cookware", *MRS Symp. LL Proc.* 805 (Quasicrystals 2003-Preparation, Properties and Applications), pp. 345-350, 2004.

[1002] S. de Palo, S. Usmani, S. Sampath, D. Sordelet and M. Besser, in: Thermal Spray (A United Forum for Scientific and Technological Advances), C. Berndt, Ed., ASM International, Materials Park, OH, USA, 1997, pp. 135.

[1003] E. Lugscheider, C. Herbst-Dederichs and H. Reimann, in: Thermal Spray: Surface Engineering via Applied Research, C. Berndt, Ed., ASM International, Materials Park, OH, USA, 2000, pp. 843.

[1004] L. Pawlowski, Science and Engineering of Thermal Spray Coatings, John Wiley and Sons, Ltd., 2008.

[1005] P. Brunet, L. Zhang, D. Sordelet, M. Besserand and J. Dubois, "Comparative study of microstructural and tribological properties of sintered, bulk icosahedral samples", *Mater. Sci. Eng.* A 294-296, pp. 74-78, 2000.

[1006] F. Schurack, J. Eckert and L. Schultz, in: Quasicrystals: Structure and Physical Properties, H. Trebin, Ed., Wiley-VCH GmbH and Co. KGaA, Weinheim, 2003, pp. 551.

[1007] Y. Kim, A. Inoue and T. Masumoto, "Ultrahigh Tensile Strengths of $Al_{88}Y_2Ni_9M_1$ (M = Mn or Fe) Amorphous Alloys Containing Finely Dispersed FCC-Al Particles", *Mater. Trans. JIM* 31, pp. 747-749, 1990.

[1008] S. Kim, A. Inoue and T. Masumoto, "Increase of Mechanical Strength of a $Mg_{85}Zn_{12}Ce_3$ Amorphous Alloy by Dispersion of Ultrafine HCP-Mg Particles", *Mater. Trans. JIM* 32, pp. 875-878, 1991.

[1009] Y. Yoshizawa, S. Oguma and K. Yamauchi, "New Fe-based soft magnetic alloys composed of ultrafine grain structure", *J. Appl. Phys.* 64, pp. 6044-6046, 1988.

[1010] S. Ishihara and A. Inoue, "Superplastic Deformation of Supercooled Liquid in Zr-Based Bulk Glassy Alloys Containing Nano-Quasicrystalline Particles", *Mater. Trans. JIM* 42, pp. 1517-1522, 2001.

[1011] Y. Kawamura, T. Shibata, A. Inoue and T. Masumoto, "Superplastic deformation of $Zr_{65}Al_{10}Ni_{10}Cu_{15}$ metallic glass", *Scripta Mater.* 37, pp. 431-436, 1997.

[1012] A. Inoue and H. Kimura, in: Novel Nanocrystalline Alloys and Magnetic Nanomaterials, B. Cantor, Ed., IoP Publishing Ltd, Bristol, Philadelphia, 2005, pp. 42.

[1013] T. Massalski, in: Physical Metallurgy, Vol. 1, R. Cahn and P. Haasen, Eds., North-Holland and Elsevier Science B.V., 1996, pp. 135.

[1014] A. Tsai and C. Gomez, in: Quasicrystals, T. Fujiwara and Y. Ishii, Eds., Handbook of Metal Physics, Elsevier B. V., 2008, pp. 75.

[1015] Y. Wang, J. Qiang, C. Wong, C. Shek and C. Dong, "Composition rule of bulk metallic glasses and quasicrystals using electron concentration criterion", *J. Mater. Res.* 18, pp. 642-648, 2003.

[1016] H. Elhor, M. Mihalkovic, M. Rouijaa, M. Scheffer, and J.Suck, in Ref. 960, pp. 382.

[1017] R. O'Handley, R. Dunlap and M. McHenry, in: Handbook of Magnetic Materials VI, K. Buschow, Ed., North-Holland, Amsterdam, 1991, p. 453.

[1018] J. Wagner, K. Wong and S. Poon, "Electronic properties of stable icosahedral alloys", *Phys. Rev.* B 39, 8091-8095, 1989.

[1019] T. Klein, C. Berger, D. Mayou and F. Cyrot-Lackmann, "Proximity of a metal-insulator transition in icosahedral phases of high structural quality", *Phys. Rev. Lett.* 66, pp. 2907-2910, 1991.

[1020] T. Eisenhammer, "Photon frequency dependent electron relaxation time in noble metals: effect of voids", *Thin Solid Films* 270, pp. 55-59, 1995.

[1021] W. Assmann, Th. Reichelt, T. Eisenhammer, H. Huber, A. Mahr, H. Schellinger and R. Wohlgemuth, "ERDA (Elastic Recoil Detection Analysis) of solar selective absorbers", *Nuclear Instruments and Methods in Physics Research, Section B: Beam Interactions with Materials and Atoms* 113, pp. 303-307, 1996.

[1022] T. Eisenhammer, A. Mahr, A. Haugeneder and W. Assmann, "Selective absorbers based on AlCuFe thin films", *Solar Energy Materials and Solar Cells* 46, pp. 53-65, 1997.

[1023] V. Demange, A. Milandri, M. de Weerd, F. Machizaud, G. Jeandel and J. Dubois, "Optical conductivity of Al-Cr-Fe approximant compounds", *Phys. Rev.* B 65, pp. 144205-15, 2002.

[1024] A. Tsai and M. Yoshimura, "Quasicrystalline Catalyst for Steam Reforming of Methanol", in: Quasicrystals: Preparation, Properties and Applications, E. Belin-Ferre, P. Thiel, A. Tsai and K. Urban, Eds., *MRS Symp. Proc.* K 643, Warrendale, 2001, pp. K16.4.1-K16.4.9.

[1025] M. Yoshimura and A. P. Tsai "Quasicrystal application on catalyst" *J. Alloys Comp.* 342, 451-454, 2002.

[1026] E. Belin-Ferré, M. Fontaine, J. Thirion, S. Kameoka, A. Tsai and J. Dubois, "Electronic structure of leached Al-Cu-Fe quasicrystals used as catalysts", *Phil. Mag.* 86, pp. 687-692, 2006.

[1027] R. Stroud, A. Viano, P. Gibbons, K. Kelton and S. Misture, "Stable Ti-based quasicrystal offers prospect for improved hydrogen storage", *Appl. Phys. Lett.* 69, pp. 2998-3000, 1996.

[1028] A. Viano, E. Majzoub, R. Stroud, M. Kramer, S. Misture, P. Gibbons and K. Kelton, "Hydrogen absorption and storage in quasicrystalline and related Ti-Zr-Ni alloys", *Phil. Mag.* A 78, pp. 131-142, 1998.

[1029] U. Köster, J. Meinhardt, S. Roos and H. Liebertz, "Formation of quasicrystals in bulk glass forming Zr-Cu-Ni-Al alloys", *Appl. Phys. Lett.* 69, pp. 179-181, 1996.

[1030] D. Zander, H. Leptien, U. Köster, N. Eliaz and D. Eliezer, "Hydrogenation of Zr-based metallic glasses and quasicrystals", *J. Non-Cryst. Solids* 250-252, pp. 893-897, 1999.

[1031] A. Tsai, A. Niikura, K. Aoki and T. Masumoto, "Hydrogen absorption in an icosahedral ZnMgY alloy", *J. Alloys Comp.* 253-254, pp. 90-93, 1997.

[1032]U. Koster, D. Zander, J. Meinhardt, N. Eliaz and D. Eliezer, in: Quasicrystals, S. Takeuchi and T. Fujiwara, Eds., World Scientific, Singapore, 1998, p. 313.

[1033]M. Kelton and P. Gibbons, "Hydrogen Storage in Quasicrystals", *MRS Bull.* 22, pp. 69-72, 1997.

[1034]A. Takasaki and K. Kelton, "High-pressure hydrogen loading in $Ti_{45}Zr_{38}Ni_{17}$ amorphous and quasicrystal powders synthesized by mechanical alloying", *J. Alloys Comp.* 347, pp. 295-300, 2002.

[1035]A. Takasaki and K. Kelton, "Hydrogen storage in Ti-based quasicrystal powders produced by mechanical alloying", *Intern. J. Hydrogen Energy* 31, pp. 183-90, 2006.

[1036]B. Liu, Y. Zhang, G. Mi, Z. Zhang and L. Wang, "Crystallographic and electrochemical characteristics of Ti-Zr-Ni-Pd quasicrystalline alloys", *Intern. J. Hydrogen Energy* 34, pp. 6925-6929, 2009.

[1037]B. Liua, G. Fan, Y. Wang, G. Mi, Y. Wu and L. Wang, "Crystallographic and electrochemical characteristics of melt-spun Ti-Zr-Ni-Y alloys", *Intern. J. Hydrogen Energy* 33, pp. 5801-5805, 2008.

[1038]A. Kocjan, P. McGuiness and S. Kobe, "Desorption of hydrogen from Ti-Zr-Ni hydrides using a mass spectrometer", *Intern. J. Hydrogen Energy* 35, pp. 259-265, 2010.

[1039]D. Zander and U. Köster, in: Quasicrystals: Structure and Physical Properties, H. Trebin, Ed., Wiley-VCH GmbH and Co. KGaA, Weinheim, 2003, pp. 551.

[1040]B. Liu, Q. Li, Z. Zhang, G. Mi, Q. Yuan and L. Wang, "Crystallographic and electrochemical characteristics of $Ti_{45}Zr_{35}Ni_{13}Pd_7$ melt-spun alloys", *Intern. J. Hydrogen Energy* 34, pp. 1890-1895, 2009.

Index

S

T

CPSIA information can be obtained
at www.ICGtesting.com
Printed in the USA
LVHW072329301121
704834LV00002B/18

* 9 7 8 1 6 0 8 0 5 5 5 0 0 *